1st Conference on Light and Particle Beams in Materials Science

(LPBMS 2013)

Journal of Physics: Conference Series Volume 502

Tsukuba, Turkey
29-31 August 2013

ISBN: 978-1-63439-282-2
ISSN: 1742-6588

Printed from e-media with permission by:

Curran Associates, Inc.
57 Morehouse Lane
Red Hook, NY 12571

Some format issues inherent in the e-media version may also appear in this print version.

Printed by Curran Associates, Inc. (2014)

For permission requests, please contact the Institute of Physics
at the address below.

Institute of Physics
Dirac House, Temple Back
Bristol BS1 6BE UK

Phone: 44 1 17 929 7481
Fax: 44 1 17 920 0979

techtracking@iop.org

Additional copies of this publication are available from:

Curran Associates, Inc.
57 Morehouse Lane
Red Hook, NY 12571 USA
Phone: 845-758-0400
Fax: 845-758-2634
Email: curran@proceedings.com
Web: www.proceedings.com

TABLE OF CONTENTS

ELECTRON SPECTROSCOPY AND MICROSCOPY

X-RAY DIFFRACTION AND SCATTERING

NEUTRON SCIENCE

Preface

From 29–31 August 2013, the 1st International Conference on Light and Particle Beams in Materials Science, LPBMS 2013, took place in the Tsukuba International Congress Center in the city of Tsukuba, Japan. The conference was a continuation of the international series Synchrotron Radiation in Materials Science (SRMS), which started in 1994. The last one, SRMS-7, was held in Oxford UK 11–14 July 2010, where the International Advisory Committee (IAC) recommended the conference be enlarged to incorporate Materials Research from Neutron, Muon, and Slow Positron Sources, as well as the science emerging from Synchrotron Light Sources.

The conference brought together contributions from academics and industrial researchers with a diverse background and experience from the physics, chemistry and engineering communities. The topics covered in the LPBMS2013 include strongly correlated electron systems, magnetism and magnetic materials, soft matter, interface and surface defects, catalysts, biomaterials, and ceramics. In the 3-day scientific program, the conference consisted of 9 plenary talks, 33 invited talks, 20 oral presentations, and 126 poster presentations.

We are pleased to publish the proceedings of the LPBMS2013 in this volume of Journal of Physics: Conference Series. This volume contains 58 papers representing the work that was presented and discussed at the conference. We hope that this volume will promote further development of this interdisciplinary materials research emerging from synchrotron light, neutron, muon, and slow positron sciences. Finally, we would like to thank the International Advisory Committee (Chair: Professor G N Greaves), sponsors, all the participants and contributors for making possible this international meeting of researchers.

Reiji Kumai & Youichi Murakami

Journal of Physics: Conference Series **502** (2014) 011001 doi:10.1088/1742-6596/502/1/011001

Program committee

Jun-ichiro Mizuki (Kwansei Gaguin University, Chair)
Katsutoshi Aoki (Tohoku University)
Taka-hisa Arima (University of Tokyo)
Kiyotaka Asakura (Hokkaido University)
Atsushi Fujimori (University of Tokyo)
Ryosuke Kadono (KEK)
Toshiji Kanaya (Kyoto University)
Ryouji Kanno (Tokyo Institute of Technology)
Shin-ya Koshihara (Tokyo Institute of Technology)
Eiichiro Matsubara (Kyoto University)
Kanta Ono (KEK)
Masaharu Oshima (University of Tokyo)
Motohiro Suzuki (JASRI)
Masaki Takata (RIKEN)

Organizing committee

Youichi Murakami (KEK, Chair)
Reiji Kumai (KEK, Secretary General)
Hitoshi Abe (KEK)
Kenta Amemiya (KEK)
Masatoshi Arai (J-PARC)
Yasuhiko Fujii (CROSS)
Ryosuke Kadono (KEK)
Kenji Kojima (KEK)
Hiroshi Kumigashira (KEK)
Jun-ichiro Mizuki (JSSRR, Kwansei University, JAEA/SPring-8)
Hironori Nakao (KEK)
Kanta Ono (KEK)
Toshiya Otomo (KEK)
Hideki Seto (KEK)
Masaki Takata (RIKEN)
Kazuyoshi Yamada (KEK)
Yuichi Yamasaki (KEK)
Junichi Yamaura (Tokyo Institute of Technology)

Peer review statement

All papers published in this volume of *Journal of Physics: Conference Series* have been peer reviewed through processes administered by the proceedings Editors. Reviews were conducted by expert referees to the professional and scientific standards expected of a proceedings journal published by IOP Publishing.

Journal of Physics: Conference Series **502** (2014) 012001 doi:10.1088/1742-6596/502/1/012001

Structural and electronic properties of anatase $Ti_{1-x}Fe_xO_{2-\delta}$ thin film prepared by RF magnetron sputtering

K. Usui[1], T. Okumura[1], E. Sakai[2], H. Kumigashira[2], and T. Higuchi[1]

[1]Department of Applied Physics, Tokyo University of Science, Tokyo 125-8585, Japan
[2]Photon Factory, KEK, Tsukuba, 305-0801, Japan

j1513601@ed.tus.ac.jp

Abstract. $Ti_{1-x}Fe_xO_{2-\delta}$ thin films have been deposited on Nb-doped $SrTiO_3$ substrate by RF magnetron sputtering. The as-deposited thin films prepared at T_{sub} = 300 ~ 500°C exhibit anatase structure. The lattice constant depends on the crystallization temperature. The substitution of Fe ions is confirmed by photoemission spectra. The Ti 2*p* X-ray absorption spectra exhibit that Ti 3*d*-band occupancy is increased as a result of the oxygen vacancies. The thin film prepared at 500°C has the Ti 3*d* -DOS at the Fermi level. These findings indicate that Ti 3*d* electrons created by the oxygen vacancies contribute to the electrical conductivity.

1. Introduction

Nb-doped TiO_2 exhibits *n*-type conductivity and becomes a metallic in the heavily doping region. Such a behavior has been also reported in $TiO_{2-\delta}$ thin film with oxygen vacancies. The *n*-type TiO_2 is expected in practical applications such as transparent electrodes and sensors. In recent years, Matsumoto *et al.* have discovered that the Co-doped TiO_2 ($Ti_{1-x}Co_xO_{2-\delta}$) thin film becomes a diluted magnetic semiconductor, which realizes practical application of spintronics device [1-3]. The ferromagnetism of $Ti_{1-x}Co_xO_{2-\delta}$ thin film is driven by the exchange interaction between the Co spins and the *n*-type carriers in the Ti 3*d* band. The oxygen vacancies of this material have an important role for magnetic properties. Yan *et al.* have reported that the oxygen vacancies induce a spin-split donor impurity band with a t_{2g} character in the band gap energy region [4]. Although similar behavior has been also expected in Fe-doped TiO_2 thin film prepared by pulsed laser deposition [5], the details have not been clarified thus far.

In this study, the authors have prepared the Fe-doped TiO_2 ($Ti_{0.993}Fe_{0.007}O_{2-\delta}$) thin films by RF magnetron sputtering and proved its electronic structure by X-ray absorption spectroscopy (XAS) and photoemission spectroscopy (PES). The oxygen vacancies of the thin film can form by controlling the deposition parameters. In this paper, the authors present that the oxygen vacancy of the thin film is closely related with the electrical and magnetic properties in terms of electronic structure.

2. Experiment

$Ti_{0.993}Fe_{0.007}O_{2-\delta}$ thin films were deposited on 0.1 wt. % Nb-doped $SrTiO_3$ (100) single crystals with metallic conductivity, which were supplied by Crystal- Base Co. Ltd. The target was sintered ceramics prepared by solid state reaction of TiO_2 and Fe_2O_3, and pressed into cylinders (ϕ 53×3 mm) at 27 MPa, then sintered in air at 1250°C for 6 h. The prepared target was examined using X-ray diffraction (XRD).

The RF magnetron sputtering system was arranged in a symmetric configuration with a rotating substrate holder for compositional uniformity. For the deposition, the Ar gas flow rate, deposition pressure and the RF power of ceramics target were fixed at 0.7sccm, 4.0×10^{-3} Torr and 50 W, respectively. The substrate temperature (T_{sub}) was changed from 300°C to 500°C. The film thickness was approximately 100 nm for the deposition time of 1 h.

The structural properties of the prepared thin films were characterized by XRD. The electronic structure was measured by XAS and PES installed at an undulator beamline BL-2C in Photon Factory at the High Energy Accelerator Research Organization in Tsukuba, Japan [6]. The energy resolutions of XAS and PES at $h\nu$=730 eV were 60 meV and 100 meV, respectively.

3. Results and Discussions

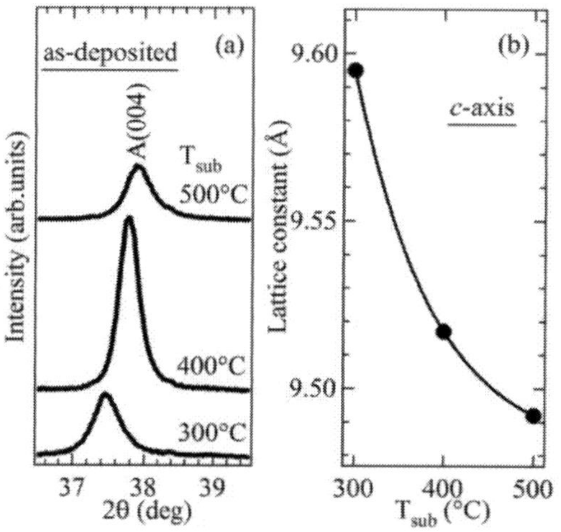

Figure 1. (a) XRD patterns and (b) lattice constants of the thin films prepared at T_{sub} = 300°C, 400°C and 500°C.

Figure 1(a) shows the XRD patterns of $Ti_{0.993}Fe_{0.007}O_{2-\delta}$ thin films prepared at various T_{sub}. The XRD patterns exhibit single phase of anatase (004). The peak position shifts to the higher 2θ side with increasing T_{sub}. The lattice constant calculated from figure 1(a) is shown in figure 1(b). The lattice constant decreases with increasing T_{sub}. These results are considered to be due to the oxygen vacancies with Fe doping.

Figure 2(a) shows the PES spectra in Fe 3*p* core level of the thin films prepared at T_{sub} = 400°C and 500°C. The Fe 3*p* peak is observed at ~65 eV, although the peak intensity is weak. This indicates that Fe ions are doped into TiO_2 Thin films. The peak position accords with Fe^{3+}. Figure 2(b) shows the Ti 2*p* XAS spectra of the thin films prepared at T_{sub} = 400°C and 500°C. The Ti 2*p* XAS spectra correspond to the transition from the Ti 2*p* core level to the unoccupied Ti 3*d* state. The spectra consist of two parts derived from the spin orbit split of L_3 ($2p_{3/2}$) and L_2 ($2p_{1/2}$) states. They are split into the t_{2g} and e_g states by the octahedral ligand field. The spectral shapes of thin films are in good agreement with anatase TiO_2 single crystal [7]. The intensity of the t_{2g} peak for L_3 state is lower in the thin film prepared at T_{sub} = 500°C. This indicates that the Ti 3*d*-band occupancy is increased as a result of the oxygen vacancies.

Figure 3 shows the PES spectra in the valence band region of the thin films prepared at various T_{sub}. The spectral shapes and peak positions do not depend on T_{sub}. The thin films prepared at T_{sub} = 300°C and 400°C have broad DOS in the band gap energy region. However, the thin film prepared at T_{sub} = 500°C has an apparent Fermi edge at E_F. The existence of the Fermi edge indicates that the thin film prepared at T_{sub} = 500°C exhibits metallic conductivity at room temperature.

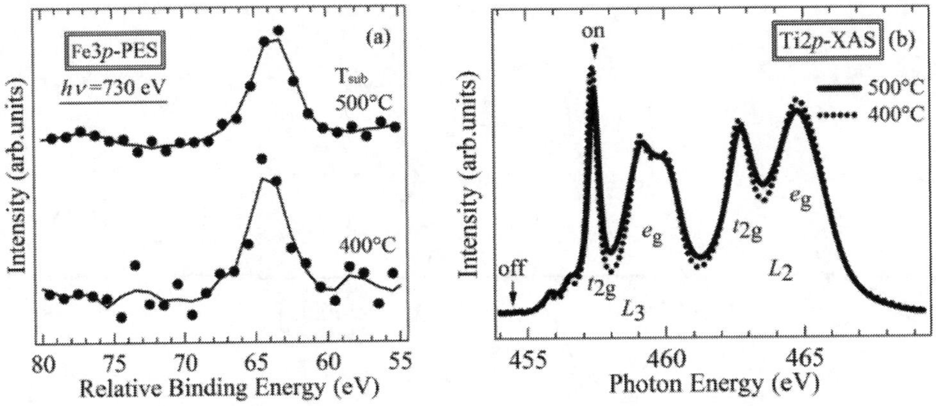

Figure 2. (a) PES spectra in the Fe 3*p* core level and (b) Ti 2*p* XAS spectra of the thin films prepared at T_{sub} = 300°C, 400°C and 500°C.

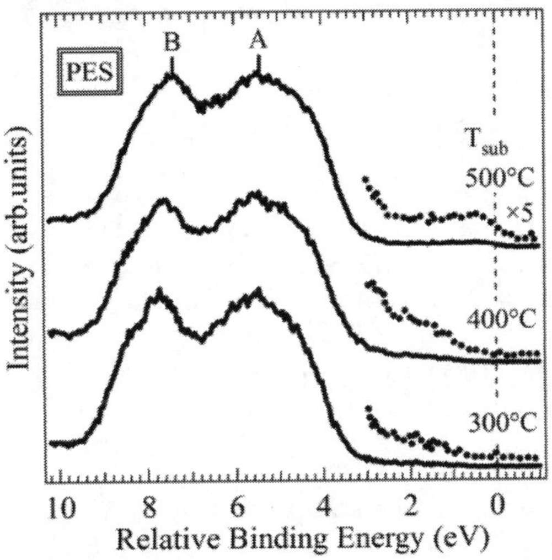

Figure 3. PES spectra in the valence band region (Solid line) and in the band gap energy region (Closed circles) of the thin films prepared at T_{sub} = 300°C, 400°C and 500°C.

Figure 4(a) show the off- and on-resonance spectra in the valence band region of the thin films prepared at T_{sub} = 500°C. The excitation energies of the off- and on-resonance spectra are shown in figure 2(b). The spectral shape is in a good agreement with O 2*p* band-DOS curve calculated by Thomas *et al.* [8]. The B peak is resonantly enhanced by the Ti 2*p* → 3*d* excitation in the on-resonance spectrum. The difference spectrum, which is obtained by subtracting from the on-resonance to off- spectra, is shown in figure 4(b). The difference spectrum is Ti 3*d* partial DOS (PDOS) in the valence band region. The PDOS distributes in the valence band region. The existence of Ti 3*d*-PDOS indicates that the Ti 3*d* state strongly hybridized with O 2*p* state in the valence band. Therefore, the peak A is the O 2*p* non-bonding state and the peak B is the O 2*p* bonding state hybridized with Ti 3*d* state [9]. On the other hand, the Ti 3*d*-PDOS is also observed at around the E_F. The Ti 3*d*-PDOS corresponds to the large Fermi edge shown in Fig. 3. When the Fe^{3+} ions are substituted into Ti^{4+} site, the thin film forms a lot of oxygen vacancies in order to keep the charge neutral condition. Therefore, the number of Ti 3*d* electron increases by Fe substitution. Thus, the electrical conductivity is mainly achieved by Ti 3*d* electron created by the oxygen vacancies during deposition, although the Fe 3*d* electron may also contribute to electrical and magnetic properties.

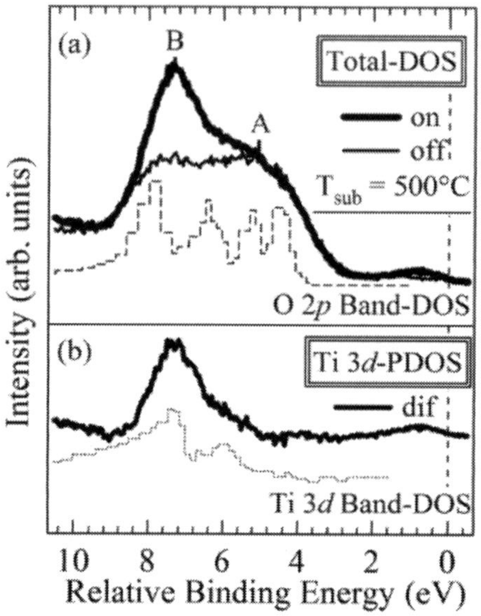

Figure 4. (a) Off- and on-resonance spectra for the thin film prepared at T_{sub} = 500°C. (b) Difference spectrum from on- to off-resonance spectra of the thin film. The dashed lines are the total-DOS and Ti 3*d*-PDOS curves calculated by Ref. [8].

4. Conclusion

The authors have studied the electronic structures of the anatase $Ti_{0.993}Fe_{0.007}O_{2-\delta}$ thin films prepared at various T_{sub} using XAS and PES. The Fe 3*p* PES spectra exhibit that Fe ions are doped into TiO_2 thin films. The Ti 2*p* XAS spectra exhibit that the Ti 3*d*-band occupancy is increased as a result of the oxygen vacancies. The PES spectrum of thin film prepared at T_{sub} = 500°C has an apparent Fermi edge, which corresponds to Ti 3*d*-PDOS, at E_F. These results indicate that Ti 3*d* electrons created by the oxygen vacancies contribute to the electrical conductivity.

Acknowledgments

This work at KEK-PF was done under the approval of the Program Advisory Committee (proposal No. 2011S2-003) at the Institute of Materials Structure Science, KEK.

References

[1] Matsumoto Y, Murakami M, Shono T, Hasegawa T, Fukumura T, Kawasaki M, Ahmet P, Chikyow T, Koshihara S and Koinuma H 2001 *Science* **291** 854
[2] Toyosaki H, Fukumura T, Yamada Y, Nakajima K, Chikyow T, Hasegawa T, Koimuma H and Kawasaki M 2004 *Nature. Material.* **3** 221
[3] Ueno K, Fukumura T, Toyosaki H, Nakano M and Kawasaki M 2007 *Appl. Phys. Lett.* **90** 072103
[4] Yan W, Sun Z, Pan Z, Liu Q, Yao T, Wu Z, Song C, Zeng F, Xie Y, Hu T and Wei S 2009 *Appl. Phys. Lett.* **94** 042508
[5] Inaba K, Hitosugi T, Hirose Y, Furubayashi Y, Kinoda G, Yamamoto Y, Won Kim T, Fujioka H, Shimada T and Hasegawa T 2006 *Jpn. J. Appl. Phys.* **45** 114
[6] Horiba K, Ohguchi H, Kumigashira H, Oshima M, Ono K, Nakagawa N, Lippmaa M, Kawasaki M and Koinuma H 2003 *Rev. Sci. Instrum.* **74** 3406
[7] Van der Laan G, 1990 *Phys. Rev.* B **41** 12366.
[8] Thomas A G, Flavell W R, Mallick A K, Kumarasinghe A R, Tsoutsou D, Khan N, Chatwin C, Rayner S and Smith G C 2007 *Phys. Rev.* B **75** 035105
[9] Tezuka Y, Shin S, Ishii T, Ejima T, Suzuki S and Sato S 1994 *J. Phys. Soc. Jpn.* **63** 347

Gigantic transverse x-ray magnetic circular dichroism in ultrathin Co in Au/Co/Au(001)

T Koide[1], K Mamiya[1], D Asakura[1], Y Osafune[2], A Fujimori[2], Y Suzuki[3,4], T Katayama[4], and S Yuasa[4]
[1]Photon Factory, IMSS, High Energy Accelerator Research Organization, Tsukuba, 305-0801, Japan
[2]Department of Physics, University of Tokyo, Bunkyo-ku, Tokyo 113-0033, Japan
[3]Graduate School of Engineering Science, Osaka University, Toyonaka, Osaka 560-8531, Japan
[4]NanoElectronics Research Institute, National Institute of Advanced Industrial Science and Technology, Tsukuba, 305-8568, Japan

E-mail: tsuneharu.koide@kek.jp

Abstract. Transverse-geometry x-ray magnetic circular dichroism (TXMCD) measurements on Au/Co-staircase/Au(001) reveal the orbital origin of intrinsic in-plane magnetic anisotropy. A gigantic TXMCD was successfully observed at the Co $L_{3,2}$ edges for Co thickness (t_{Co}) in the 2-monolayer regime. A TXMCD-sum-rule analysis shows a remarkable enhancement of an orbital-moment anisotropy (Δm_{orb}) and of an in-plane magnetic dipole moment ($m_T^{\parallel}$). Both Δm_{orb} and $m_T^{\parallel}$ exhibit close similarity in t_{Co} dependence, reflecting the in-plane magnetic anisotropy. These observations evidence that extremely strong, intrinsic, in-plane magnetic anisotropy originates from the anisotropic orbital part of the wave function, dominating the dipole-dipole-interaction-derived, extrinsic, in-plane magnetic anisotropy.

1. Introduction

Magnetic anisotropy (MA) in two-dimensional systems has recently attracted much scientific and technological attention, because MA provides a physically intriguing issue of its intrinsic origin and high potential for spintronic applications. Most of previous studies have been focused on perpendicular magnetic anisotropy (PMA) in magnetic ultrathin films [1], superlattices [2], and nano-clusters [3-5]. This is mainly because PMA could have potential applications for high-density magnetic recording media. PMA is also physically of interest, since it is observed only in systems involving special magnetic interaction that overcomes dipole-dipole interaction which favors in-plane magnetic anisotropy (IMA).

IMA has attracted less interest than PMA, partly due to its little potential for spintronic applications and partly due to its extrinsic origin. Physically, IMA could originate also from the intrinsic, microscopic, magnetic states. But IMA is normally dominated by the extrinsic anisotropy arising from macroscopic dipole-dipole interaction. The intrinsic IMA is thus difficult to separately observe. The intrinsic IMA may be scientifically of interest and of importance, since separately revealing intrinsic IMA provides much information about the microscopic, electronic and magnetic states of low-dimensional systems [6,7]. Observing the intrinsic IMA requires special experimental techniques that allow probing the masked, microscopic electronic states. X-ray magnetic circular dichroism in transverse geometry (TXMCD) has such an ability [8, 9]. Along with the T-geometry sum rules [8], it allows a determination of the orbital-moment anisotropy ($\Delta m_{orb} = m_{orb}^{\parallel} - m_{orb}^{\perp}$) and the anisotropic magnetic dipole moments ($m_T^{\parallel}$=-1/2 $m_T^{\perp}$). Importantly, Δm_{orb} is directly related with IMA and $m_T^{\parallel}$ should be closely connected with Δm_{orb}.

The origin of intrinsic IMA is not easy to approach by the usual longitudinal XMCD (LXMCD) alone, because the spin magnetic moment (m_{spin}) dominates LXMCD but m_{spin} is essentially isotropic, being not responsible for MA. By contrast, in TXMCD the magnetic field (***B***) is applied to the sample perpendicularly to the photon helicity (h) and fully aligns m_{spin} along the field, resulting in m_{spin} perpendic-

Journal of Physics: Conference Series **502** (2014) 012002 doi:10.1088/1742-6596/502/1/012002

ular to h. The contribution of m_{spin} to TXMCD is thus effectively "killed." In contrast, m_{orb} and m_T are not completely perpendicular to h even in T geometry. Thus, TXMCD allows a determination of the anisotropic m_{orb} and m_T, giving information about the intrinsic IMA. Here we report on a TXMCD study of ultrathin Au/Co(t_{Co})/Au(001).

2. Experiment

An Au(2nm)/Co-staircase($2 \leqq t_{Co} \leqq 9$ML)/Au(001)(100nm) sample was prepared by molecular-beam epitaxy with an MgO(001) substrate. A 100-nm-thick Au(001) seed layer was prepared on the 40-nm Cr buffer layer. The Co thickness (t_{Co}) ranged from 2 to 9 monoatomic layers (ML) with each step height of 1 ML. A layer-by-layer growth of Co during deposition was confirmed *in-situ* by reflection high-energy electron diffraction. The Co layer was protected from oxidation with a 2-nm-thick Au capping layer. TXMCD measurements were made at the Co $L_{3,2}$ edges at room temperature using circularly polarized synchrotron radiation on bending-magnet beam line BL-11A at the Photon Factory. The degree of circular polarization (P_C) was chosen to be P_C= +78±3% [3]. Polarization-dependent x-ray absorption spectra (XAS) were taken using the total electron-yield method with h fixed and $\boldsymbol{B}$ reversed. The incident-light intensity was monitored with the photocurrent from an Au-coated mirror. The T geometry was set up by rotating the whole superconducting-magnet XMCD apparatus [10] around the vertical axis. The ±5-T field was applied to the sample perpendicular to h at a light-incident angle of θ = 45° with an accuracy of ±0.3°.

3. Results and discussion

Figure 1 shows the flux-normalized, polarization-dependent Co $L_{3,2}$-edge XAS (μ_+ and μ_-) and TXMCD ($\Delta\mu = \mu_+ - \mu_-$) spectra for t_{Co} = 7 ML. Here μ_+ and μ_- represent the absorption coefficients with their sign defined by the direction of $\boldsymbol{B}$ with respect to the surface normal ($\boldsymbol{n}$), as denoted by an inset of Fig.1. The spectra were taken at room temperature at an incidence angle of θ = 45° under a high field of $\boldsymbol{B}$ = ±5T. Correction for P_C was made in XAS and TXMCD. The energy integral of the TXMCD spectrum is shown by a red curve in Fig. 1(b). It is crucial to assure no artifact arising from mixing of a strong LXMCD into a weak TXMCD due to possible alignment errors. We measured conventional LXMCD at the Co $L_{3,2}$ edges on a bcc Co 6-ML film with an MgO(001) substrate. A strong, normal-incidence LXMCD was observed at the Co $L_{3,2}$ edges but no TXMCD signal was detected on the identical sample. This shows no artifact in the TXMCD.

The TXMCD spectrum is similar in line shape to the separately measured LXMCD. The TXMCD is naturally much weaker in amplitude than the LXMCD. The small TXMCD is reasonable, since m_{spin} does not contribute to TXMCD; m_{spin} is fully aligned along the high magnetic field ($\boldsymbol{B}$) and is perpendicular to h. The observed TXMCD is genuine and indicates the anisotropic orbital moment ($\Delta m_{orb}= m^{\parallel}_{orb} - m^{\perp}_{orb}$) and magnetic dipole moment ($m^{\parallel}_T = -1/2\, m^{\perp}_T$). This result leads us to interest in the t_{Co} dependence of TXMCD.

Figure 2(a) shows the Co $L_{3,2}$-edge TXMCD spectra as functions of t_{Co} ranging from 9 to 2 ML. It is evident that the TXMCD rapidly increases in amplitude with decreasing t_{Co}. Surprisingly, the negative L_3-edge TXMCD amplitude amounts to ≳15% at the thinnest limit of t_{Co} = 2 ML. This is even comparable

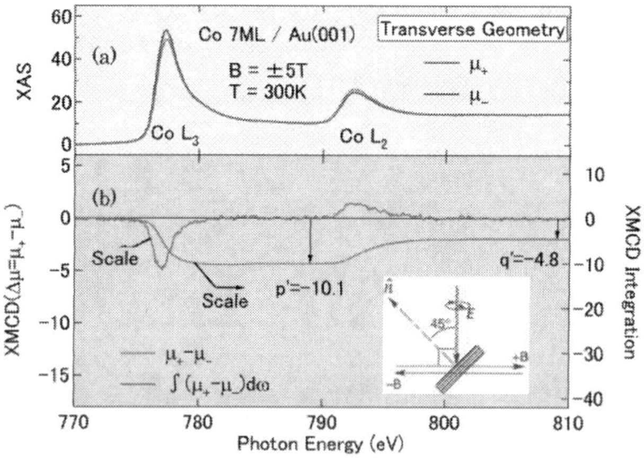

Figure 1 (a) 45°-incidence Co $L_{3,2}$-edge XAS (μ_+ and μ_-) and (b) T-geometry TXMCD ($\Delta\mu = \mu_+ - \mu_-$) for t_{Co}= 7ML (green curve). The red curve in (b) denotes the energy integral of TXMCD. The spectra were taken at room temperature and for $\boldsymbol{B}$ = ±5T. The inset denotes the definition of μ_+ and μ_- by the relative orientation of $\boldsymbol{B}$ with respect to the surface normal ($\boldsymbol{n}$). p' and q' denote the energy integrals of TXMCD over the L_3 and $L_3 + L_2$ regions, respectively

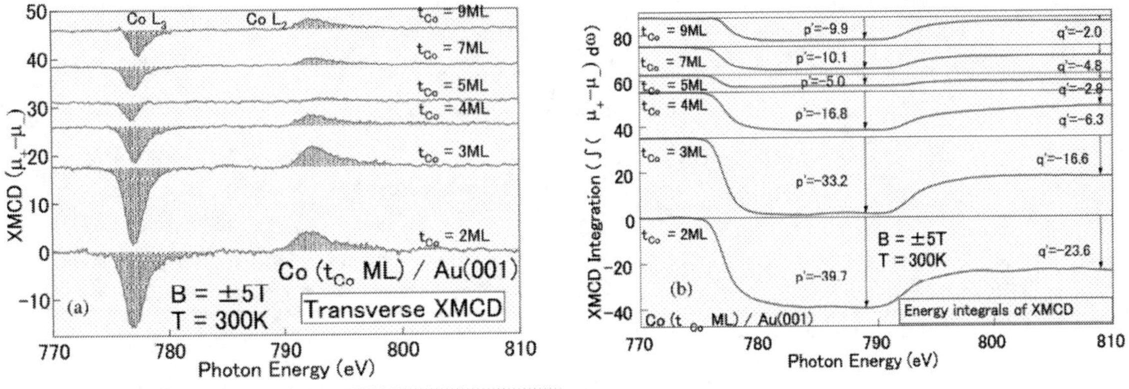

Figure 2(a) Co thickness (t_{Co}) dependence of Co $L_{3,2}$-edge TXMCD with t_{Co} ranging from 9 to 2 ML. Note the remarkably enhanced TXMCD for t_{Co} = 2 ML. (b) Energy integrals of Co $L_{3,2}$-edge TXMCD as functions of t_{Co}. Note the remarkably enhanced energy-integrated TXMCD for t_{Co} = 2 ML.

to a usual LXMCD amplitude in ferromagnetic bulk substances. The observation of the gigantic TXMCD indicates extremely strong IMA arising from the increased low dimensionality. Figure 2(b) shows the t_{Co} dependence of the energy integral of the Co $L_{3,2}$-edge TXMCD over a range $2 \le t_{Co} \le 9$ ML. The notations p' and q' denote the energy integrals of the TXMCD over the L_3 and $L_3 + L_2$ edges. They are used below for the sum rule analyses. Corresponding to the TXMCD behavior, the energy-integrated TXMCD remarkably increases in amplitude with lowering t_{Co}, showing enhanced Δm_{orb} and $m_T^{\parallel}$.

To quantify magnetic anisotropy, we determine anisotropic magnetic moments using the TXMCD sum rules [8, 11, 12]. The T sum rules [8], for the present definition of μ_+ and μ_-, state for $\theta = 45°$,

$$\Delta m_{orb} = -8/3 \cdot \left[\Delta A_3 + \Delta A_2\right]^{\theta=45°} \cdot n_h / \left[A_3 + A_2\right], \quad (1)$$

$$m_T^{\parallel} = -4/21 \cdot \left[\Delta A_3 - 2\Delta A_2\right]^{\theta=45°} \cdot n_h / \left[A_3 + A_2\right]. \quad (2)$$

Here $\Delta m_{orb} = m_{orb}^{\parallel} - m_{orb}^{\perp}$ is the orbital-moment anisotropy, $m_{orb}^{\parallel}$ and $m_{orb}^{\perp}$ ($m_T^{\parallel}$ and $m_T^{\perp}$) are the in-plane and out-of-plane orbital (magnetic dipole) moments, ΔA_3^{θ} and ΔA_2^{θ} (A_3^{θ} and A_2^{θ}) are the energy integrals of θ-directed L_3- and L_2-edge TXMCD (XAS), and n_h is the $3d$ hole number. There are additional relationships, $m_{orb}^{\theta} = -<L_{\theta}> \mu_B / \hbar$ and $m_T^{\theta} = -<T_{\theta}> \mu_B / \hbar$ with $<T_{\theta}>$ being the expectation value of the intra-atomic magnetic dipole operator, $\boldsymbol{T} = \boldsymbol{S} - 3\boldsymbol{r}(\boldsymbol{r} \cdot \boldsymbol{S})/r^2$ [12, 13]. Using eqs. (1) and (2) and a theoretical $3d$ hole number of $n_h = 2.49$ for Co [14], we have determined $\Delta m_{orb} = m_{orb}^{\parallel} - m_{orb}^{\perp}$ and $m_T^{\parallel}$.

Figure 3 shows the results for Δm_{orb} and $m_T^{\parallel}$ as functions of t_{Co}. Both Δm_{orb} and $m_T^{\parallel}$ increase steeply with decreasing t_{Co}. Most interestingly, Δm_{orb} and $m_T^{\parallel}$ exhibit almost the same t_{Co} dependence. Since Δm_{orb} is directly related with the IMA energy and $m_T^{\parallel}$ is closely connected with Δm_{orb}, our finding indicates that this system is of strong intrinsic IMA with the IMA energy increasing with decreasing t_{Co}. The similar behavior of Δm_{orb} and $m_T^{\parallel}$ with t_{Co} is reasonable if one takes into account the theoretical prediction that non-zero values of those quantities are the manifestation of the anisotropic orbital part of the electron wave function.

We have obtained for t_{Co} = 2 ML the numerical results of Δm_{orb} = +0.35 μ_B and $m_T^{\parallel}$ = +0.075 μ_B. It should be recalled that both Δm_{orb} and $m_T^{\parallel}$ are almost zero under high-symmetry crystal fields in bulk materials. The extremely large Δm_{orb} shows an enhanced in-plane orbital moment, indicating strong IMA of Au/Co/Au(001) at t_{Co} = 2 ML. The present value of Δm_{orb} = +0.35 μ_B sharply contrasts in sign with and is very similar in amplitude to the reported values of Δm_{orb} = -0.12 μ_B for 4-ML Co in Au/Co ultrathin film/Au(111) [1] and Δm_{orb} = -0.11μ_B for 2-ML-height Co cluster in Au/Co nano-clusters/Au(111) [3]. Interestingly, the present Co/Au(001) is of strong IMA whereas the previous Co/Au(111) showed strong PMA, indicating the strong effect of a substrate orientation on the thin-film magnetic anisotropy.

We estimate an intrinsic IMA energy ΔE_{SO} caused by spin-orbit (SO) interaction. We take an approximation incorporating an effective renormalization factor α [14],

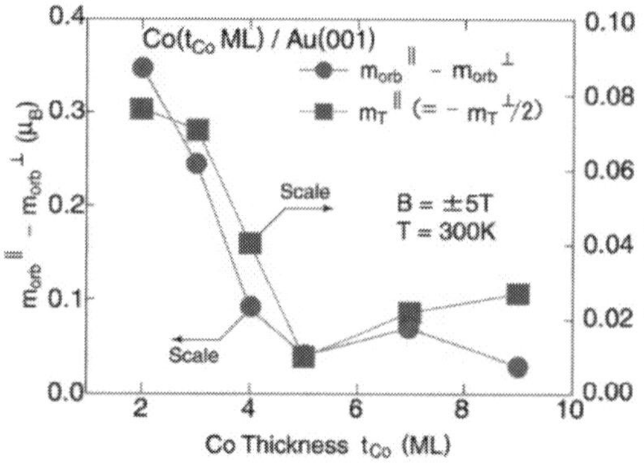

Figure 3 Co thickness (t_{Co}) dependence of T-geometry-derived orbital-moment anisotropy ($\Delta m_{orb} = m_{orb}^{\parallel} - m_{orb}^{\perp}$) and in-plane magnetic dipole moment ($m_T^{\parallel}$). Note that Δm_{orb} and $m_T^{\parallel}$ exhibit very similar t_{Co} dependence.

$$\Delta E_{so} \approx -\alpha \cdot \xi_{3d} (m_{orb}^{\parallel} - m_{orb}^{\perp}) / (4\mu_B), \tag{4}$$

where ξ_{3d} (≈70meV) is the 3d SO interaction constant in Co. Using a TXMCD-derived value of Δm_{orb} = +0.35 μ_B and a reported value of $\alpha \approx$ 0.2 for Co [1,14], we obtain an extremely large value of $\Delta E_{SO} \approx -1.2 \times 10^{-3}$ eV . This is a very large, *intrinsic*, electronic-state-derived IMA energy. In contrast, a dipole-dipole interaction energy, which favors extrinsic IMA, is estimated to be $-2\pi M_s^2 \times 0.93 = -0.86 \times 10^{-4}$ eV/Co with a 2-ML-thickness correction in the Co film. We have $|-\Delta E_{SO}| \gg |-2\pi M_s^2 \times 0.93|$, which shows that the electronic-state-derived, intrinsic IMA is much stronger than the extrinsic IMA originating from dipole-dipole interaction.

We consider the physical implication of the enhanced $m_T^{\parallel}$ with the x axis taken to be in the layer plane. It holds that $m_T^{\parallel}$ $(= m_T^x)$ $= - < T_{\parallel} > \mu_B / \hbar$ and $< T_{\parallel} >$ $(=< T_x >)$ $= 7\sum_i \alpha^i < S^i > \cdot < Q_{xx}^i >$ $\approx 7 < S > \cdot < Q_{xx} >$ [13]. Here S is the spin quantum number of the total system, i denotes the symmetry of the d orbitals ($i = x^2$-y^2, $3z^2$-r^2, xy, yz, zx), $< Q_{xx}^i > = \langle i | Q_{xx} | i \rangle$ is the quadrupole electron-density distribution of orbital i and $< Q_{xx} > = \sum_i \alpha^i \langle Q_{xx}^i \rangle = \sum_i \alpha^i < i | Q_{xx} | i >$ represents the $< Q_{xx}^i >$ averaged over the occupied five d_i orbitals with α^i denoting the occupation probability of orbital i. The extremely large $m_T^{\parallel}$ indicates a net appearance of $<Q_{xx}>$ of the total Co 3d states. Using the values of the LXMCD-derived S =0.855$\hbar$ (m_{spin}=1.71μ_B) and the TXMCD-derived $m_T^{\parallel}$ = +0.075 μ_B, we obtain a 3d-state-averaged value of $<Q_{xx}>$ = 0.0125. The present non-zero $<Q_{xx}>$ implies that the four-lobe nature of the out-of-plane orbitals partially manifests itself when observed from the in-plane x axis because of the relative high occupation of the out-of-plane orbitals.

4. Conclusion

We have successfully observed a giant TXMCD at the Co $L_{3,2}$ edges in thin Co layers in Au/Co staircase/Au(001). TXMCD sum-rule analyses have given a remarkably enhanced orbital-moment anisotropy and an enhanced in-plane magnetic dipole moment. Those anisotropic moments were shown to originate from the non-equally populated, anisotropic 3d states and to produce strong intrinsic IMA.

References

[1] Weller D *et al.* 1995 *Phys. Rev. Lett.* **75** 3752.
[2] Nakajima N *et al.* 1998 *Phys. Rev. Lett.* **81** 5229.
[3] Koide T *et al.* 2001 *Phys. Rev. Lett.* **87** 257201.
[4] Koide T *et al.* 2004 *J. Electron Spectrosc. and Relat. Phenom.* **136** 107.
[5] Gambardella P *et al.* 2003 Science **300** 1130.
[6] Honolka J *et al.* 2012 *Phys. Rev. Lett.* **108** 256811.
[7] Du C, Adur R, Wang H, Hauser A J, Yang F and Hammel P C 2013 *Phys. Rev. Lett.* **110** 147204.
[8] Dürr H A, Guo G Y, van der Laan G, Lee J, Lauhoff G and Bland J A C 1997 Science **277** 213.
[9] Mamiya K *et al.* 2006 *Rad. Phys. and Chem.* **75** 1872.
[10] Koide T, Shidara T and Fukutani H 1992 *Rev. Sci. Instrum.* **63** 1462.
[11] Thole B T, Carra P, Sette F and van der Laan G 1992 *Phys. Rev. Lett.* **68** 1943.
[12] Carra P, Thole B T, Altarelli M and Wang X 1993 *Phys. Rev. Lett.* **70** 694.
[13] Stöhr J and König H 1995 *Phys. Rev. Lett.* **75** 3748.
[14] Bruno P 1989 *Phys. Rev.* B **39** 865.

Journal of Physics: Conference Series **502** (2014) 012003 doi:10.1088/1742-6596/502/1/012003

Microscopic origin of ferrimagnetism of a double perovskite Sr_2FeMoO_6: *An x-ray magnetic circular dichroism study*

T Koide[1], T Sekine[1], H Miyauchi[1], H Manaka[1], D Asakura[1], A Fujimori[2], K -I Kobayashi[3], Y Tomioka[3], T Kimura[4] and Y Tokura[3, 4]

[1]Photon Factory, IMSS, High Energy Accelerator Research Organization, Tsukuba, Ibaraki 305-0801, Japan

[2]Department of Physics, University of Tokyo, Bunkyo-ku, Tokyo 113-0033, Japan

[3] CERC, National Institute of Advanced Industrial Science and Technology, Tsukuba, Ibaraki 305-8568, Japan

[4]Department of Applied Physics, University of Tokyo, Bunkyo-ku, Tokyo 113-0033, Japan

E-mail: tsuneharu.koide@kek.jp

Abstract. We have studied the microscopic magnetic state of a double perovskite Sr_2FeMoO_6 by x-ray magnetic circular dichroism (XMCD). An Fe $L_{3,2}$-edge XMCD reveals that the formal Fe^{3+} ions have an appreciable orbital moment (m_{orb}) parallel to their large spin moment (m_{spin}), showing the charge-transfer-induced, minority-spin (↓) Fe t_{2g} electrons. A weak but undoubted O K-edge XMCD was successfully detected, showing an O 2p-hole m_{orb} induced by 2p-3d hybridization. Along with information obtained from the observed Mo $M_{3,2}$-edge XMCD, our findings verify the partially filled Fe 3d t_{2g}↓-O 2p_π↓-Mo 4d t_{2g}↓ hybridized bands at the Fermi level, which mediates ferrimagnetism.

1. Introduction

Tunneling magnetoresistance (TMR) under low fields even at room temperature in ordered double perovskite, Sr_2FeMoO_6 [1], offers the possibility of a new class of magnetoresistive oxides. Despite intrinsically non-magnetic Mo atoms being located between Fe atoms [1], the perovskite has been reported to be a ferrimagnet with the Mo spin antiparallel to the Fe spin [2] and with a high Curie temperature of $T_C \gtrsim 400K$. The oxide is theoretically predicted to be half metallic, with only the minority-spin (↓) density of states being at the Fermi level (E_F) [1]. This nature is thought to be the origin of the TMR effect. These properties are partly similar to those of manganites [3], thereby being expected to provide a key clue to underlying physics of colossal MR in manganites [3]. Despite many studies of Sr_2FeMoO_6, fundamental issues for understanding the origin of its unique magneto-transport properties still remain controversial.

The key issues to be experimentally clarified in Sr_2FeMoO_6 include: (i) What is the real valence or 3d-electron configuration of the nominal Fe^{3+} ions? Do the formal Fe^{3+} ions have a finite orbital magnetic moment (m_{orb}) or not ?; (ii) Is the O 2p state magnetically polarized? What role and how do the formal O^{2-} ions play in mediating ferrimagnetic interaction? Theories predict the reported ferrimagnetism ascribable to the Mo 4d-Fe 3d hybridization by implicitly considering the O 2p states [4-7]. This effect should induce p-d charge transfers and strongly affect the valence and the m_{spin} and m_{orb} values of constituent ions. X-ray magnetic circular dichroism (XMCD) [2, 8], neutron-diffraction [9, 10], and Mössbauer spectroscopy [11, 12] studies have reported contradictory results. Most importantly, no experiment has yet been reported on the O 2p state, though it should play a crucial role in mediating the magnetic interaction. Thus, the issues on the microscopic electronic and magnetic states of Sr_2FeMoO_6 are yet unsettled. XMCD experiments [13] with sum-rules [14,15] could allow directly approaching the issues through an element-specific, separate determination of m_{spin} and m_{orb}. Here we report on a microscopic study of the element-specific, valence-orbital-selective spin and orbital states of Sr_2FeMoO_6 by high-precision XMCD measurements at the Fe $L_{3,2}$ and O K edges, resolving issues (i) and (ii).

2. Experiment

Polycrystalline samples of Sr_2FeMoO_6 were used. The Curie temperature (T_C) was determined to be $T_C \approx$ 430 K by magnetization measurements. The details of sample growth and characterization were reported elsewhere [1]. XMCD experiments were made by detecting x-ray absorption spectra (XAS) using the total electron-yield method with circularly polarized synchrotron radiation on helical-undulator (AR-NE1B) and bending-magnet (BL-11A) beam lines at the Photon Factory. The sample temperature was $\approx$ 10-30 K much lower than T_C. The sample surface was diamond-filed *in-situ* in a preparation chamber. Magnetic fields of $\boldsymbol{B}=\pm(1.1\text{-}2.5)$T were applied to the sample parallel and antiparallel to the photon helicity (h) using a superconducting magnet [16] and a permanent-magnet flipper-type XMCD apparatus. The field direction was switched with h fixed on BL-11A. With the flipper on NE1B, the field was reversed at every photon energy without any derivative-like XMCD artifact, and the h was also reversed. The degree of circular polarization (P_C) was evaluated to be $\pm95\pm3$ % on NE1B and chosen to be +78$\pm$3% on BL-11A [13].

3. Results and discussion

Figures 1(a) and 1(b) show polarization-dependent Fe $L_{3,2}$-edge XAS (μ_+ and μ_-) and P_C-corrected XMCD ($\Delta\mu = \mu_+ - \mu_-$) taken at $T \approx$30 K under $\boldsymbol{B} = \pm2.5$ T. The energy integrated XMCD is also displayed in Fig.1(b). A linear background has been subtracted in XAS. Here μ_+ and μ_- stand for the absorption coefficients with h parallel and antiparallel to the Fe 3d majority-spin (↑) direction. A strong XMCD was

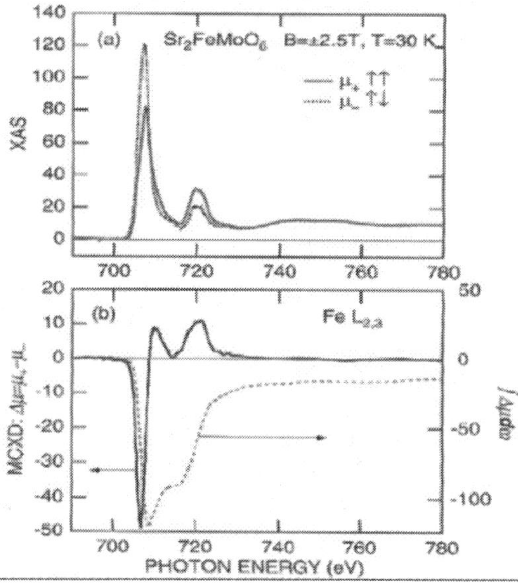

Figure. 1 (a) Polarization-dependent XAS and (b) XMCD (solid) and its energy-integrated (dashed) spectra at the Fe $L_{3,2}$ edges in Sr_2FeMoO_6. The spectra were taken at T = 30 K and $\boldsymbol{B} = \pm2.5$ T. A linear background has been subtracted in (a). Correction for P_C has been made in (a) and (b). A normalization was made in (a) such that the sum of the L_3 XAS peaks was equal to 200.

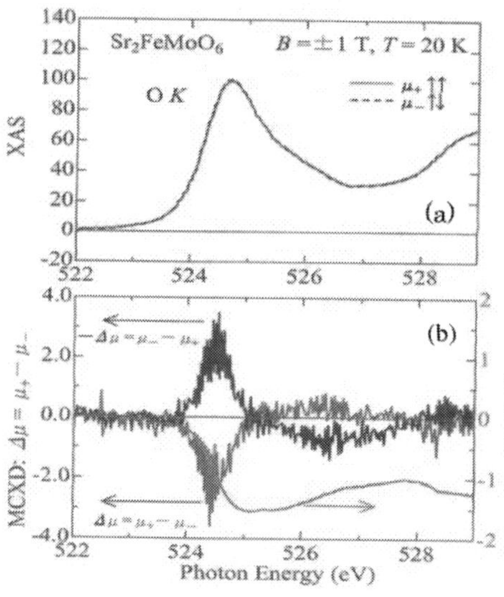

Figure. 2 (a) Polarization-dependent XAS and (b) XMCD (green and blue curves) and its energy-integrated (red curve) spectra at the O K edge in Sr_2FeMoO_6. The spectra were taken at T = 20 K and $\boldsymbol{B} = \pm1.1$ T. The XMCD spectrum taken by a reversal of h (blue curve:$-\Delta\mu = \mu_- - \mu_+$) is displayed for comparison. A linear background has been subtracted in (a). Correction for P_C has been made.

observed at the Fe $L_{3,2}$ edges, which indicates the high-spin state of the Fe ion. In addition, our analytical result, $\Delta A_{L_3} + \Delta A_{L_2} < 0$, shows that m_{orb}(Fe) (> 0) is *appreciable* and parallel to m_{spin}(Fe) (> 0) on the basis of the XMCD orbital sum rule [14]. This result is in sharp contrast with the previous XMCD studies [2, 8] which reported that $\Delta A_{L_3} + \Delta A_{L_2} \approx 0$, i.e., $m_{orb}(Fe) \approx 0$, being important for elucidating the electronic states of Sr_2FeMoO_6. Since the nominal high-spin Fe^{3+} ion should take the $t_{2g}\uparrow^3 e_g\uparrow^2$ configuration with a vanishing m_{orb}, the present result demonstrates the existence of the charge-transfer-induced Fe $3d$ $t_{2g}\downarrow$ electrons. We thus conclude that the Fe ion is in the O $2p\downarrow$-to-Fe $3d\downarrow$ charge-transferred ($\delta > 0$), more-than-half, $Fe^{+3-\delta}$ $t_{2g}\uparrow^3 e_g\uparrow^2 t_{2g}\downarrow^{\delta}$ configuration. This result is consistent with the Mössbauer result [11,12]. Although there is a possibility of coexistence of the charge-transferred state and the mixed-valence state, the resolution of the present measurements is not high enough for judging it from the line shape.

Figure 2 shows the polarization-dependent O *K*-edge XAS and P_C-corrected XMCD in the first XAS sharp-peak region. The measurements were made at $T \approx 20$ K and for $\boldsymbol{B} = \pm 1.1$ T. Green and blue curves in Fig. 2(b) represent $\Delta\mu$ (= $\mu_+ - \mu_-$) and $-\Delta\mu$ (= $\mu_- - \mu_+$), respectively, taken by reversing the helicity (h) of helical undulator radiation. Although μ_+ and μ_- in Fig. 2(a) show no appreciable difference, a small but undoubted XMCD was successfully observed and confirmed by its sign reversal on reversing h in addition to the flipper's assurance of no differential XMCD artifact.

The O *K*-edge XAS shows the existence of O $2p$ holes. The result that $\int_K \Delta\mu d\omega < 0$ demonstrates the magnetic polarization of the O $2p$ holes with a finite m_{orb}(O) (> 0) which is parallel to m_{spin}(Fe) (> 0) by the orbital sum rule [14]. From the fact that m_{orb}(Fe) (> 0) and m_{orb}(O) (> 0), we infer that the observed m_{orb}(O) should be induced by $2p$-$3d$ hybridization and a resultant $2p$-to-$3d$ charge transfer. The O *K*-edge XMCD gives no information about m_{spin}(O) because of the absence of spin-orbit (SO) interaction in the O $1s$ core hole. However, the O $2p\downarrow$-to-Fe $3d\downarrow$ charge transfer indicates that the holes should reside in the O $2p\downarrow$ state with a resulting net up-spin ($\uparrow$) polarization of the O $2p$ level, i.e., $m_{spin}(O) > 0$ with the sign definition of $m_{spin} > 0$ for $S_z > 0$. The substantial O *K*-edge XMCD (~2%) indicates that m_{orb}(O) should arise from the orbital current directly transferred from m_{orb}(Fe) via the Fe $3d$ $t_{2g}\downarrow$-O $2p\downarrow$ wave-function overlap but cannot be ascribed to the two-step mechanism by the O $2p$-hole SO interaction under hybridization-induced $2p$ spin polarization.

We now determine m_{spin} and m_{orb} using the XMCD sum rules [14,15]. The orbital [14] and angle-averaged spin [15,17] sum rules are given for the $L_{3,2}$ and $M_{3,2}$ edges by

$$m_{orb} = -\frac{4}{3}\frac{\Delta A_{j_3} + \Delta A_{j_2}}{A_{j_3} + A_{j_2}} n_h , \quad (1)$$

$$m_{spin} = -\frac{2}{3}\sum_{\alpha=x,y,z}\frac{(\Delta A_{j_3} - 2\Delta A_{j_2})^{\alpha}}{(A_{j_3} + A_{j_2})} n_h , \quad (2)$$

where m_{orb} and m_{spin} are in units of μ_B/atom, n_h is the $3d$ or $4d$ hole number, ΔA_{j_k} and A_{j_k} ($j = L, M$; $k = 2, 3$) are the j_k-edge-integrated XMCD and XAS intensities, respectively. For the present polycrystalline sample, m_{spin} can be directly determined using the angle-averaged spin sum rule (2) with allowed neglect of the magnetic dipole moment [17]. We shall employ for m_{spin} a correction factor (C.F.) given by theory [18] that takes into account the Coulomb interaction between electrons based on the full atomic multiplet and O_h crystal field. Using a $3d$ hole number of $n_h(3d) = 4.4 \pm 0.3$ ($\delta = 0.6 \pm 0.3$) supported by Mössbauer spectroscopy studies [11, 12] and a C.F. as averaged from theoretical ones for the Fe^{+2} and Fe^{+3} high-spin ions [18], we obtain $m_{spin}(Fe) = 2.80 \pm 0.30\ \mu_B$, $m_{orb}(Fe) = 0.093 \pm 0.01\ \mu_B$, and $m_{tot}(Fe) = 2.89 \pm 0.30\ \mu_B$.

We have observed the XAS and XMCD at the Mo $M_{3,2}$ edges (not shown here). For Mo, we first determine the ratio of the orbital to spin moments. Relying only on the Mo-edge XMCD integral, we get $m_{orb}(Mo)/m_{spin}(Mo) = -0.105 \pm 0.020$. Using a zeroth-order $4d$ hole number, $n_h(4d) = 9.0$, we obtain $m_{spin}(Mo) = -(0.36 \pm 0.03)\ \mu_B$, $m_{orb}(Mo) = +(0.037 \pm 0.015)\ \mu_B$, and $m_{tot}(Mo) = -(0.32 \pm 0.03)\ \mu_B$. The nonzero m_{orb}(Mo) which is antiparallel to m_{spin}(Mo) is reasonable for the present less-than-half Mo system.

We next estimate m_{orb}(O). The XAS and XMCD at the Mo $M_{3,2}$ edges are much weaker than those at the Fe $L_{3,2}$ edges. This fact indicates that the contribution to the O *K*-edge XAS and XMCD of the Mo $4d$

states via $2p$-$4d$ hybridization is negligibly small even for charge transfer (δ') $\neq 0$ compared with that of the Fe $3d$ states through $2p$-$3d$ hybridization. Hence, approximating the O $2p$ hole number, n_h(O), to that arising only from the $2p$-to-$3d$ charge transfer ($\delta \approx 0.6$) leads to $n_h(O) = \delta/6 \approx 0.1$. This gives an estimate of m_{orb}(O) from the XMCD using the K-edge orbital sum rule as $m_{orb}(O) = -2/3 \cdot (\Delta A_K / A_K) \cdot n_h(O) \approx (0.3 \pm 0.1) \times 10^{-3}$ μ_B, where ΔA_K and A_K are the XMCD and XAS energy integrals at the O K edge. This value is as small as 1.5% of $m_{orb}(O) \approx 0.02$ μ_B in $SrFeCoO_3$ [19], where the O $2p$-Fe, Co $3d$ hybridization and the resultant XMCD are gigantic. In contrast to m_{orb}(O), m_{spin}(O) is difficult to estimate, since an accurate n_h(O) value is unknown and m_{spin}(O) does not contribute to the O K-edge XMCD. However, the existence of holes in the down spin indicates the O $2p$ spin polarization, $m_{spin}(O) > 0$. Neglecting m_{spin}(O) and m_{orb}(O), we get $m_{tot} \approx m_{tot}(Fe) + m_{tot}(Mo) \approx 2.57 \pm 0.20$ μ_B. The XMCD-derived m_{tot} value of the system agrees with a magnetization-derived m_{tot} value of 3.0 μ_B [1].

The present m_{orb}(Fe) agrees with theories [5,6] and Mössbauer-spectroscopy studies [11,12] which claimed the intermediate $Fe^{+2.5}$ valence for the nominal Fe^{3+} ions. The result that $m_{spin}(Fe) > 0$, $m_{spin}(O) > 0$, and $m_{spin}(Mo) < 0$ shows that the nominal O^{2-} ion makes a ferromagnetic interaction with the formal Fe^{3+} ions and an antiferromagnetic interaction with the Mo^{5+} ions, leading to total ferrimagnetic order. Our results negate the results of previous XMCD studies [2,8], which reported an undetectable O K-edge XMCD, a very small m_{orb}(Fe) or even a small negative $m_{orb}(Fe)$ (= -0.036 μ_B), antiparallel to m_{spin}(Fe), despite for the more-than-half system.

We give here a schematic picture of the element-specific, spin-dependent density of states and of the representative Fe $3d_{xy}$ $t_{2g}{\downarrow}$-O $2p_\pi{\downarrow}$-Mo $4d_{xy}$ $t_{2g}{\downarrow}$ hybridized states near E_F, which is deduced from the present observations. Our results show that unpaired electrons exist in the Fe $3d$ $t_{2g}{\downarrow}$-O $2p_\pi{\downarrow}$-Mo $4d$ $t_{2g}{\downarrow}$ hybridized bands crossing E_F and that the magnetically polarized O $2p$ states should play an essential role in mediating ferrimagnetic interaction. This finding is crucial for uncovering the magnetic and electronic states of Sr_2FeMoO_6, since the O $2p$-mediated Fe $3d$-Mo $4d$ hybridization is a theoretical prerequisite for the driving force of ferrimagnetism with a high T_C [4-7]. The picture shows half-metallic nature of Sr_2FeMoO_6, giving firm support for theoretical predictions. Our result also implies that charge carriers are not holes but are minority-spin electrons and agrees with the sign of the normal Hall effect that shows the electron-like nature of charge carriers.

4. Conclusion

We have presented evidence for substantial m_{orb}(Fe) and m_{orb}(O) which are parallel to m_{spin}(Fe) in Sr_2FeMoO_6. We have observed finite antiparallel m_{spin}(Mo) and m_{orb}(Mo), the former being antiparallel to m_{spin}(Fe). Combined information of these quantities resolve controversial issues (i)-(ii), verifying theoretically predicted Fe $3d$ $t_{2g}{\downarrow}$-O $2p_\pi{\downarrow}$-Mo $4d$ $t_{2g}{\downarrow}$ hybridized bands across E_F.

References

[1] Kobayashi K -I, Kimura T, Sawada H, Terakura K and Tokura Y 1998 *Nature* **395** 677.
[2] Besse M *et al* 2002 *Europhys. Lett.* **60** 608.
[3] Tokura Y (ed.) 2000 *Colossal Magnetoresistive Oxides* (Gordon and Breach Science).
[4] Sarma D D, Mahadevan P, Saha-Dasgupta T, Ray S and Kumar A 2000 *Phys. Rev. Lett.* **85** 2549.
[5] Fang Z, Terakura K and Kanamori J 2001 *Phys. Rev.* B **63** 180407(R).
[6] Kanchana V, Vaitheeswaran G, Alouani M and Delin A 2007 *Phys. Rev.* B **75**, 220404(R).
[7] Erten O, Meetei O N, Mukherjee A, Randeria M, Trivedi N and Woodward P 2011 *Phys. Rev. Lett.* **107** 257201.
[8] Ray S, Kumar A, Sarma D D, Cimino R, Turchini S, Zennaro S and Zema N 2001 *Phys. Rev. Lett.* **87** 097204.
[9] García-Landa B *et al* 1999 *Solid State Commun.* **110** 435.
[10] Moritomo Y *et al* 2000 *J. Phys. Soc. Jpn.* **69** 1723.
[11] Lindén J, Yamamoto T, Karppinen M, Yamauchi H and Pietari T 2000 *Appl. Phys. Lett.* **76** 2925.
[12] Lindén J, Yamamoto T, Nakamura J, Yamauchi H and Karppinen M 2002 *Phys. Rev.* B **66** 184408.
[13] Koide T *et al.* 2001 *Phys. Rev. Lett.* **87** 246404.
[14] Thole B T, Carra P, Sette F and van der Laan G 1992 *Phys. Rev. Lett.* **68** 1943.
[15] Carra P, Thole B T, Altarelli M, and Wang X 1993 *Phys. Rev. Lett.* **70** 694.
[16] Koide T, Shidara T, and Fukutani H 1992 *Rev. Sci. Instrum.* **63** 1462.
[17] Stöhr J and König H 1995 *Phys. Rev. Lett.* **75** 3748.
[18] Teramura Y, Tanaka A and Jo T 1996 *J. Phys. Soc. Jpn.* **65** 1053.
[19] Koide T *et al.*, manuscript in preparation.

Electronic structure of $Li_2Fe_{1-x}Mn_xP_2O_7$ for lithium-ion battery studied by resonant photoemission spectroscopy

Koji Horiba[1,2,6], Shota Ito[1], Shodai Kurosumi[1], Naoka Nagamura[1,2], Satoshi Toyoda[1,2], Hiroshi Kumigashira[3,4], Masaharu Oshima[1,2], Naoya Furuta[5], Shin-ichi Nishimura[5], Atsuo Yamada[5] and Noritaka Mizuno[1]

[1] Department of Applied Chemistry, The University of Tokyo, 7-3-1, Hongo, Bunkyo-ku, Tokyo 113-8656, Japan
[2] Synchrotron Radiation Research Organization, The University of Tokyo, 7-3-1, Hongo, Bunkyo-ku, Tokyo 113-8656, Japan
[3] Photon Factory, Institute of Materials Structure Science, High Energy Accelerator Research Organization, 1-1 Oho, Tsukuba, Ibaraki 305-0801, Japan
[4] Precursory Research for Embryonic Science and Technology, Japan Science and Technology Agency, Gobancho Building, 7, Gobancho, Chiyodaku, Tokyo 102-0076 Japan
[5] Department of Chemical System Engineering, The University of Tokyo, 7-3-1, Hongo, Bunkyo-ku, Tokyo 113-8656, Japan

E-mail: horiba@post.kek.jp

Abstract. In order to clarify changes in the electronic structures, especially Fe partial density of states (DOS), of $Li_2Fe_{1-x}Mn_xP_2O_7$ with Mn substitution, we have performed x-ray absorption spectroscopy and resonant photoemission spectroscopy (RPES) experiments for $Li_2Fe_{1-x}Mn_xP_2O_7$. Using RPES teqniques, we have succeeded in extracting the Fe^{2+} partial DOS. We have found the systematic shift to higher binding energy and broadening of Fe 3d t_{2g} down-spin states accompanying with the Mn substitution. The peak shift of the Fe 3d t_{2g} down-spin states is matched very well to the change of Fe^{3+}/Fe^{2+} redox potential, suggesting that the origin of high Fe^{3+}/Fe^{2+} redox potential in $Li_2Fe_{1-x}Mn_xP_2O_7$ is the shift of the Fe 3d t_{2g} down-spin states to the higher binding energy with Mn substitution.

1. Introduction

A pyrophosphate iron compound $Li_2FeP_2O_7$ is a newly designed polyanionic-cathode material for a Li-ion battery [1]. Compared to olivine fomula $LiFePO_4$ [2, 3], $Li_2FeP_2O_7$ has some superior characteristics such as high redox voltage of 3.5 eV, reversible electrochemical activity without any particle-downsizing/carbon coating, a two-dimensional channel for Li-diffusion, and the potential of a 2-electron redox reaction. In addition, recently it has been reported that the Fe^{3+}/Fe^{2+} redox potential upshifts to 4 V with partial substitution of Mn for Fe sites in $Li_2FeP_2O_7$ [4]. Since the possibility of Mn^{3+}/Mn^{2+} redox reaction below 4 V has been ruled out,

[6] Present address: Photon Factory, Institute of Materials Structure Science, High Energy Accelerator Research Organization, 1-1 Oho, Tsukuba, Ibaraki 305-0801, Japan

the upshift of the redox potential in $Li_2Fe_{1-x}Mn_xP_2O_7$ is thought to be derived from changes in the Fe states with Mn substitution. In order to clarify changes in the electronic structures, especially Fe partial density of states (DOS), of $Li_2Fe_{1-x}Mn_xP_2O_7$ with Mn substitution, we have performed x-ray absorption spectroscopy (XAS) and resonant photoemission spectroscopy (RPES) experiments for $Li_2Fe_{1-x}Mn_xP_2O_7$.

2. Experimental

The mixed-metal pyrophosphate $Li_2Fe_{1-x}Mn_xP_2O_7$ ($x = 0$, 0.25, 0.5, and 0.75) powders were synthesized by conventional solid state reactions [4] using Li_2CO_3, $FeC_2O_4{\cdot}2H_2O$, MnC_2O_4, and $(NH_4)_2HPO_4$. Stoichiometric amounts of precursors and optional 10 wt% Ketjen black carbon additives for improvement of electron conductivity were thoroughly mixed and then sintered at 600 °C for 12 h under Ar gas flow. The formation of $Li_2Fe_{1-x}Mn_xP_2O_7$ crystals were confirmed by powder x-ray diffraction measurements.

Before soft x-ray spectroscopic measurements, we performed electrochemical charge/discharge cycles for characterization and control of Li concentration. Working electrodes were formulated with mixing the $Li_2Fe_{1-x}Mn_xP_2O_7$/C composites, additional carbon black, and polyvinylidene fluoride binder in a minimal amount of N-methylpyrrolidone solvent. The cells were charged and discharged over the potential range from 2.0 V to 4.5 V in an Ar-filled glove box and we confirmed that galvanostatic cycling profiles were the same as those in the previous reports [4]. After the several charge/discharge cycles, each cell was charged up to 2.0 V for full lithiation.

Spectroscopic measurements were performed at BL-2C of the Photon Factory in High Energy Accelerator Research Organization. In order to avoid moisture air exposure, we enclosed the electrode samples in an Ar-filled transfer chamber at the inside of the glove box and connected the transfer chamber to a sample entry of measurement chamber without any exposure to air. The total energy resolution of XAS and RPES measurements was set to 200 meV. The binding energies were calibrated using an Au 4f core-level peak on a gold foil reference and a C 1s core-level peak on carbon black additives in the samples.

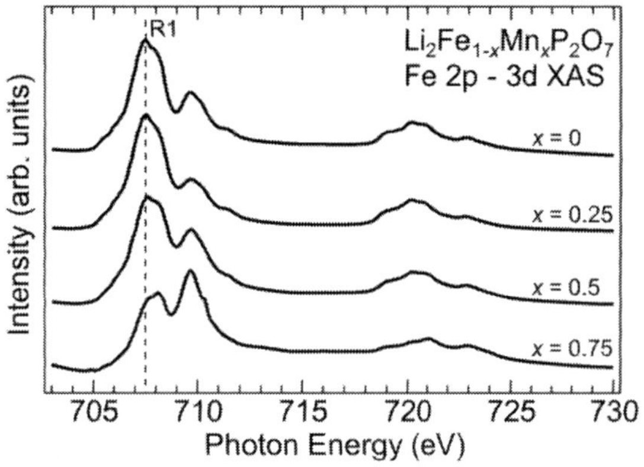

Figure 1. Fe $2p$ - $3d$ XAS spectra for $Li_2Fe_{1-x}Mn_xP_2O_7$. A dashed line labeled R1 corresponds to the photon energy used for RPES measurements in Fig. 2.

3. Results and Discussion

Figure 1 shows Fe $2p$ - $3d$ XAS spectra for $Li_2Fe_{1-x}Mn_xP_2O_7$. The XAS spectra on $Li_2FeP_2O_7$ and $Li_2Fe_{0.75}Mn_{0.25}P_2O_7$ are very similar to those on olivine-type $LiFePO_4$ [5] and $LiFe_{0.5}Mn_{0.5}PO_4$ [6] with Fe^{2+} states. With increasing Mn concentration, Fe^{3+} components slightly increase [5]. Since the Mössbauer spectra, which are completely sensitive to bulk conditions, show only small signal of Fe^{3+} impurity without any Mn-concentration dependence [4], the origin of the Fe^{3+} components is probably a slight desorption of Li ions at the surface.

By the slight Li desorption, the valencies of Fe ions are more strongly affected for samples with less amount of Fe such as $x = 0.75$. because the valencies of Mn ions hardly change. In order to extract the Fe^{2+} contribution in the valence band, we have performed RPES measurements at the excitation energy R1 (a dashed line in Fig. 1) corresponding to the absorption edge of Fe^{2+} states.

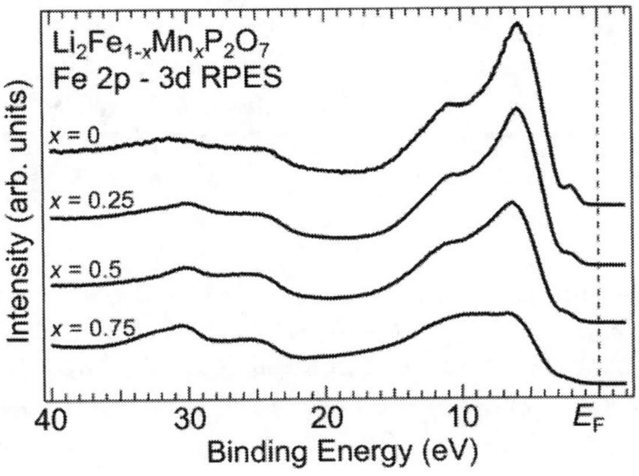

Figure 2. Fe $2p$ - $3d$ RPES spectra for $Li_2Fe_{1-x}Mn_xP_2O_7$ measured at the excitation energy R1 (a dashed line in Fig. 1). The spectral intensity is normalized by the area of O 2s core-level peaks around 25 eV.

Figure 2 shows Fe $2p$ - $3d$ RPES spectra for $Li_2Fe_{1-x}Mn_xP_2O_7$ measured at the excitation energy R1. The RPES spectrum on $Li_2FeP_2O_7$ is also very similar to that for olivine-type $LiFePO_4$ [5] with Fe^{2+} states, suggesting that Li-Fe pyrophosphate and olivine materials have similar electronic structures. A distinctive feature around 2 eV corresponding to the Fe 3d t_{2g} down-spin states in Fe^{2+} [7] is clearly observed. With decreasing Fe concentration, although the intensity of resonant valence band spectra decreases, Fe 3d t_{2g} down-spin states still remain observable.

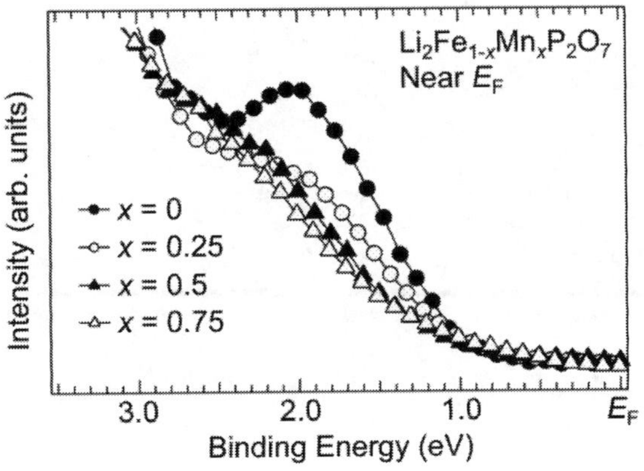

Figure 3. The expanded RPES spectra near E_F for $Li_2Fe_{1-x}Mn_xP_2O_7$.

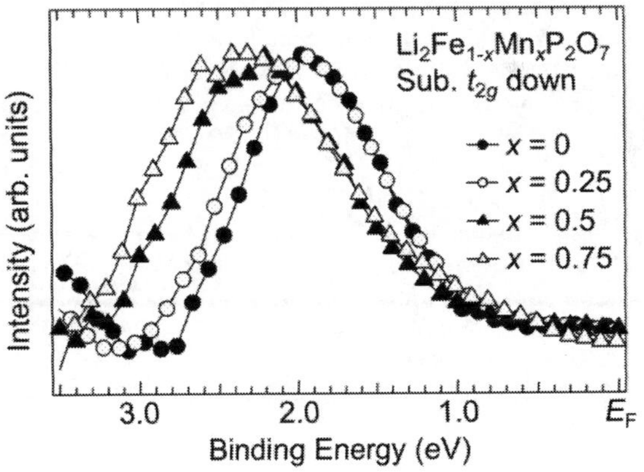

Figure 4. The extracted Fe 3d t_{2g} down-spin states obtained by subtracting the contribution from the other states from the RPES spectra for $Li_2Fe_{1-x}Mn_xP_2O_7$.

The expanded RPES spectra near E_F are shown in Fig. 3. We note that the Fe 3d t_{2g} down-spin states gradually shift to higher binding energy with increasing Mn concentration. In order to analyze the shift of the Fe 3d t_{2g} down-spin states in detail and quantitatively, we extract the Fe 3d t_{2g} down-spin states by subtracting the contribution from the other states, as shown in Fig. 4. The systematic peak shift and broadening accompanying with the Mn substitution is observed more clearly in the extracted Fe 3d t_{2g} down-spin states. Thus, we have determined that the center of the peak position at $x = 0$, 0.25, 0.5, and 0.75 is 1.9 eV, 2.0 eV, 2.3 eV, and 2.4 eV, respectively. The peak shift between 1.9 eV at $x = 0$ and 2.4 eV at $x = 0.75$ is matched very well to the difference of Fe^{3+}/Fe^{2+} redox potential between 3.5 eV for $Li_2FeP_2O_7$ and around 4 eV for $Li_2Fe_{1-x}Mn_xP_2O_7$ with high Mn concentration [4]. These results suggests that the origin of high Fe^{3+}/Fe^{2+} redox potential in $Li_2Fe_{1-x}Mn_xP_2O_7$ is the change in the electronic structures of Fe^{2+} partial DOS, namely the shift of the Fe 3d t_{2g} down-spin states to the higher binding energy with Mn substitution. Possible explanations for the Mn-induced change in the Fe 3d electronic structure are enhancements of Fe - O bond length and Fe ionicity resulting from the lattice volume expansion [8], and development of destabilization and higher Gibbs free energy due to strong electrostatic repulsion between Li^+ and Fe^{3+} (or, lithium vacancy and Mn^{2+}) in oxidized Fe^{3+} and unreacted Mn^{2+} surrounding the remaining Li^+ ions [9].

In conclusion, we have investigated the changes in the electronic structures, especially Fe partial DOS, of $Li_2Fe_{1-x}Mn_xP_2O_7$ with Mn substitution using XAS and RPES techniques. We have found the systematic shift to higher binding energy and broadening of Fe 3d t_{2g} down-spin states accompanying with the Mn substitution. The peak shift of the Fe 3d t_{2g} down-spin states is matched very well to the change of Fe^{3+}/Fe^{2+} redox potential, suggesting that the origin of high Fe^{3+}/Fe^{2+} redox potential in $Li_2Fe_{1-x}Mn_xP_2O_7$ is the shift of the Fe 3d t_{2g} down-spin states to the higher binding energy with Mn substitution.

Acknowledgments

This research is supported by the Japan Society for the Promotion of Science (JSPS) through its "Funding Program for World-Leading Innovative R&D on Science and Technology (FIRST Program)."

References

[1] Nishimura S, Nakamura M, Natsui R and Yamada A 2010 *J. Am. Chem. Soc.* **132** 13596
[2] Padhi A K, Nanjundaswamy K S and Goodenough J B 1997 *J. Electrochem. Soc.* **144** 1188
[3] Yamada A, Kudo Y and Liu K Y 2001 *J. Electrochem. Soc.* **148** A1153
[4] Furuta N, Nishimura S, Barpanda P and Yamada A 2012 *Chem. Mater.* **24** 1055
[5] Kurosumi S, Nagamura N, Toyoda S, Horiba K, Kumigashira H, Oshima M, Furutsuki S, Nishimura S, Yamada A and Mizuno N 2011 *J. Phys. Chem. C* **115** 25519
[6] Kurosumi S, Horiba K, Nagamura N, Toyoda S, Kumigashira H, Oshima M, Furutsuki S, Nishimura S, Yamada A and Mizuno N 2013 *J. Power Sources* **226** 42
[7] Augustsson A, Zhuang G V, Butorin S M, Osorio-Guillen J M, Dong C L, Ahuja R, Chang C L, Ross P N, Nordgren J and Guo J H 2005 *J. Chem. Phys.* **123** 184717
[8] Yamada A, Takei Y, Koizumi H, Sonoyama N, Kanno R, Itoh K, Yonemura T and Kamiyama T 2006 *Chem. Mater.* **18** 804
[9] Padhi A K, Nanjundaswamy K S, Masquelier C, Okada S and Goodenough J B 1997 *J. Electrochem. Soc.* **144** 1609

An automated HAXPES measurement system with user-friendly GUI for R4000-10 keV at BL46XU in SPring-8

H Oji[1,2], T Matsumoto[1,2], Y-T Cui[1], and J-Y Son[1,2]

[1] Japan Synchrotron Radiation Research Institute (JASRI), 1-1-1 Kouto, Sayo, Hyogo 679-5198, JAPAN
[2] SPring-8 Service Co., Ltd., 1-20-5 Kouto, Shingu, Tatsuno, Hyogo 679-5165, Japan

E-mail: oji-h@spring8.or.jp

Abstract. We have developed a noble HAXPES measurement system, where the controls for sample manipulation and VG Scienta R4000 work together. The main part of the system is developed by spec. R4000 is controlled by sending commands from spec to the TCP server programed by LabVIEW with SESWrapper library. The command line interface of spec is wrapped by a graphical user interface (GUI) which can be easily operated by users. The sample can be aligned in the intuitive way by clicking the position to be measured on the sample image. Fine alignment can be done by scanning the sample position to maximize the count rate of photoelectrons. After setting measurement positions and conditions on the GUI, the automatic measurements through the multiple measurement positions can be performed in this system.

1. Introduction

In the Engineering Science Research III beamline, BL46XU, of SPring-8, a hard X-ray photoemission spectroscopy (HAXPES) system equipped with VG-Scienta R4000-10 keV analyzer has been opened to users since 2008. Utilizing a large escape depth of photoelectrons excited by hard X-ray, the R4000 system has been served as a powerful tool for elucidating the electronic states deep inside the material, such as electrode/dielectric interfaces buried in gate stack structures and the electrodes in Li-ion batteries covered by solid electrolyte interfaces (SEI) layers with considerable thickness (several tens of nm). Nowadays, an increasing number of users are applying HAXPES analysis to solve their problems. Thus the improvement in the efficiency and the usability of the HAXPES system is highly needed.

The efficiency and the usability will be improved if the analyzer control and sample manipulation work together, since it makes the automatic measurements through different measurement positions possible. In the SpecsLab2 software for the SPECS Phoibos analyzer [1], the ObjectServer integrates the various controls including the sample manipulation through CORBA (Common Object Request Broker Architecture) interfaces. The ProCSA software for FOCUS HV-CSA analyzer [2] works as the TCP server to be controlled by other systems externally. Although the genuine program for R4000, SES, itself does not have such functions, the so-called SESWrapper library for LabVIEW of National Instruments distributed by VG Scienta [3] can be used to build the external control system for R4000.

In the conventional HAXPES system of BL46XU, R4000 was controlled by SES, and spec of Certified Scientific Software was employed for the sample position control. Since they were not linked together, the automatic measurements through multiple measurement positions and the scanning of a sample position with monitoring the count rate of the photoemission were impossible. To realize these, we have developed a novel HAXPES measurement system, in which the control for R4000 analyzer and that for sample manipulation are integrated. The main part of the system was coded by spec. An external control system for R4000 which can be controlled by spec via TCP/IP protocol was developed by using LabVIEW and the SESWrapper library. For improve the usability, the command line interface of spec is wrapped by the graphical user interface (GUI).

2. Instrumentation

2.1. Details of new HAXPES measurement system

Figure 1 shows the comparison between the conventional and the new HAXPES measurement systems.

In the conventional system (Figure 1(a)), users manipulate samples by executing commands at a terminal emulator (TeraTerm) on Windows PC connected to spec running on Linux PC. On the other hand, the setting of the kinetic energy, the monitoring of the count rate of photoemission, and the measurement of the spectra are done by SES. For the measurements, users have to do the following procedures for each measurement position. Firstly, they align the sample by spec monitoring the sample image from the CCD cameras mounted on the measurement chamber. Secondly, they adjust finely the sample position on spec so that the count-rate of photoemission from the sample monitored by SES maximizes. Finally, they set the measurement conditions and run the measurements on SES. Since spec and SES are independent of each other, the automatic measurement through multiple positions is impossible in the conventional system.

In order to integrate the control of R4000 and that of sample manipulation, we constructed new measurement system depicted in Figure 1(b). The main part of the system was coded by spec, which controls both sample position and R4000. In order to control R4000 from spec, a TCP server has been developed by using LabVIEW 8.6. We adopted this design because of the following reasons. (i) Development of the system by spec is simple and fast. In addition, most of the basic commands of spec to manipulate the sample in the conventional system can be also used in the new system as they are. (ii) The part for controlling R4000 is readily constructed by LabVIEW, for which the library to

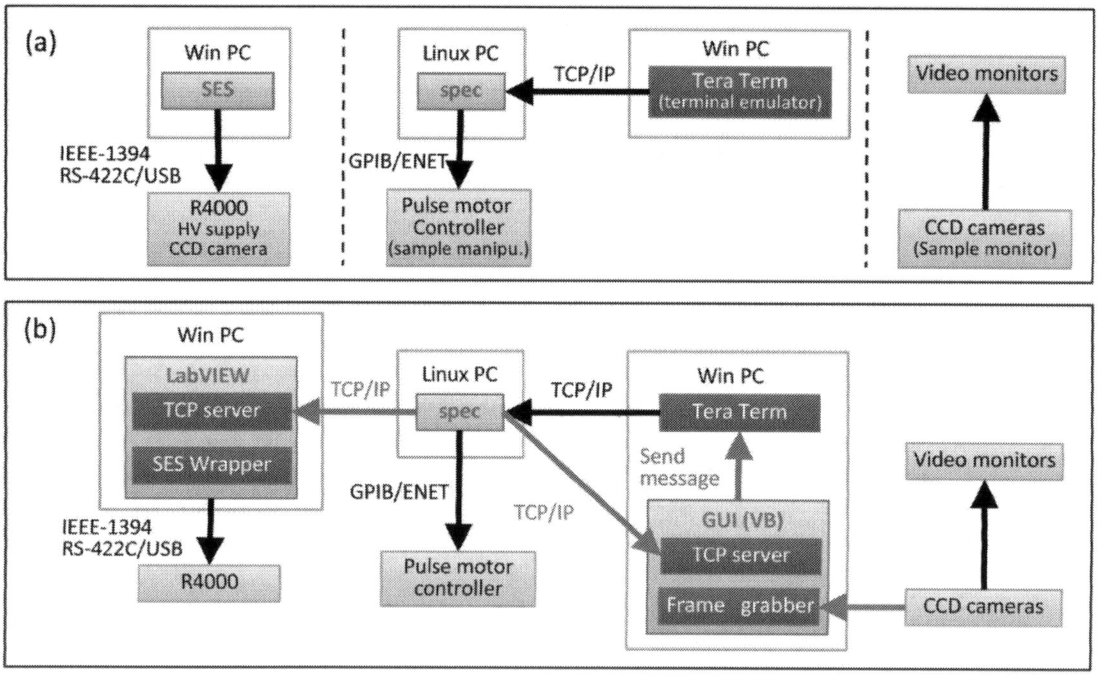

Figure 1. A schematic diagram illustrating the HAXPES measurement systems installed in BL46XU: (a) Conventional system. (b) New system.

control R4000 (SESWrapper) is distributed by VG Scienta [1].

The TCP server receives the commands to control R4000 from spec and returns values to it, if any, via TCP/IP. Thus the external control of R4000 by spec has become possible. The TCP server program also acts as the real time monitors for the detector image and the count rate, and the spectrum of an ongoing measurement (Figures 2(a) and (b), respectively). On the other hand, a number of spec commands necessary for the automatic measurements have been newly produced, *i.e.*, the commands for setting voltages of the electrodes, getting photoemission count, setting the measurement conditions, starting a measurement, saving the data, storing a sample position to be measured, etc. With the system described up to here, the automatic measurements are basically possible by making a macro file with properly arranged commands and loading it to spec.

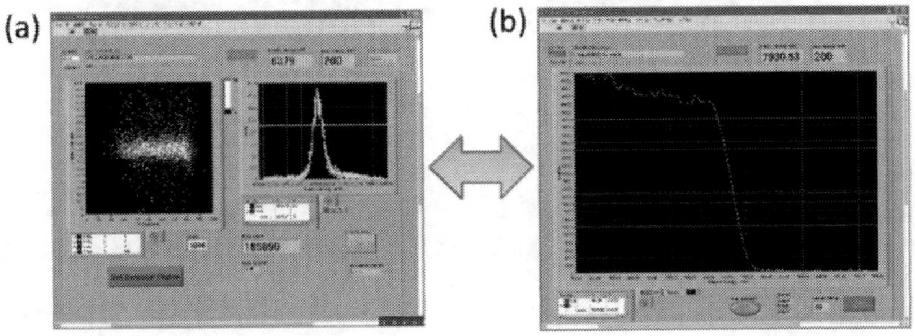

Figure 2. TCP server developed for the new system. (a) Detector image and count rate monitor mode. (b) Spectrum monitor mode.

For easier use of the system, we developed a GUI by Visual Basic 2010 of Microsoft. It throws spec commands to the terminal emulator according to the actions (events) made on it using the SendMessage function of Visual Basic. As shown in Figure 3, there are the sample alignment mode (a) and the measurement condition setting mode (b) in the GUI, which are switchable by the tabs in the upper part of the GUI.

In the sample alignment mode (Figure 3(a)), the real-time image of the sample from one of the CCD cameras (Cam-1 and Cam-2 in the Figure 3) is displayed at the center of the GUI. The images of the CCD cameras are captured into PC by the frame grabber board, DFG/MC4/PCIe, of The Imaging Source. Cam-1 is mounted 45° above the X-ray beam indicated by the yellow line in the image, and Cam-2 is on the lens axis of the analyzer indicated by the gray crosshairs in the image. Thus the images of Cam-1 and Cam-2 are used for the sample alignment with respect to the X-ray beam and analyzer lens axis, respectively.

The sample alignment is usually performed as follows. (i) The number of the face (1 – 6, since the sample holder has six faces for sample mounting) and a take-off angle are set to align the samples mounted on the designated face roughly to the measurement position. (ii) Then the sample is aligned on the sample image shown in the middle part of the GUI. By clicking a certain point on the image, the sample manipulator moves so that the clicked position on the sample aligns on the X-ray axis (the yellow line) when Cam-1 is selected, or on the axis of analyzer lens (the gray crosshairs) when Cam-2 is selected. (iii) Fine alignment is done by scanning the sample position with respect to the photoemission counts detected by R4000. (iv) After the alignment is finished, the position (x, y, z, θ coordinates of the manipulator) is stored with an arbitrary position name. By repeating the procedures (i) – (iv), users can store multiple measurement positions in the system.

After setting all the sample positions to be measured, users proceed to the measurement condition setting mode (Figure 3(b)). A measurement region with the information of the energy range, energy step, and dwell time can be created with an arbitrary region name by opening the dialog box for editing the regions. The conditions for a measurement, i.e., a sample position, a measurement region, and a number of sweeps is specified in a row of the table in the window. Sample positions and regions can be selected easily from the position and region names in the dropdown list which emerges by clicking a cell. The rows to be measured are checked at the second column of the table. When all the procedures described above are finished, an automatic measurement can be started by pressing the start button.

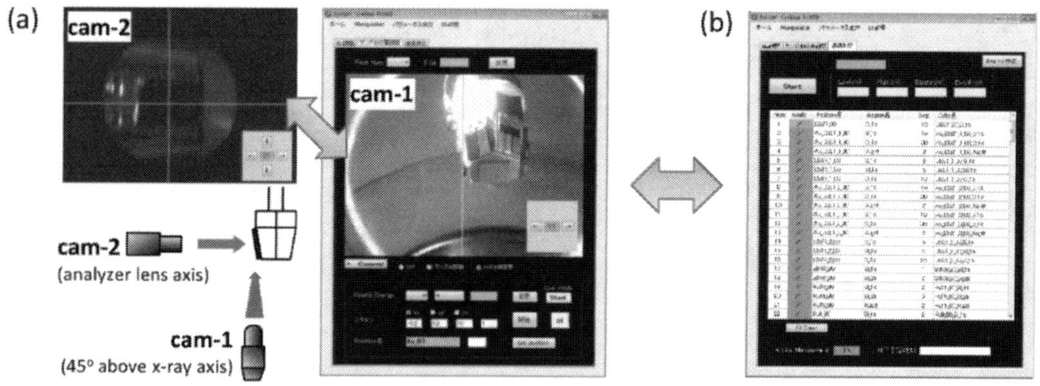

Figure 3. GUI for new HAXPES measurement system. (a) Sample alignment mode. (b) Measurement condition setting mode.

2.2. Characteristics of the new system

The new measurement system has been opened to users since Oct. 2012 and has been widely used by them. The system has several advantageous characteristics as described below.

First of all, automatic measurements through a lot of measurement positions are available. This feature is quite advantageous when users want to execute the measurements of many samples and/or at many take-off angles in a routine way. In some cases, the automatic measurements for several hours or even longer are possible. This feature is guaranteed by the beam stability of BL46XU. However there is a concern about the beam abort of the storage ring. We cope with it by checking the ring current during the measurement. If the ring current falls below the critical value, the system stops a measurement and sends the voice mail alerting the failure to the PHSs for users.

Secondly, the system is easy to operate. The design of GUI is simple and easy to use even for unaccustomed users. Especially the sample alignment system using the frame grabber board made us possible to align samples intuitively. Note that the concept of the sample alignment with frame-grabber board may be applicable to other experiment systems.

Thirdly, the system is flexible and easy to customize. In some customized experiments, the automatic measurement using the GUI will not be possible. It will take long time to develop sophisticated GUIs for customized experiments. However, the base system developed by spec is easy to customize. For example, we utilized this property to the bias applied HAXPES measurements. Firstly, we made the spec macro to change the voltage of the DC power source (ADC 6240A) via GPIB. This is easy, since spec has GPIB interface and command-sets to control GPIB instruments. Then we created a macro file where the commands to change the sample position, to measure spectrum, and to change of bias voltage are simply arranged. Finally the file was loaded to spec, and the automatic measurements defined in the file was executed. Thus the new system is advantageous not only for routine experiments but also for customized ones.

Acknowledgements

We would like to thank Dr. T. Sugiyama of Kyusyu University for his kind advice on the programing using the SESWrapper library. We are also grateful to Dr. I. Hirosawa of JASRI for the careful reading of the manuscript and helpful suggestions. The system was developed in the beam time approved by JASRI (Proposal Nos. 2011B2088 and 2012B1464).

References

[1] http://www.bessy.de/rglab/doc/Manual-SpecsLab2.pdf
[2] http://www.focus-gmbh.com/page2/page48/page8/page20/page20.html
[3] http://www.vgscienta.com/productlist.aspx?MID=287.

A HAXPES measurement system up to 15 keV developed at BL46XU of SPring-8

H Oji[1,2], Y-T Cui[1], T Koganezawa[1], N Isomura[3], K Dohmae[3] and J-Y Son[1,2]

[1] Japan Synchrotron Radiation Research Institute (JASRI), 1-1-1 Kouto, Sayo, Hyogo 679-5198, Japan
[2] SPring-8 Service Co., Ltd., 1-20-5 Kouto, Shingu, Tatsuno, Hyogo 679-5165, Japan
[3] Toyota Central R&D Labs., Inc., 41-1, Yokomichi, Nagakute, Aichi 480-1192, Japan

E-mail: oji-h@spring8.or.jp

Abstract. In order to achieve much larger probing depth than the conventional HAXPES system of BL46XU, a HAXPES measurement system equipped with a cylindrical sector analyzer, Focus HV-CSA 300/15 has been developed, by which photoelectrons with the kinetic energy up to 15 keV can be analyzed. The Si 1s peak which comes from the buried Si wafer underneath the 60 nm SiO_2 thin films can be clearly identified in the spectra excited by the photon energy of 14 keV, indicating the much larger probing depth than the conventional HAXPES measurement with 8 keV X-ray. The total energy resolution estimated from the Au Fermi edge spectra was ~0.5 eV, which is sufficient for the chemical state analysis of materials.

1. Introduction

In BL46XU of SPring-8, a hard X-ray photoemission spectroscopy (HAXPES) measurement system equipped with an electron energy analyzer, VG Scienta R4000-10 keV, has been installed since 2008, by which photoelectrons up to 10 keV can be analyzed [1]. Since then, the system has been opened to industrial researchers and has served them as a powerful tool for directly exploring the electronic structure deep inside the material, such as electrode/dielectric interfaces buried in gate stack structures, which are not accessible by conventional XPS.

To achieve much larger probing depth than R4000, we developed a new HAXPES measurement system equipped with Focus HV-CSA 300/15 analyzer [2], by which the electron kinetic energy up to 15 keV can be analyzed. There are only a few literatures reporting the HAXPES spectra excited by photon energy more than 10 keV [2]. In the present paper, the details of the system and the HAXPES spectra excited by 14 keV with the energy resolution sufficient for practical analysis are reported.

2. Instrumentation

The system is illustrated in Figure 1. It consists of main and sub chambers. The four-axis (x, y, z, θ) sample manipulator, the electron analyzer, the neutralizer (PHI 04-091), and the electron gun (RDEC RDA-003G) are installed in the main chamber. The incident X-ray is introduced from the beryllium window (^{t}100 nm) attached at the front of the main chamber. A large-sized sample holder (78 mm × 28 mm × ^{t}2 mm) has adopted to load large number of samples at once (Figure 1(b)). From a large quick access window, the samples on the sample holder are loaded to the sub chamber and are transferred to the main chamber for the measurement. Three CCD cameras are attached for monitoring the samples.

The HV-CSA 300/15 analyzer consists of a cylindrical sector field analyzer, a pre-retarding lens system, and a 2-D event-counting detector equipped with a multi-channel plate (MCP), a phosphor screen, and a CMOS camera. The entrance slit of the analyzer has a rectangular shape with the three selectable sizes in the energy dispersive direction (4.5 mm, 1.5 mm, and 0.5 mm) and the fixed size (12 mm) in the non-dispersive direction. The lens mode can be selected from the magnification lists from 3 to 60, stepwise. The pass energies are continuously changeable from 1 to 500 eV. The HV-CSA has the advantage of increasing luminosity by the second order focusing property along the dispersive direction [2]. In addition, it has much less volume and weight than the other commercial hemispherical HAXPES analyzers.

The manipulator and the base frame can be controlled by spec programs of Scientific Certified Software through the pulse step motors. The analyzer is controlled by the genuine software ProCSA supplied by Focus, which can act as a TCP server for the external control by other systems. Using this function, several scan commands of spec were developed for the sample alignment easily.

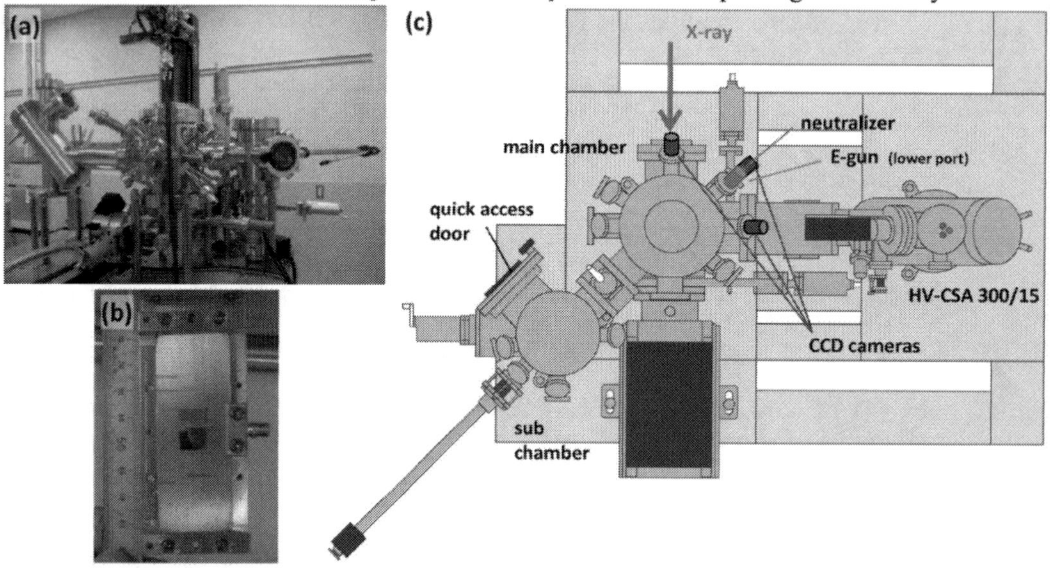

Figure 1. The HAXPES measurement system with Focus HV-CSA 300/15 analyzer up to 15 keV developed in this study. (a) Photograph of the system. (b) Sample holder. (c) Illustration of overall system from top view.

3. Experimental

The experiments were performed at the undulator beamlines, BL46XU and BL33XU, of SPring-8.

In BL46XU [1], the white light from the undulator is monochromatized by a Si double crystal monochromator (DCM). The X-ray beam was horizontally focused by Rh-coated 700-mm-long double bend mirrors with their incident angles set to 3.15 mrad. The size of the front-end slit was set to 0.7 mm (horizontal) × 0.7 mm (vertical), and the other slits were set wide enough to avoid cutting the X-ray beam. The beam size at the sample was ~0.25 mm (h) × ~0.5 mm (v).

In BL33XU [3], the white light is monochromatized by a Si channel-cut monochromator (CCM), and was horizontally and vertically focused by Rh-coated 1000-mm-long double upstream mirrors (M1, M2) and 700-mm-long double downstream mirrors (M3, M4), respectively. The incidence angles of the mirrors are set to 1.5 mrad for M1, M2 and 3.5 mrad for M3, M4. The size of the front-end slit was set to 0.5 mm (h) × 0.5 mm (v), and the other slits are set so that the beam size at the sample was shaped into ~0.8 mm (h) × ~0.15 mm (v).

In the both beamlines, Si(111) and Si(333) reflections were used to obtain monochromatized X-ray of 14 keV for the excitation of photoelectrons. The voltages for screen and MCP of the analyzer were set to 4500 V and 2600 V, respectively. The base pressure of the main chamber was ~ 2×10^{-8} Pa.

Journal of Physics: Conference Series **502** (2014) 012006 doi:10.1088/1742-6596/502/1/012006

4. Results and Discussion

Figure 2 shows the Si 1s spectra of Si(100) substrate excited by 14 keV at different analyzer slit sizes, pass energies, and Si reflection plane for the monochromator. The peak widths by means of full width at half maximum (FWHM) are indicated in the inset. The peak width of Si 1s excited by X-ray of Si(333) DCM is considerably narrower than that of Si(111) DCM, indicating the difference in the X-ray energy resolution between the two reflection planes (theoretical energy resolution, $\Delta E/E$, for Si(111) and Si(333) are 1.1×10^{-4} and 8.5×10^{-6}, respectively [4]). The Si 1s peak width becomes smaller as the slit width gets narrower, indicating the energy resolution of the analyzer increases by narrowing down the slit width. The width is narrowest (0.95 eV in FWHM) at the slit size of 0.5 mm × 12 mm and the pass energy of 50 eV, which is comparable to the results measured by R4000 with 8 keV X-ray (0.87 eV, not shown in this report).

Figure 3 shows the Si 1s spectra of 60 nm SiO_2 thin film formed on Si(111) substrate by thermal oxidization measured with 14 keV X-ray at different monochromator reflection planes, slit sizes, and take-off angles (TOAs). The Si 1s peak from the bulk-Si under the 60 nm-thick SiO_2 layer overlaps with the Si-oxide component in the spectrum with X-ray monochromatized by Si(111) DCM, while the bulk peak is clearly distinguished at ~12173 eV and well separated from the oxide component in that with X-ray monochromatized by Si(333) DCM. Thus the usage of Si(333) DCM reflection for 14 keV X-ray not only realizes the much deep probing depth, but also increases the energy resolution of X-ray and thus the total energy resolution of the measurement. At TOA = 30°, the peak of the bulk-Si is not clearly distinguished, since the probing depth at TOA = 30° is about a half of that at TOA = 80°.

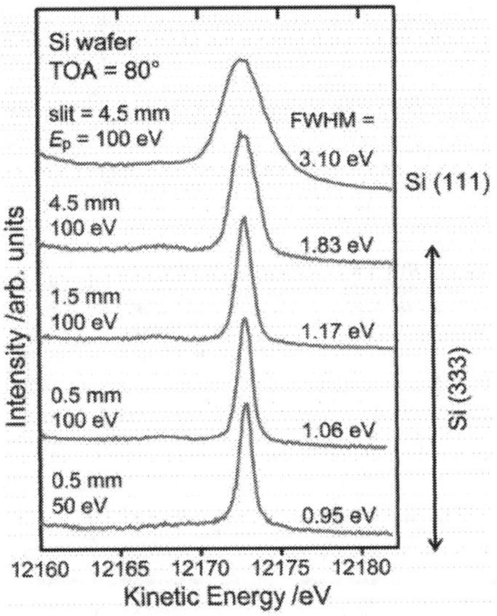

Figure 2. Si 1s spectra of Si wafer taken at $h\nu = 14$ keV as a function of the reflection planes of monochromator, analyzer slit widths, and pass energies.

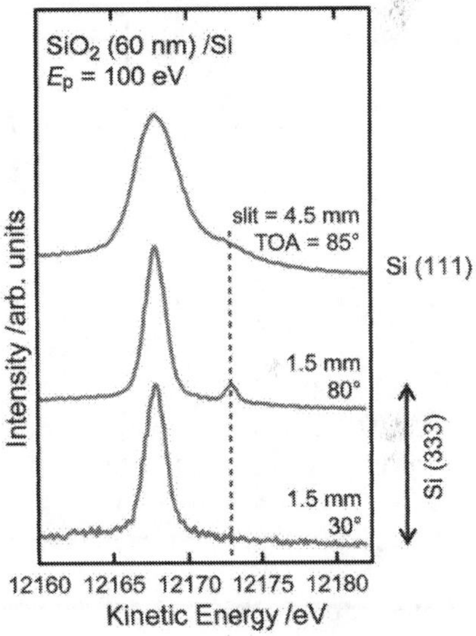

Figure 3. Si 1s spectra of SiO_2 (60 nm) / Si at $h\nu = 14$ keV as a function of the reflection planes of monochromator, analyzer slit widths, and TOAs.

The HAXPES spectra of 100 nm Au film sputtered on Si substrate by $h\nu = 14$ keV are shown in Figure 4. Unlike the conventional XPS, the relative intensity of Au 4f peaks are significantly low compared to Au 5s and 5p peaks, because the photoionization cross-section for the state with higher angular momentum quantum number decreases much faster than those with lower one as the excitation energy increases. The doublet peaks of Au 4f are well resolved, as shown in Figures 4 (a) and (b), where the peaks become much sharper as the slit width becomes narrower. Rubio-Zuazo *et al.*, reported the HAXPES spectra of Au (21 nm) on Cu polycrystal with $h\nu = 15$ keV, while 4f peak are not well resolved in their data, due to the worse total resolution by Si(111) DCM reflection [2]. As shown in the Figure 4(c), one can see the Fermi edge and detailed valence band structures clearly in the spectrum. The total energy resolution estimated by fitting the Fermi-Dirac distribution function (the solid line in the figure) to the experimental curve is 0.50 eV.

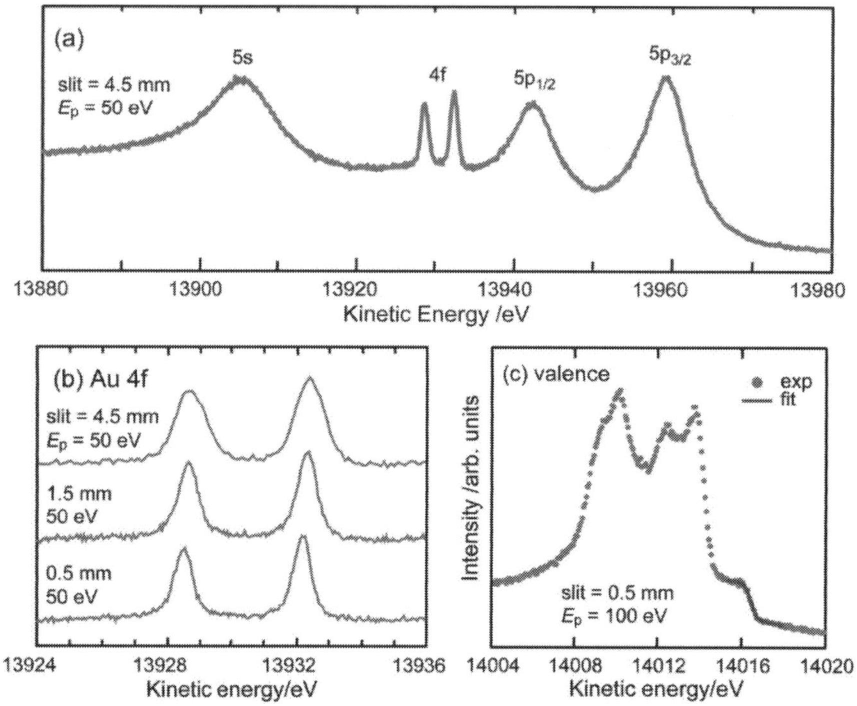

Figure 4. The HAXPES spectra of Au (100 nm) / Si taken at $h\nu = 14$ keV. (a) Survey spectrum from Au 5s to 5p regions. (b) Au 4f region. (c) Valence band spectrum. Experimental spectrum (filled red circles) is fitted with the Fermi-Dirac distribution function (solid blue line).

5. Conclusion

The HAXPES measurement system equipped with Focus HV-CSA analyzer with the electron kinetic energy up to 15 keV has been developed. With the X-ray monochromatized by Si(333) reflection and the proper setup of the analyzer, the energy resolution sufficient for the chemical state analysis has been achieved. Furthermore, the large probing depth of the system has been confirmed.

Acknowledgements

The authors would like to express our thanks to Drs. Y. Hayashi, T. Nonaka, and Y. Hirose of Toyota Central R&D Laboratory for their kind help during the experiments at BL33XU. We also thank Mr. N. Shimizu of Omicron Nanotechnology Japan for participating in the starting up experiments and helping us a lot. The system was developed in the beam time approved by JASRI (Proposal Nos. 2011A1746, 2011A1850, 2011B1971, 2012A1340, 2012A7019, 2012B7019, and 2013A1819).

References

[1] http://www.spring8.or.jp/wkg/BL46XU/instrument/lang-en/INS-0000001484/instrument_summary_view
[2] Rubio-Zuazo J, Escher M, Merkel M and Castro G R 2010 *Rev. Sci. Instr.* **81** 043304
[3] Nonaka T, Dohmae K, Araki T, Hayashi Y, Hirose Y, Uruga T, Yamazaki H, Mochizuki T, Tanida H and Goto S 2012 *Rev. Sci. Instr.* **83** 083112
[4] Ishikawa T, Tamasaku K and Yabashi M 2005 *Nucl. Instr. Method. Phys. Res.* A **547** 42

Investigation of group IVA elements combined with HAXPES and first-principles calculations

Y-T Cui[1,*], G-L Li[2], H Oji[1,3], J-Y Son[1,3]

[1] Industrial Application Division, JASRI / Spring-8, 1-1-1 Kouto, Sayo-cho, Hyogo 679-5198, Japan
[2] School of Physics and Engineering, Henan University of Science and Technology, Luoyang 471023, China
[3] SPring-8 Service Co., Ltd., 1-20-5 Kouto, Shingu, Tatsuno, Hyogo 679-5165, Japan

*E-mail: yitaocui@spring8.or.jp

Abstract. The core level and valence band spectra of group IVA elements were investigated with hard x-ray photoemission spectroscopy (HAXPES) photon energy of 7.939 keV by bulk sensitive manner. The survey and valance band spectra were presented, relative peaks intensity are discussed by thinking about inelastic mean free path (IMFP) and photoionization cross section of photoelectrons (PICS). In order to understand bulk band structures, valence bands are compared with the calculated ones by considering PICS, IMFP and total energy resolution. The calculated results by GGA, HSE06 and GW0 methods are simply discussed by comparing with experiment spectra.

1. Introduction

The group IVA elements of the periodic table (C, Si, Ge, Sn and Pb) were shown great interest in past years not only due to the important applications on semiconductor industries, but also their crystal structure changes against pressures and atomic numbers. Such as in group IVA above Sn, the elements C, Si, and Ge tend to form a diamond structure with strong covalent bonds, while below Sn the stable phase for Pb is metallic and fcc. It turns out that Sn is border line between covalent and metallic bonding, below 286 K, Sn crystallizes into a grey material with a diamond cubic structure and with a zero band gap and a metallic and tetragonal crystal structure (β-Sn) above this temperature. Even though, there are so many works have been done about the investigation of electronic states by means of photoemission spectroscopy (PES). But due to the much surface sensitive of UPS and XPS, the electronic states for true bulk information are still untouched so far. The newly developed high-energy, high-resolution HAXPES at third generation synchrotron light sources, improves greatly the photoelectron IMFP (~3 nm for lab XPS at $h\nu \sim 1486$ eV, and 10 ~ 20 nm for HAXPES at $h\nu \sim 7.939$ keV) [1]. It indicates that HAXPES can be used to detect the bulk information, interface or buried phases, even for the samples without surface treatment [2], which usually has to be done for XPS or UPS methods.

In this work, by combination of HAXPES and first-principles calculations, the bulk electronic states of group IVA elements (except C) were investigated to deepen our understanding of real bulk information for these elements and their compounds.

2. Experimental and method of calculation

Hard x-ray photoemission measurements were carried out at the undulator beamline BL46XU of SPring-8. The excitation energy used for HAXPES was monochromatized with Si (111) double crystal and Si (444) channel-cut monochromators, which provides an x-ray energy resolution of 40 meV at around 7939 eV, a focal spot size of 0.02 mm (vertical) × 0.2 mm (horizontal), and a photon flux of 2×10^{11} photons/s. A VG Scienta R4000 electron analyser for high electron kinetic energies of up to 10 keV was used. The electrons emitted at direction of 80 degree respect to the sample surface were detected with an acceptance angle of $\pm8^{\circ}$. The curved slit with the width of 0.5 mm is used and pass energy is fixed to 200 eV for sufficient counting rates, which result in a total energy resolution of 230 meV at room temperature evaluated by Au Fermi edge. The experiment was carried out at room temperature with the vacuum better than $\sim5\times10^{-6}$ Pa in the analyser chamber. The single crystal samples for Si and Ge, as well as high purity polycrystalline samples for Sn and Pb are used. Si samples (111 and 100 plane, n and p type doping) were measured without surface cleaning. Ge, Sn and Pb samples were cleaved or scraped to get clean surface before measurements.

The calculations are done using the projector-augmented wave method as implemented in the Vienna *Ab Initio* Simulation Package (VASP) [3, 4] with the Perdew, Bourke and Ernzerhof (PBE) generalized gradient approximation (GGA) exchange-correlation functional [5]. For the hybrid functional calculations, the Heyd-Scuseria-Ernzerhof (HSE06) hybrid functional is used [6]. For the GW0 calculations [7], the initial wave functions and eigenvalues in G were determined using the HSE06 functional. Enough bands have to be included, and for the update of G and W, four steps are necessary to find band gap energies converged up to 10^{-3} eV. The valence configurations are treated as s^2p^2 for Si, Ge, Sn, and Pb. The cut-off energy for plane-wave basis functions is 400 eV. The experimental structural parameters are taken for the elemental materials. For all the calculations, energy convergence and k-point grids are carefully checked.

For our analyser set-up, the photoelectron intensity can be described as (after refs. [8, 9]):

$$= F_x\frac{\sigma_i}{4\pi}[1+\beta_i]\Delta\Omega A_0 DTf\rho\lambda_e\sin\theta \quad (1)$$

Where, F_x is the average flux of x-ray on the sample, $\Delta\Omega$ is the acceptance solid angle of the analyzer, A_0 is the area of the sample irradiated by x-rays that is viewed by the analyzer for photoemission normal to the surface, D is the efficiency of the detector, T is the transmission function of analyzer, f is the fraction of photoemitted electrons that appears in the main or no loss peak, ρ is the total atomic density of the sample, $\lambda_e$is the IMFP for the photoelectrons in the sample with certain kinetic energy, σ_i is the photoionization cross section [10], β_i is the dipole angular distribution [10]; i is the index of subshell, θ is the mean emission direction of photoelectron with respect to the sample surface. Eq. (1), will be used for both the quantitative analysis of core levels and the comparison of theoretically calculated results and experimental ones.

3. Results and discussion

The survey spectra of Si, Ge, Sn, and Pb samples were presented in figure 1. Strong plasmon peaks can be found in Si, Ge, Sn samples, and relative weak one in Pb sample. The inset of figure 1 shows the comparison of the standard spectra of lab XPS (by Al Kα monochromatic x-ray) and that of HAXPES for Sn metal sample. The relative peak intensity is much different for lab XPS and HAXPES due to the change of PICS against excitation photon energy [10] as well as variation of IMFP with kinetic energy of phtoelectrons [1], since the photoelectron intensity is strongly dependent with PICS and IMFP with respect to Eq. (1). For example, the intensity of Sn 3d is much larger than that of 3p for spectra by Al Kα, while the intensity is opposite for HAXPES ($h\nu$~7939 eV). Also, some deep core levels such as Si 1s, Sn 2p, and Pb 3d, etc. can be probed by HAXPES. Additionally, in HAXPES spectra, the Sn MNN Auger spectra move out of the low binding energy regions where they appear in lab XPS, owing to the changes of photoelectrons in kinetic energy.

The O and C 1s core levels are checked for all samples, their contribution is too weak to be seen except Si sample, since Si sample is measured without surface cleaning, while other samples are cleaved or scraped to get clean surface before measurements. It is confirmed that Si valance band spectra obtained by x-ray of 7939 eV show negligible difference with and without thin natural oxidized layer, due to the small contribution (<5%, by fitting the Si 2p core levels), which can also be easily understood since the thickness of natural oxidized layer (~1nm) is so thin comparing with the total probing depth of HAXPES (~39 nm) [11].

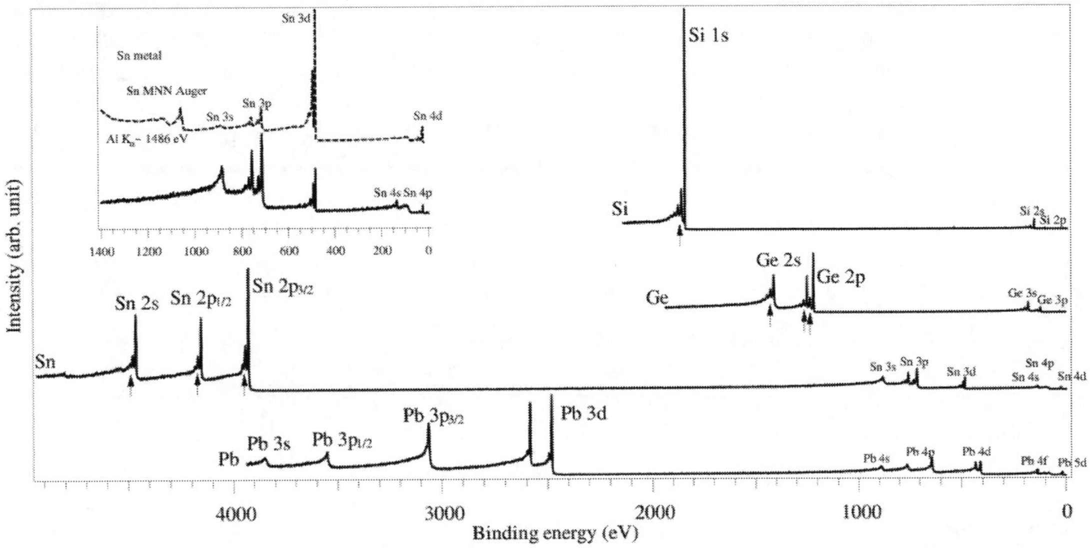

Figure 1 Survey spectra of group IVA elements with hard x-ray (*hv* ~ 7939 eV). The inset shows the comparison of survey spectra of Sn metal with Al Kα (dotted line) and hard (solid line) x-rays. Arrows mark the position of plasmons.

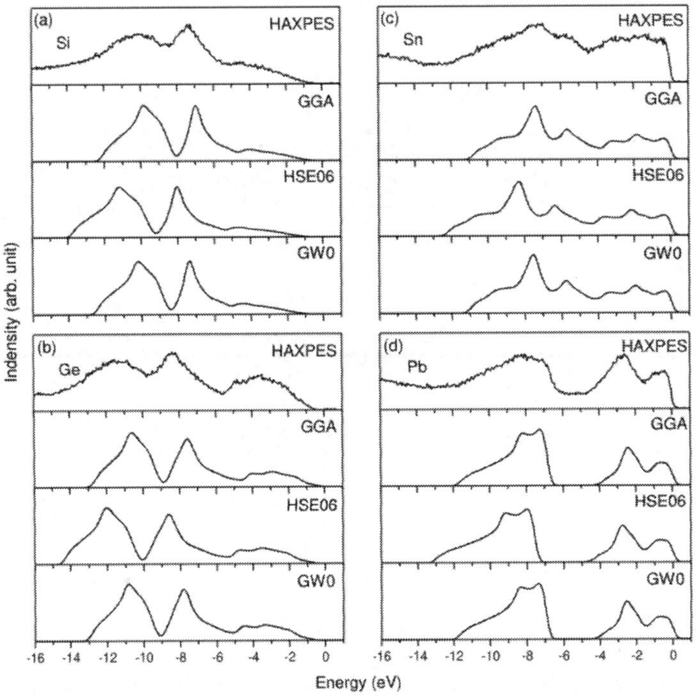

Figure 2 The valence band structures and the calculated density of states after consideration of photoionization cross section, Fermi-Dirac distribution and convolution with the experimental total energy resolution determined Gaussian function.

The valence band structures of Si, Ge, Sn and Pb measured with HAXPES at room temperature are shown in Figure 2. The occupied states of Si and Ge show 3 dominated structures due to the strong sp3 hybridization in diamond crystal structure. Sn, and Pb show different electronic structures, since Sn shows the tetragonal crystal structure (β-Sn) and Pb is a fcc crystal structure at room temperature. In order to understand the experimental structures, the calculated spectra by GGA, HSE06 and GW0 were presented in Figure 2 after consideration of photoionization cross section (basing on Eq. (1)), Fermi-Dirac distribution and convolution with the experimental total energy resolution determined Gaussian function. The calculated band gaps and occupied band widths are list in Table 1. The different calculation methods give the same band shapes and relative intensities except some differences in some peak positions in other words band gaps and band widths. By comparing the experimental and calculated results, one finds that the GGA and GW0 give proper band widths for all of samples and show nearly the same peak position as experimental ones, while the HSE06 provides the relative large band widths than that of experiment. If considering the bang gaps of Si, Ge and α-Sn (1.11 eV, 0.67 eV and 0.08 eV respectively) from experiments [12, 13], the HSE06 and GW0 provide the reasonable band gaps value relative to that of experiment, while the GGA seems under evaluation of the band gap values. Basing on the discussion above, the GW0 is the best method to give proper band structures and band gap of group IVA elements. If just the band width is cared, the GGA can give reasonable results for much reduction of calculation time comparing with that of GW0.

Table 1 The calculated band gaps E_g (occupied band widths E_w) of Si, Ge, Sn, Pb (in eV)

	Si	Ge	α-Sn	β-Sn	Pb
GGA	0.58 (11.96)	0.07 (12.78)	0.00 (10.64)	(11.15)	(11.80)
HSE06	1.21 (13.28)	0.75 (14.07)	0.01 (11.79)	(12.54)	(12.54)
GW0	1.28 (11.91)	0.80 (12.62)	0.05 (11.02)	(11.30)	(11.30)

4. Conclusions

In conclusion, the group IVA elements Si, Ge, Sn, Pb were investigated with HAXPES by bulk sensitive manner. The relative peaks intensity is discussed by thinking about IMFP and PICS, which will provide the principles to quantitative analysis of these elements with HAXPES. Calculated valence bands are combined well with the experimental ones by considering PICS, IMFP and total energy resolution. Amount of calculated results, GW0 method is the best one by comparing with experiment spectra and band gaps.

5. Acknowledgements

The experiments were performed at the BL46XU of SPring-8 (Proposal Nos. 2011B2085, 2012A1358, 2012A1749, 2012B1244 and 2013A1400). The calculations were financially supported by Henan University of Science and Technology (Nos. 09001638, 13000908) and National Natural Science Foundation of China (Nos.11204066, 11147166). The computational resources were provided by Shanghai Supercomputer Centre.

References

[1] Tanuma S, et al., Surf. Interf. Anal., **21**, 165 (1994).
[2] Kobayashi K, et al., Appl. Phys. Lett. **83**, 1005 (2003).
[3] Blochl P E, Phys. Rev. B **50**, 17953 (1994).
[4] Kresse G, et al., Phys. Rev. B **47**, R558 (1993) and **54**, 11169 (1996); Comput. Mater. Sci. **6**, 15 (1996).
[5] Perdew J P, et al., Phys. Rev. Lett. **77**, 3865 (1996).
[6] Krukau A V, et al., J. Chem. Phys. **125**, 224106 (2006).
[7] Stroppa A, et al., Phys. Rev. B **79**, 201201(R) (2009).
[8] Powell C J, et al. J. Electron Spectronsc. Relat. Phenom., **178-179**, 331 (2010).
[9] Kunz C, et al., Nucl. Instrum. Method. Phys. Res. A **547**, 257 (2005).
[10] Yeh J J, et al., Atom. Data and Nucl. Data Tables, **32**, 1 (1985); Trzhaskovskaya M, et al., Atom. Data Nucl. Data Tables,**77**, 97, (2001); **82**, 257 (2002); **92**, 345 (2006)
[11] Here, the probing depth (PD) is defined as 95% of the detacted signal originates, where PD ~ $3\lambda_e$. For $h\nu$ ~ 7939 eV, λ_e~ 13nm [1], $3\lambda_e$~ 39nm.
[12] Streetman B G, et al., Solid State electronic Devices (5th ed.). New Jersey: Prentice Hall (2000).
[13] Heine V, et al., J. Phys. C, **2**, 719 (1969).

Micromanipulation and Pick-Up System for X-Ray Diffraction Characterization of Micrometer-Sized Single Particles

Y Takeichi[1], N Inami[1], T Ueno[2], K Saito[1], H Otori[1], R Sagayama[1], R Kumai[1] and K Ono[1]

[1] Photon Factory, Institute of Materials Structure Science, High Energy Accelerator Research Organization, 1-1 Oho, Tsukuba, Ibaraki 305-0801, Japan
[2] National Institute for Material Science, 1-2-1 Sengen, Tsukuba, Ibaraki 305-0047, Japan

E-mail: yasuo.takeichi@kek.jp

Abstract. We describe a micromanipulation and pick-up system for preparing a micrometer-sized single particle for X-ray diffraction characterization. Combining a microgripper based on microelectromechanical systems, piezo-motor-driven linear stages, and a gamepad, the system provides precise and intuitive handling of the object. Single-crystal X-ray diffraction measurements of Sm-Fe-N permanent magnet were performed using this system. We also describe a method to distinguish crystallographically homogeneous particles found in powder-form samples.

1. Introduction

Powder X-ray diffraction measurement using synchrotron radiation provides fruitful information about the crystal structure and chemical composition of materials [1]. Rietveld analysis of the obtained diffraction pattern is the common method to determine lattice structures of the samples. In some cases, however, the Rietveld analysis does not converge well; for example, the Debye-Scherrer rings become discontinuous and spot-like if the number of particles included in the investigated volume is insufficient. It becomes more difficult to obtain a well converged structure when the sample material has a preferred orientation. Powder-form ferromagnetic samples with strong magnetic anisotropy are a case in point. To deduce the lattice structure of those materials, single-crystal X-ray diffraction measurement is necessary.

When the particle size is of 100 μm or more, one can separate and prepare a single sample specimen for measurement by hand using micro-handling tools [2] while viewing them with a stereoscopic microscope. For smaller particles, however, direct manipulation by hand is no longer applicable and micromanipulation requires the help of electronic devices.

Recently, the commercial use of micrometer-sized fabrication and motion technology has been widely extended; microelectromechanical systems (MEMS) are available for sample handling, as well as piezo-motor-driven stages for precise motion of the tool and/or the sample. Today, they are commercially available at reasonable prices. We have combined some devices to develop a micromanipulation and pick-up system. For use in routine X-ray diffraction measurements, easy and intuitive control of the system is desirable.

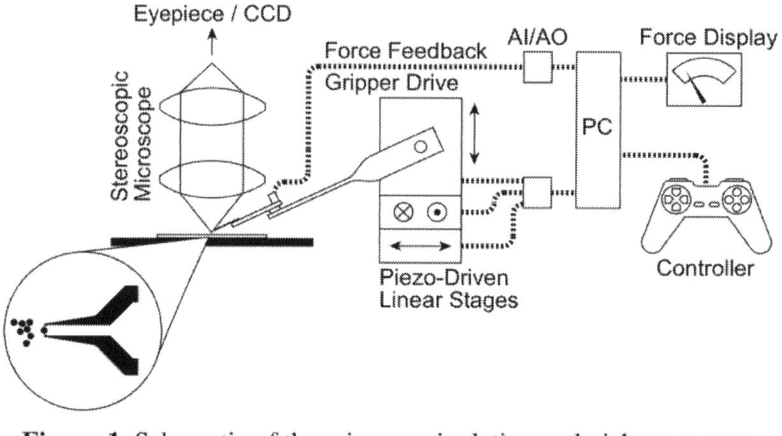

Figure 1. Schematic of the micromanipulation and pick-up system.

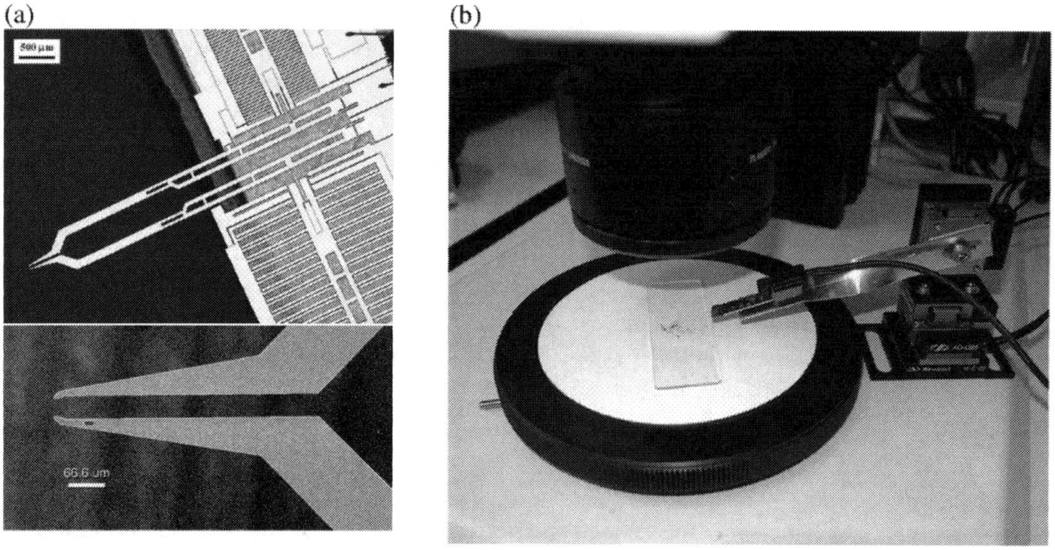

Figure 2. (a) The upper panel shows a microscopic image of the microgripper FT-G32; The scale bar is 500 μm. The lower panel shows a SEM image of the gripper tip; the scale bar is 66.6 μm. (b) Photograph of the micromanipulation and pick-up system. The sample powder is dispersed on the microscope slide.

2. Micromanipulation and pick-up system

Figure 1 shows a schematic of our micromanipulation and pick-up system. MEMS-based microgrippers (FemtoTools FT-G series [3]) are used for sample handling. The microgripper can grasp a particle up to 30 μm (FT-G32, see figure 2(a)) or up to 100 μm (FT-G102). Both the models of the gripper are equipped with gripping force sensors. The gripper is mounted on an XYZ stage composed of Newport Agilis piezo-motor-driven linear stages. They are driven at a wide range of speeds, up to 0.5 mm/s and down to incremental sub-micrometer motions. The operating program, based on National Instruments LabVIEW, controls the analog in/out for the gripper drive and force sensor, and the XYZ stage via a USB connection. All motions of the gripper and the stages are intuitively controlled using a commercial gamepad. The motion of the gripper is monitored with a stereoscopic microscope (Leica M205 C) with magnification of up to 320 times. The microscope image is also monitored with a CCD camera connected to a PC via a Gigabit Ethernet.

To pick up a single particle from the powder-form sample, the powder should be first dispersed on a microscopic slide, as seen in figure 2(b). An organic solvent is often useful in sparsely dispersing dense powder. Next the gripper is slid carefully up to the object particle by moving the XYZ stage. Steric observation through the eyepiece is essential to control the gripper precisely. After gripping the particle, a constant-force gripping mode is available using force feedback control. This helps in holding soft or fragile materials such as biological cells. Finally, the particle can be mounted on polymer MicroMounts [2] or onto the tip of a glass needle for X-ray diffraction measurements.

3. X-ray diffraction measurement

A single-particle X-ray diffraction measurement of powdered samples was performed using the system described above. The sample was a Sm-Fe-N rare-earth permanent magnet [4] which has strong uniaxial magnetic anisotropy. First, a powder X-ray diffraction measurement of this sample was performed. However, the Rietveld analysis of the obtained two-theta diffraction patterns (not shown here) did not converge well, possibly owing to the preferred orientation caused by magnetostatic interactions between the particles.

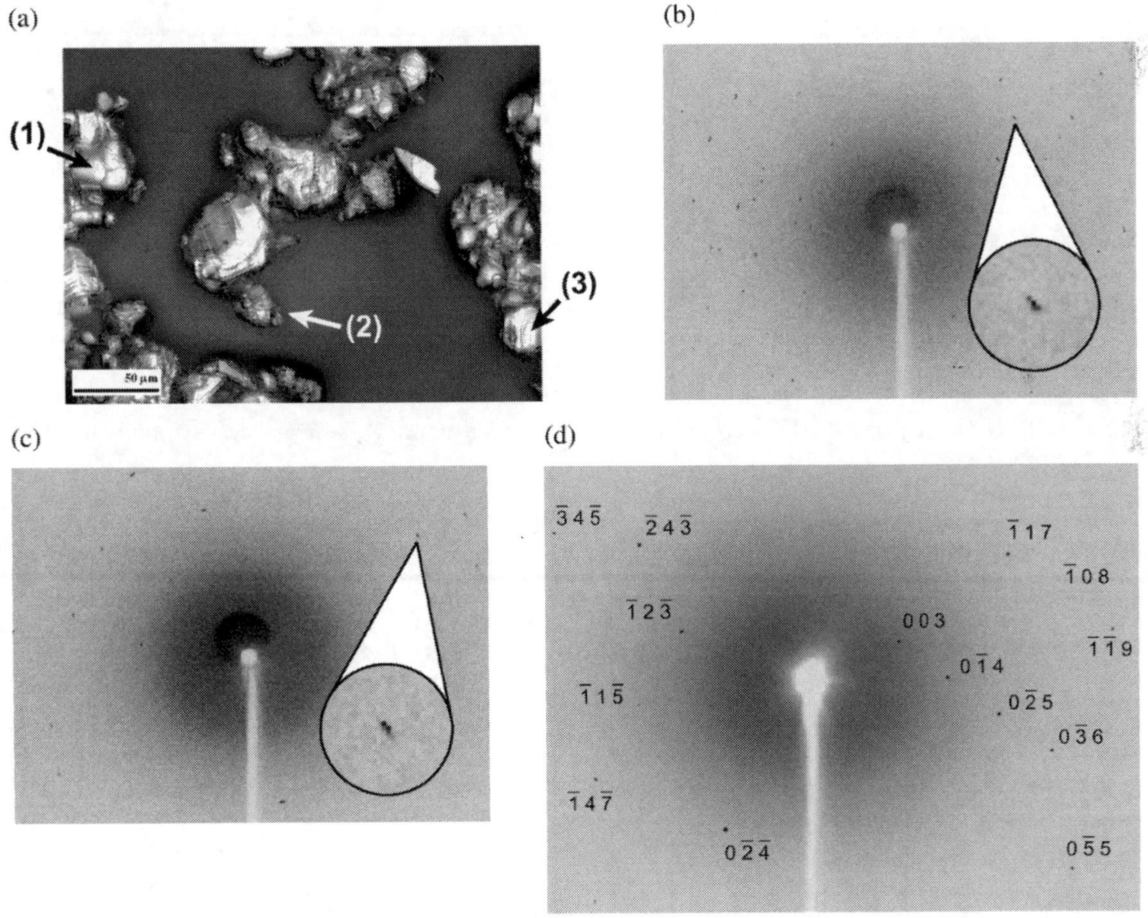

Figure 3. (a) Morphology of Sm-Fe-N powder observed with optical microscope. The scale bar is 50 μm. (b–d) are the X-ray diffraction patterns of (b) a polycrystalline Sm-Fe-N particle including subspots, (c) a particle consisting of a twin of single crystal grains, and (d) a single crystal particle.

Figure 3(a) shows a microscopic image of the sample. The particle size mainly ranges from 20 to 30 μm, but most of the main particles are accompanied with smaller ones of size ~10 μm. To sparsely disperse the particles sticking to each other, vacuum grease was used as solvent. After successful pick-up the sample using the FT-G102 gripper, it was stuck onto a glass needle.

X-ray diffraction measurements were performed at BL-8A/B of the Photon Factory, Japan. The X-ray energy was set to 18 keV. A careful examination of the diffraction pattern on the imaging plate reveals crystallographic homogeneity of the measured particle. Figure 3(b) shows the diffraction spots accompanied with weaker subspots obtained from the large particle (> 30 μm) marked as region (1) in figure 3(a). The particle marked as region (2) exhibited arrays of two or more distinct spots, as shown in figure 3(c). Those particles were deduced to be polycrystalline.

Figure 3(d) shows the diffraction pattern of a particle of size ~20 μm (region (3) in figure 3(a)) after repeated pick-up and measurement trials. All the observed spots were indexed in an analysis using results of successive measurements from rotating the sample. We therefore concluded that the single-crystal X-ray diffraction measurement of this material was performed successfully. The method described above enables us to analyze crystallographic structure more precisely than that reported in previous literature using the powder diffraction method [5, 6]. The crystal structure obtained from the present results will be described elsewhere [7].

4. Conclusion

We have developed a micromanipulation and pick-up system for X-ray diffraction measurements. The system enables us to pick up a single particle from powder-form samples of varying sizes from several to 100 μm. We have succeeded in distinguishing the crystallographic homogeneity of the ferromagnetic particles. The first results of single-crystal X-ray diffraction measurements of Sm-Fe-N permanent magnet was presented.

5. Acknowledgement

The part of this work is supported by the Elements Strategy Initiative Center for Magnetic Materials under the outsourcing project of MEXT. The authors thank Sumitomo Metal Mining Co. Ltd. for providing the samples.

References

[1] Will G 2006 *Powder Diffraction: The Rietveld Method and the Two-Stage Method* (Berlin: Springer)
[2] MiTeGen LLC, http://www.mitegen.com
[3] Vijayasai A P, Sivakumar G, Mulsow M, Lacouture S, Holness A and Dallas T E 2010 *Rev. Sci. Instrum.* **81** 105114
[4] Coey J M D and Sun H 1990 *J. Mag. Mag. Mat.* **87** L251
[5] Sun H, Coey J M D, Otani Y, Hurley D P F 1990 *J. Phys.: Condens. Matter* **2** 6465
[6] Katter M, Wecker J, Kuhrt C, Schultz L, Grössinger R 1992 *J. Mag. Mag. Mat.* **117** 419
[7] Inami N, Takeichi Y, Ueno T, Saito K, Sagayama R, Kumai R, Ono K, accepted to *J. Appl. Phys.*

Development of a Compact Scanning Transmission X-Ray Microscope

Y Takeichi[1], N Inami[1], H Suga[2], T Ueno[3], S Kishimoto[1], Y Takahashi[2] and K Ono[1]

[1] Photon Factory, Institute of Materials Structure Science, High Energy Accelerator Research Organization, 1-1 Oho, Tsukuba, Ibaraki 305-0801, Japan
[2] Department of Earth and Planetary Systems Science, Hiroshima University, 1-3-1 Kagamiyama, Higashi-Hiroshima, Hiroshima 739-8526, Japan
[3] National Institute for Material Science, 1-2-1 Sengen, Tsukuba, Ibaraki 305-0047, Japan

E-mail: yasuo.takeichi@kek.jp

Abstract. We report a compact scanning transmission X-ray microscope newly designed and developed at the Photon Factory. The microscope has very compact size and is equipped with fully digitized control electronics to realize high stability, precise positioning and fast data acquisition. The hardware design of the microscope is described in detail. Results of measurement using test samples are also presented.

1. Introduction

Recent developments in microscopic measurement techniques using focused X-rays from a synchrotron radiation source opened a wide field of research in physics, chemistry, biology and environmental science. These methods, which are often referred to as X-ray spectromicroscopy or microspectroscopy, enable elemental composition or chemical state mapping, and the local X-ray absorption spectroscopy (XAS) of materials. Scanning transmission X-ray microscopy (STXM) [1] using soft X-rays is a powerful tool with features of high spatial resolution down to nanometer scale, high sensitivity to chemical properties, and relatively small radiation damage to the sample. It also allows a wide variety of advanced measurement techniques such as magnetic-domain imaging by using magnetic circular dichroism [2], time-resolved experiments [3], and luminescence detection [4].

In this report, we describe our STXM installed at the soft X-ray undulator beamline of the Photon Factory, Japan. The STXM is a newly designed microscope with very compact size that (i) improves the vibration behavior by increasing the resonance frequency of the microscope components, (ii) realizes a fast response of the stages driven by piezo motors, and (iii) can be used with a multi-purpose beamline that is already in operation.

2. Design of the compact STXM

Figure 1 shows the optics of the present STXM. The undulator radiation is first monochromatized and focused on the four-way aperture slit. A Fresnel zone plate (FZP) with outermost zone width of 30 nm is placed at 1 m distant from the slit, and is designed to have focal distance of 0.7–5 mm. An order sorting aperture (OSA) is placed between the FZP and the sample to stop the zeroth and higher order diffraction. A sample image is obtained by scanning the sample position and detecting the transmitted X-rays. All components from the aperture slit to the detector are mounted on a single optical table to

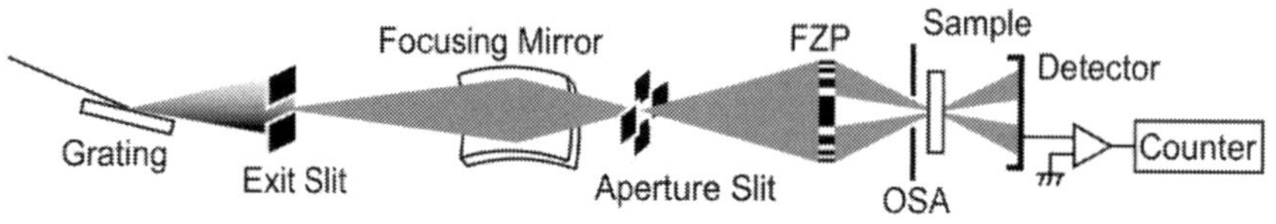

Figure 1. Optics of the compact STXM.

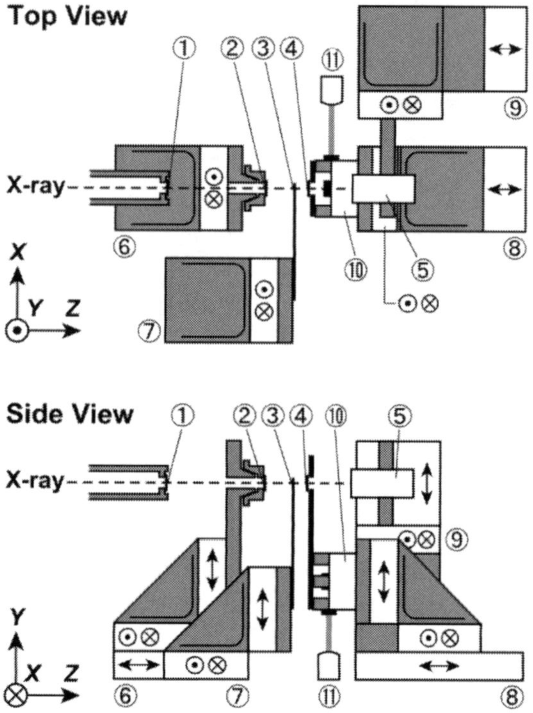

Figure 2. Schematic diagrams of the compact STXM. 1. Vacuum window. 2. FZP. 3. OSA. 4. Sample. 5. Detector. 6. FZP stage. 7. OSA stage. 8. Sample coarse stage. 9. Detector stage. 10. Sample scanner. 11. Laser interferometric sensor head.

reduce the relative motion of the optics by vibration.

Figure 2 shows the scheme of the STXM components, which is in principle the same as the scheme of the common design [1]. In designing the components, computational simulation of the vibration behavior employing the finite element method was performed to keep the resonant frequency higher than 200 Hz. Using piezo-motor-driven linear stages (Attocube ECS series) for all of the coarse stages, the whole system shown in figure 2 is contained in a small vacuum chamber with interior dimensions of 220×310×200 mm^3. The ECS stages are capable of positioning with nanometer precision by means of integrated optical encoders. The coarse stages thus enable fast and precise positioning in the optics adjustment and rough imaging of the sample. The sample position is monitored with laser interferometric position sensors (Attocube FPS3010) with sensitivity of 25 pm. The position values are read out at a rate of 52 kHz for feedback positioning by the sample scanner and on-the-fly measurement.

To realize transmitted X-ray detection with good linearity in a wide range of intensity, fast pulse counting using a silicon avalanche photodiode (APD) [5] is implemented. The APD is capable of pulse counting up to 10^7 cps, which corresponds to the maximum intensity of focused X-rays expected at the sample. The APD is also suitable for our system because it is much smaller and lighter than the scintillator–photomultiplier assembly commonly used in a commercial STXM.

The pulse counting function, scanner drive and position read-out from the interferometric sensor are all integrated in a field-programmable-gated-array (FPGA) circuit. The FPGA implementation and control program of the STXM are developed on a National Instruments LabVIEW platform.

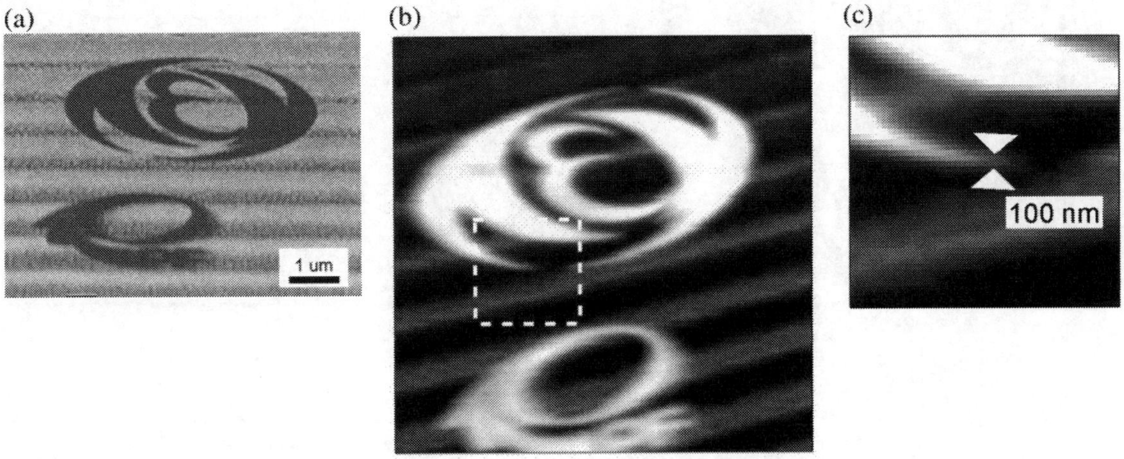

Figure 3. (a) Oblique-view SEM image of KEK and PF logos fabricated with an FIB. (b) STXM image of the logomarks. (c) Enlarged image of the area indicated in (b).

3. Experiment

3.1. Microscopic image

The experiment was performed at BL-13A and BL-16A of the Photon Factory. Energy resolution was set to $E/\Delta E \sim 5000$. Figure 3(a) shows an oblique-view scanning electron microscopy (SEM) image of a test sample fabricated with a focused ion beam (FIB). The logos are 200 nm deep and patterned on ≈300-nm-thick W film deposited on SiN membrane. Figure 3(b) shows the STXM image taken with photon energy of 350 eV. Figure 3(c) is an enlarged image of the area indicated by the dashed square in figure 3(b). A structure with width of 100 nm was resolved. The spatial resolution of the microscope is still being improved by optimizing the focusing optics and vibration isolation according to the circumstances. Using the present FZP, Rayleigh resolution up to 30 nm is fairly possible. To further improve spatial resolution, we also plan to introduce FZPs with a more precise outermost zone width in the future.

At BL-16A, we examined the stability of the image while changing the polarization mode of the undulator. By taking images of a PtIr standard sample (not shown here), the lateral image shift by resulting from changing the helicity of the circular polarized radiation was estimated to be 80 nm in the horizontal direction, which is acceptable for magnetic-domain imaging.

3.2. Microspectroscopy

To perform microspectroscopy with our STXM, the Na *K* edge XAS spectrum of NaCl was measured. A solution of NaCl was dropped onto SiN membrane and then air-dried to obtain plate crystals of various sizes. Figure 4(a) and (b) show the oblique-view SEM image and STXM image of the measured sample. Absorption spectra of the area indicated in figure 4(b) were obtained in a line-profile manner; the sample position was scanned at each point of the photon energy. Figure 4(c) shows the spectrum obtained from the region on (solid line) and away from the NaCl solid (dashed line). Using the out-of-the-sample spectrum as a reference (I_0), the XAS spectrum of NaCl was obtained as shown in figure 4(d). The spectrum is identical to the XAS spectrum at room temperature reported in the literature [6].

4. Concluding remarks

We designed and developed a compact STXM using piezo-motor-driven linear stages. Structures of 100 nm were shown to be resolved. We also obtained local XAS spectra of NaCl plate crystal. At the beamlines used here, the STXM covers a photon energy range of 250–1500 eV, *i.e.*, from the C *K*

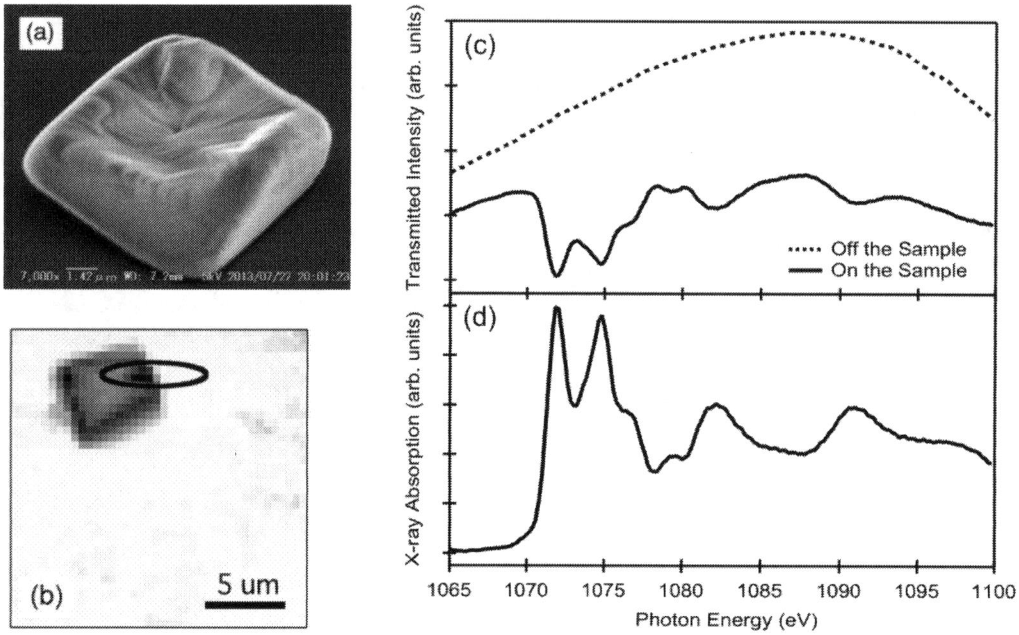

Figure 4. (a) Oblique view SEM image of NaCl dried on SiN membrane. (b) STXM image of the dried NaCl. (c) Transmitted X-ray intensity measured on and off the NaCl sample at the designated area in (b). (d) Na *K* edge XAS spectrum obtained from (c).

edge to the *M* edges of rare earths. Our STXM will be used for the magnetic domain imaging and nanometer-scale chemical characterization of rare-earth permanent magnets using circular polarized radiation [2] and the chemical mapping of materials in environmental sciences [7].

5. Acknowledgements

This work is partly supported by the Japan Science and Technology Agency (JST) under Collaborative Research Based on Industrial Demand "High Performance Magnets: Towards Innovative Development of Next Generation Magnets". The experiment was performed with the approval of the Photon Factory Program Advisory Committee (Proposal No. 2013S2-003). YT thanks J. Raabe and B. Watts for fruitful discussion. C. David is also acknowledged for providing the Fresnel zone plate.

References

[1] Kilcoyne A L D, Tyliszczak T, Steele W F, Fakra S, Hitchcock P, Franck K, Anderson E, Harteneck B, Rightor E G, Mitchell G E, Hitchcock A P, Yang L, Warwick T and Ade H 2003 *J. Synchrotron Rad.* **10** 125

[2] Ono K, Araki T, Yano M, Miyamoto N, Shoji T, Kato A, Manabe A, Nozaki H, Kaneko Y and Raabe J 2011 *IEEE Trans. Magn.* **47** 2672

[3] Raabe J, Quitmann C, Back C H, Nolting F, Johnson S and Buehler C 2005 *Phys. Rev. Lett.* **94** 217204

[4] Vaz C A F, Moutafis C, Quitmann C and Raabe J 2012 *Appl. Phys. Lett.* **101** 083114

[5] Baron A Q R, Kishimoto S, Morse J and Rigal J M 2006 *J. Synchrotron Rad.* **13** 131

[6] Murata T, Matsukawa T and Naoe S 1988 *Solid State Commun.* **66** 787

[7] Kikuchi S, Makita H, Mitsunobu S, Terada Y, Yamaguchi N, Takai K and Takahashi Y 2011 *Chem. Lett.* **40** 680

Journal of Physics: Conference Series **502** (2014) 012010 doi:10.1088/1742-6596/502/1/012010

Visualization of magnetic dipolar interaction based on scanning transmission X-ray microscopy

Hiroyuki Ohtori[1], Kaoru Iwano[1], Chiharu Mitsumata[2], Yasuo Takeichi[1], Masao Yano[3], Akira Kato[3], Noritaka Miyamoto[3], Tetsuya Shoji[3], Akira Manabe[3], Kanta Ono[1]

[1] Photon Factory, IMSS, KEK, Tsukuba, Ibaraki 305-0801, Japan
[2] NIMS, Tsukuba, Ibaraki 305-0047, Japan
[3] TOYOTA Motor Corporation, Toyota, Aichi 471-8571, Japan

E-mail: hiroyuki.otori@kek.jp

Abstract. Using scanning transmission X-ray microscopy (STXM), in this report we visualized the magnetic dipolar interactions in nanocrystalline Nd-Fe-B magnets and imaged their magnetization distributions at various applied fields. We calculated the magnetic dipolar interaction by analyzing the interaction between the magnetization at each point and those at the other points on the STXM image.

1. Introduction

Nd-Fe-B magnets have recently attracted much attention as they have become increasingly used [1] and intensively examined [2, 3, 4, 5, 6]. In nanocrystalline Nd-Fe-B magnets, whose particle size is less than the single-domain size, magnetic dipolar interactions become essential in magnetization reversal processes. However, the significance of dipolar energy in these magnets is not yet understood, because a quantitatively measuring dipolar energy in magnetic materials is difficult. Therefore, we consider that it is important to visualize the magnetic dipolar interaction for revealing the nature of the magnetization reversal processes, and demonstrate a visualization of the magnetic dipolar interaction in permanent magnets using scanning transmission X-ray microscopy (STXM).

In STXM, a Fresnel zone plate (FZP) focuses the X-rays to a diffraction-limited focus, and images are acquired by raster-scanning the sample through the focus while measuring the transmitted intensity with a relatively large X-ray detector. In our experiment, we used a FZP with an outermost zone width of 25 nm and a focus size (or spatial Resolution) of about 30 nm. Additionally, we used a combination of a central stop and an order-sorting aperture to block zero-order light not diffracted by the zone plate and higher-order light. The samples, mounted on TEM grids, are first held in the focal position of the beam. The sample is then scanned with respect to the focused soft X-ray beam using an interferometer-controlled piezo stage. Transmitted light is measured using a photomultiplier with a scintillator. Essentially, STXM is a microscope that takes advantage of the X-ray magnetic circular dichroism (XMCD) effect. Because the XMCD effect is proportional to the magnetization of a specific element and is sensitive to the projection of the magnetization along the direction of photon propagation, the STXM image of the magnetic contrast shows the real space distribution of element-specific magnetization with high spatial resolution. Therefore, STXM is a powerful tool to investigate the magnetic microstructure of Nd-Fe-B magnets [7].

In this paper, we analyzed the STXM image of the magnetization distribution with a special attention to the direction perpendicular to the surface. We calculated the potential energy of the magnetic dipolar interaction between the magnetization at each point and those at the other points on the STXM image in order to obtain the distribution of the dipolar energy. Finally, we visualized the magnetic dipolar interaction in the nanocrystalline Nd-Fe-B magnets at remanent states after applying 0, 1, 2, 4, 6, and 10 kOe.

2. Formalism

To visualize the magnetic dipolar interaction energies, we must calculate them from the magnetization distribution, which can be obtained from the STXM image [8]. The potential energy of the magnetic dipolar interaction U is given by

$$U = -\int \frac{\nabla \cdot \boldsymbol{M}(\boldsymbol{r}')}{|\boldsymbol{r}-\boldsymbol{r}'|} dV' + \int \frac{\boldsymbol{n} \cdot \boldsymbol{M}(\boldsymbol{r}')}{|\boldsymbol{r}-\boldsymbol{r}'|} dS'. \tag{1}$$

In this study, we assume the sample is thin enough to calculate the magnetization as $\boldsymbol{M} = (0, 0, m(x, y))$. By substituting $\boldsymbol{M} = (0, 0, m(x, y))$ into eq.(1), we obtain

$$\begin{aligned} U &= -\int dV' \frac{\frac{\partial}{\partial z'} m(x', y')}{|\boldsymbol{r}-\boldsymbol{r}'|} + \int dS' \int dz' \frac{\partial}{\partial z'} \frac{m(x', y')}{\sqrt{(x-x')^2 + (y-y')^2 + (z-z')^2}} \\ &= \int_{-a}^{a} dx' \int_{-b}^{b} dy' \int_{-c}^{c} dz' \frac{(z-z')m(x', y')}{((x-x')^2 + (y-y')^2 + (z-z')^2)^{\frac{3}{2}}}, \end{aligned} \tag{2}$$

where $\boldsymbol{r} = (x, y, z)$ is the coordinate of the observation point, $\boldsymbol{r}' = (x', y', z')$ is the coordinate of the source point, $2a$ is the width of the sample in the x direction, $2b$ is the width of the sample in the y direction and $2c$ is the thickness of the sample. The relation between the magnetic dipolar interaction energy E_d and the magnetic field $\boldsymbol{H}$ is given by

$$E_d = -\frac{1}{2} \int \boldsymbol{M} \cdot \boldsymbol{H} dV = -\frac{1}{2} \int m(x, y) H_z dV, \tag{3}$$

where H_z is the z component of $\boldsymbol{H}$. On the other hand, the relation between U and $\boldsymbol{H}$ is given by $\boldsymbol{H} = -\nabla U$. Therefore, we can obtain E_d as the function of U as follows:

$$\begin{aligned} E_d &= \frac{1}{2} \int dx \int dy\, m(x, y) \int dz \frac{\partial}{\partial z} U(x, y, z) \\ &= \frac{1}{2} \int dx \int dy\, m(x, y)(U(x, y, c) - U(x, y, -c)). \end{aligned} \tag{4}$$

From eq.(2), the term of U in eq.(4) is given by

$$\begin{aligned} U(x, y, c) &- U(x, y, -c) \\ &= \int dx' \int dy'\, m(x', y') \int dz' \left[\frac{c - z'}{(r_{xy}^2 + (c-z')^2)^{\frac{3}{2}}} - \frac{-c - z'}{(r_{xy}^2 + (-c-z')^2)^{\frac{3}{2}}} \right] \\ &= \int dx' \int dy'\, m(x', y') 2 \left[\frac{1}{r_{xy}} - \frac{1}{\sqrt{r_{xy}^2 + (2c)^2}} \right], \end{aligned} \tag{5}$$

where $r_{xy} = \sqrt{(x-x')^2 + (y-y')^2}$ is the length between the observation point and the source point. By using eq.(4) and eq.(5), we can visualize the distribution of the dipolar energy as shown in Fig.1(b) from the STXM image of the magnetization distribution as shown in Fig.1(a).

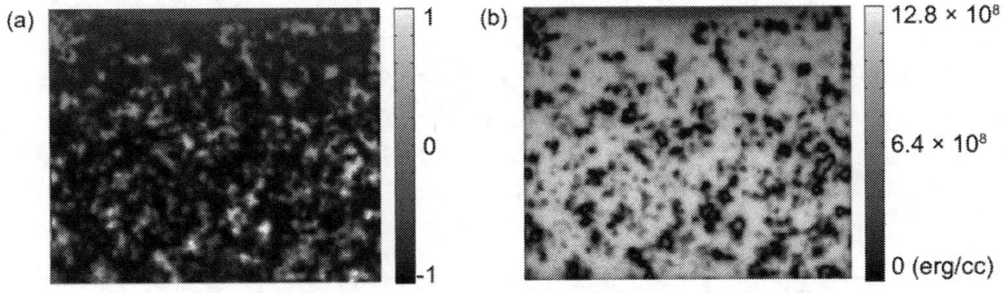

Figure 1. (a) STXM image of the magnetization distribution in the nanocrystalline Nd-Fe-B magnet. The scale unit is normalized to be equal to 1 at the maximum value. (b) Distribution of the magnetic dipolar interaction energy obtained from (a). The thickness of the sample is 60 nm.

3. Results

By using the method shown in Fig.1, we visualize the magnetic dipolar interaction in the nanocrystalline Nd-Fe-B magnet from the STXM image. We can obtain the two-dimensional magnetic dipolar interaction energy $E_{d_{xy}}$, which also represents the density of the magnetic dipolar interaction energy E_d at (x, y), by substituting eq.(5) into eq.(4)

$$E_{d_{xy}}(x, y) = \frac{1}{2}\frac{m(x, y)}{2c}\int\int dx'dy'g(\boldsymbol{x}, \boldsymbol{x}')m(x', y'), \tag{6}$$

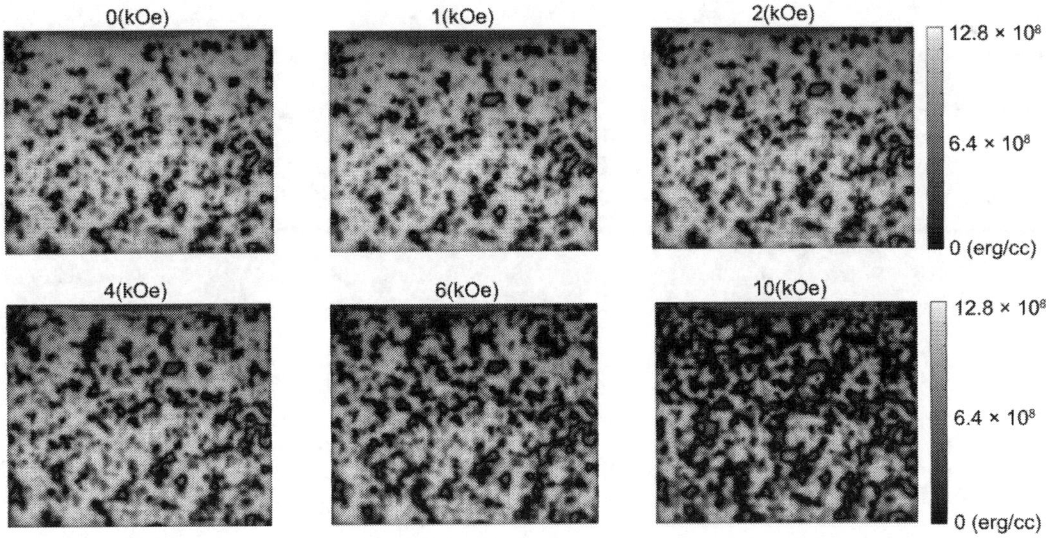

Figure 2. Distribution of the dipolar energy for applied fields of 0, 1, 2, 4, 6 and 10 kOe in the nanocrystalline Nd-Fe-B magnet. The thickness of the sample is 60 nm. The dipolar energy is high in white areas and small in black areas.

where

$$g(\boldsymbol{x}, \boldsymbol{x}') \equiv 2\left(\frac{1}{r_{xy}} - \frac{1}{\sqrt{r_{xy}^2 + (2c)^2}}\right), \tag{7}$$

$\boldsymbol{x} = (x, y)$ is the coordinate of the observation point on the x-y plane, and $\boldsymbol{x}' = (x', y')$ is the coordinate of the source point on the same plane. In Fig.2, we show the visualized magnetic dipolar interaction in nanocrystalline Nd-Fe-B magnet at remanent states after applying 0, 1, 2, 4, 6, and 10 kOe. When the applied field approaches the coercivity field (~ 10kOe), the total dipolar energy decreases. This behavior suggests that the decreased dipolar energy contributes to the stability of permanent magnets against the applied field.

4. Summary

We demonstrated a visualization of the magnetic dipolar interaction in a nanocrystalline Nd-Fe-B magnet using STXM. Using those images, we calculated the potential energy of the magnetic dipolar interaction between the magnetization at each point and those at the other points on the STXM image. Therefore, we obtained the distribution of the dipolar energy and found that the total dipolar energy decreases when the applied field approaches the coercivity field (~ 10kOe). This behavior suggests that the decreased dipolar energy contributes to the stability of permanent magnets against the applied field. Our results show that visualizing the magnetic dipolar interaction is important for observing the development of dipolar energy in magnetization reversal processes and helps to reveal the nature of those processes.

Acknowledgment

This research was supported by the Japan Science and Technology Agency (JST) under Collaborative Research Based on Industrial Demand "High Performance Magnets : Towards Innovative Development of Next Generation Magnets". Sample fabrication was partly performed under the Magnetic Materials for High-Efficiency Motors (MagHEM) project.

References

[1] Gutfleisch O, AWillard M, Bruck E, Chen C H, Sankar S G and Liu J P 2011 *Adv. Mater.* **23** 821
[2] Livingston J D 1985 *J. Appl. Phys.* **57** 4137
[3] Lemke H, Goddenhenrich T, Heiden C and Thomas G 1997 *IEEE Trans. Magn.* **33** 3865
[4] Zueco E, Rave W, Schafer R, Hubert A and Schultz L 1998 *J. Magn. Magn. Mater.* **190** 42
[5] Zhu Y and McCartney M R 1998 *J. Appl. Phys.* **84** 3267
[6] Szmaja W 2006 *J. Magn. Magn. Mater.* **301** 546
[7] Ono K, Araki T, Yano M, Miyamoto N, Shoji T, Kato A, Manabe A, Nonaki H, Kaneko Y and Raabe J 2011 *IEEE Trans. Magn.* **47** 2672
[8] Ono K, Yano M, Araki T, Manabe A, Miyamoto N, Shoji T, Kato A, Kaneko Y, Nonaki H and Raabe J *22nd International Workshop on Rare-Earth Permanent Magnets and Their Applications (REPM' 12)*

Real-time motion control and data acquisition system for scanning X-ray microscopy using programmable hardware

N Inami, Y Takeichi, and K Ono

Photon Factory, Institute of Materials Structure Science, High Energy Accelerator Research Organization (KEK), Tsukuba, Ibaraki 305-0801, Japan

E-mail: nobuhito.inami@kek.jp

Abstract. We developed a real-time motion control and measurement system for a scanning transmission X-ray microscope. The system was designed with an FPGA (field-programmable gate array), which is a programmable, reconfigurable circuit for real-time feature modification. This system enabled specimen stage control driven by a piezoelectric element and precise stage position measurement with a laser interferometer. Programmable hardware processing with FPGA realized high-speed control and measurement that cannot be handled by software processing. This system enables very high-speed control and measurement such as "on the fly" measurement.

1. Introduction

A scanning transmission X-ray microscope (STXM) [1] using synchrotron radiation is an attractive apparatus for magnetic domain observation of rare earth magnets and local component analysis for rare metal resources. [2] STXM uses an X-ray optic element called Fresnel zone plate (FZP) to converge radiated light into nanometer spot sizes and scan specimens to obtain chemical and magnetic images. To obtain high-resolution images, a large volume of sampling data is required. When a measurement takes a long time, it is difficult to use limited beam time effectively. Moreover, data deterioration by external factors such as drift of the specimen becomes more pronounced over time. For quick measurements, high counting rate X-ray detectors, real-time high-speed position control, and a capable measurement and data processing system is crucial. In addition, for nanometer scale X-ray absorption spectrum measurements, very precise position control on the order of nanometers, including beam stability and feedback control, is necessary. For scanning microscope systems in particular, real-time position control and measurement data acquisition is essential. PC-driven control systems and software can't guarantee real-time functionality, so it is essential to develop hardware-based real-time control and measurement systems. We developed a real-time control and measurement system for a STXM using a programmable, reconfigurable FPGA (field-programmable gate array), capable of producing images with 100 nm resolution.

2. Methods

The block diagram for a real-time motion control and data acquisition system for an STXM is shown in figure 1. We adopted two ways to detect X-ray photons. A photodiode (PD) IRD AXUV (Opto Diode Corporation) is employed in the detection for higher intensity X-rays above approximately 10 MHz. The PD signal to a real-time controller was amplified and converted from current to voltage by an electric current amplifier (Amp.). An avalanche photodiode (APD) is also used as the X-ray photon

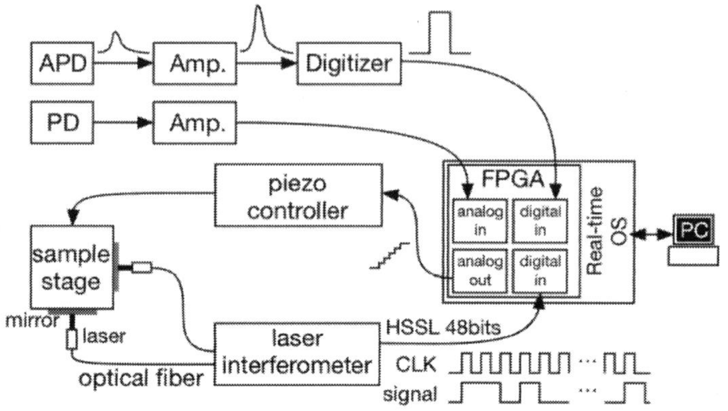

Figure 1. Block diagram of real-time motion control and data acquisition system for STXM.

pulse detector [4] below the 10 MHz rate for measurement of a wide dynamic range. The APD signal to the real-time controller with a pulse counter by an FPGA circuit was amplified by a high frequency amplifier and rectified by a digitizer.

A sample stage was moved by piezoelectric positioners (ECS3030, attocube systems AG) for large movements in the XYZ directions, and by a piezoelectric scanner (ANSxy100, attocube systems AG) for fine movement in the XY directions. The positioner range was 20 mm with a 50 nm minimum step, and the scanner range was 50 μm with sub nm resolution. The positioner was controlled by an ANC 350 multi-functional controller (attocube system AG) with a standard PC, and the scanner was controlled by an ANC300 (attocube system AG) with an FPGA controller. The positioner had a built-in optical encoder, whereas the scanner had not. The scanner performed the fine control of the sample stage position using a laser interferometer (FPS3010, attocube systems AG). The laser interferometer had a position resolution of 30 pm and output bandwidth of 10 MHz. It supported two types of outputs: quadrature signal with low voltage differential signaling (LVDS) and high-speed serial link (HSSL) with low voltage TTL (LVTTL). In this study we used HSSL to record absolute position.

It is usually necessary that a customized printed-circuit board equipped with FPGA programs be used for building a control and measurement system. In particular, designing a data transmission system to connect with a PC for control and data collection can be difficult. However, with a National Instruments CompactRIO (NI cRIO) 9075, this communication system was easily implemented. The NI cRIO was employed as the FPGA-based system for real-time control and measurement. It was equipped with a Spartan6 LX25 FPGA and a built-in 40 MHz on-board clock. It also supported real-time OS and could process signal without the PC. The NI cRIO 9075 was connected to a PC equipped with a 100Base-TX through a network hub. Each network port was assigned a private IP address. The PC and the NI cRIO were connected via TCP/IP. However, since LabVIEW on a PC communicates via an FPGA I/O node, we didn't need to consider handling TCP/IP protocol or string processing, making data transmission easier.

The NI cRIO series was equipped with slots for modules such as digital I/O and analog I/O. The NI cRIO 9075 had four module slots. In this study, we installed a NI 9264 analog output module to drive the piezoelectric positioner, a NI 9402 high-speed digital I/O module to handle the digital pulse from the laser interferometer, a NI 9205 analog input module to acquire measurements, and a NI9402 high-speed digital I/O module to count X-ray pulses.

LabVIEW FPGA running on PC was used for the FPGA programming of the NI cRIO. Although the available functions were limited due to the hardware circuitry, such as clock timing, LabVIEW FPGA could handle most of the same functions as LabVIEW for PC. It was a big advantage of LabVIEW FPGA that it could handle hardware circuit in the same manner as LabVIEW software processing.

Journal of Physics: Conference Series **502** (2014) 012011 doi:10.1088/1742-6596/502/1/012011

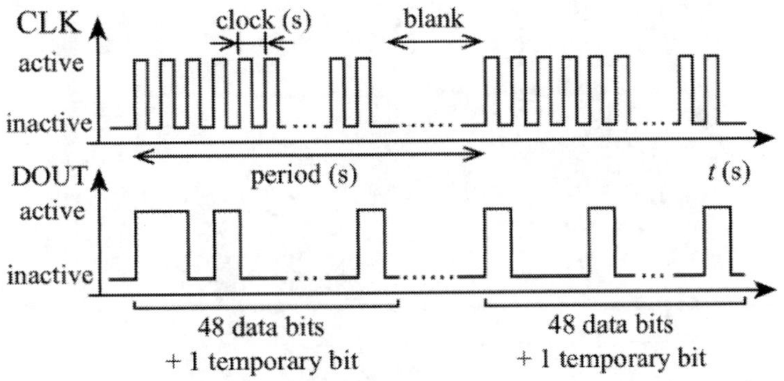

Figure 2. Schematic of clock HSSL signal from FPS3010 laser interferometer.

3. Results

3.1. Analog output for stage control

To move the sample stage, an analog voltage signal was sent to the ANC350 that controlled the piezoelectric positioner and drove the piezoelectric scanner ANSxy100. A voltage spike of 0 to +10 V was sent to the ANC350 that drove the piezoelectric scanner. The laser interferometer controlled the position with high accuracy. LabVIEW on the PC was controlled via the LabVIEW FPGA I/O node.

3.2. Digital input for HSSL signal

High precision position measurement with laser interferometer was achieved using digital input to the FPS3010. From FPS3010, clock signals in batches of 48 pulses (48 bits) and HSSL position information signal were output (figure 2 [3]), and synchronized with the clock signal. When the clock

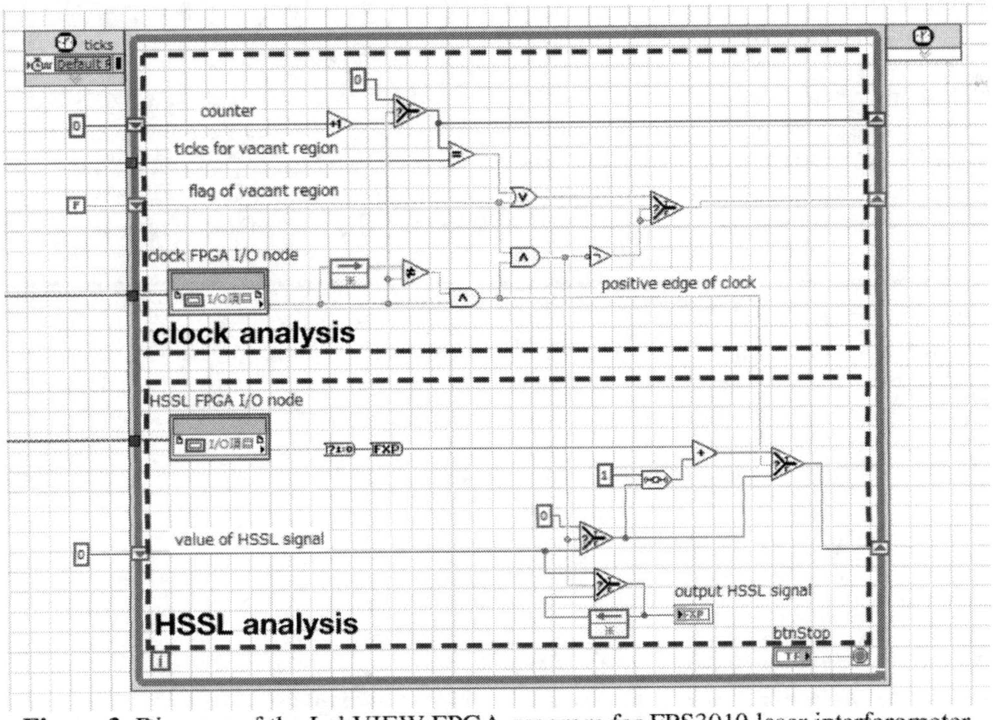

Figure 3. Diagram of the LabVIEW FPGA program for FPS3010 laser interferometer.

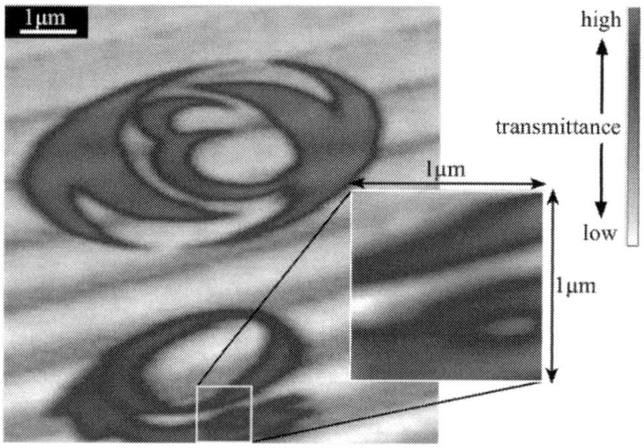

Figure 4. Scanning transmission X-ray microscope image of the KEK and Photon Factory logos.

signal was inactive, no HSSL pulse was output. The clock signal consisted of two parts; pulse signals at even intervals and intervals without pulse signals for a defined period. In this study we set the clock signal width to 200 ns and the clock signal frequency to 5 MHz for a pulse signal duration of 200 ns × 48 bit = 9.6 μs. We adjusted the period such that the interval without pulse signals equaled the signal duration.

Figure 3 shows the LabVIEW FPGA diagram for HSSL signal analysis. The main processing routine processed a loop synchronized with an on-board clock. The clock signal processing routine monitored and counted periods without clock pulse signal. When a clock pulse was input, the counter was reset and deserialization of the HSSL signal was initiated to convert the serial signal of HSSL to 48 bits value. The HSSL signal handling routine continued deserialization while the clock pulse was input. When the clock signal went into the blank period and the next clock pulse was input, the position value from HSSL was sent to the output HSSL signal. The pulse processing of the clock HSSL signal by this LabVIEW FPGA program was of the positive edge type.

3.3. Microscopy application

The STXM observation program was integrated in LabVIEW on a PC with the above elements. Figure 4 shows an STXM image of the KEK and Photon Factory logos patterned on a W thin film by a focused ion beam system. The X-ray beam energy used for the observation was 350 eV. The logos on the microscope image have a resolution of about 100 nm, and were obtained using the FZP with an outer most zone width of 30 nm.

4. Summary

We developed a real-time motion control and data acquisition system for STXM using the FPGA-based system for real-time control and measurement. It processed the position control of a specimen stage through a piezoelectric positioner and stage position measurement by laser interferometer. The performance of this system is sufficient to control and measure the stage position for STXM.

This work is partly supported by the Japan Science and Technology Agency (JST) under Collaborative Research Based on Industrial Demand "High Performance Magnets: Towards Innovative Development of Next Generation Magnets".

References

[1] Kirz J, Burg R and Rarback H 1980 *Ann. N. Y. Acad. Sci.* **342** 135
[2] Ono K *et al.* 2011 *IEEE Trans. Magn.* **47** 2672
[3] Attocube system AG 2012 AttoFPSX010 User Manual Ver. 1.2 p 25
[4] Kishimoto S, Taniguchi T and Tanaka M 2010 *AIP Conf. Proc.* **1234** 819

Positron and Ion Migrations and the Attractive Interactions between like Ion Pairs in the Liquids: Based on Studies with Slow Positron Beam

I. Kanazawa, T. Sasaki, K. Yamada, E. Imai

Department of Physics, Tokyo Gakugei University, Nukuikita-machi 4-1-1, koganei-shi, Tokyo, 184-8501, Japan

E-mail: kanazawa@u-gakugei.ac.jp

Abstract. We have discussed positron and ion diffusions in liquids by using the gauge-invariant effection Lagrange density with the spontaneously broken density (the hedgehog-like density) with the internal non-linear gauge fields (Yaug-Mills gauge fields), and have presented the relation to the Hubbard-Onsager theory.

1. Introduction

The Investigation for ion mobility in the liquid phase has been one of the central area of physical chemistry. The electrohydrodynamic theory by Hubbard and Onsager [1,2] and the stochastic theory [3,4] for ionic conductivity in the liquid phase have been proposed.Since fluctuation from the equilibrium medium is preferable in liquid phase, localization of ions such as positrons is highly probable [5,6]. Free energy density functional theories [7,8] provide self-trapping as solution of the ions in a given host liquid. The sensitivity of positrons to changes induced by melting has already been reported by the positron annihilation angular correlation [9-11] and lifetime [12,13] studies for liquids. Gramsh et al. [14,15] have observed very different behavior of the diffusion length L_+ of positrons in liquid and solid metals. That is, on melting, L_+ decreases remarkably, and in the liquid phase, L_+ increases with temperature. Kanazawa [16-20] has introduced the effective Lagrangian in the gauge-invariant formula with spontaneous symmetry breaking for liquids and glasses, and has discussed the origin of the boson peak, the glass transition, and the viscosity of the supercooled liquids. Kanazawa and coworkers [21-24] proposed a qualitative explanation for the increase of the positron diffusion length L_+ with temperature in the liquid phase, by using the theoretical formula, which is based on the gauge-invariant effective Lagrangian with spontaneously broken density (the hedgehog like fluctuation) and the massive internal gauge fields. In addition Kanazawa et al. [28] have suggested one origin of the attractive interaction between like ion pairs in liquids, extending the theoretical formula [20-23]. Recently Yamada et al. [29] have explained the origin of the attractive interaction between the Cl^- - Cl^- pair in water, which is introduced by the integral equation method [1], by using the theoretical formula [28]. In this study, we have analyzed the positron diffusion in the solid and liquid Pb, and have discussed the relation to the Hubbard-Onsager's dielectric theory for positrons and ions migrations in liquids.

2. The hedgehog-like density fluctuation and the positron diffusion

We shall introduce a field-theoretical model to treat three-dimensional liquids by using excited cluster (solitons). The connection on gauge field describes how the orientation of the fiber changes as one goes along a path in base space, or how the path in the base space is lifted into a path in the fiber bundle. We introduce a field-theoretical model to treat the problem of a charged particle in a fluid host in three spatial dimensions. It has been proposed that the parameter $\rho(t,r,u) = \rho(t,x,y,z,u)$ in three dimensional liquids is specified by the rotation, which is related to the non-linear gauge fields (Yang Mills gauge fields) A^a_μ of SO(4) symmetry of S^3. The curvature can be represented by using a component, u, in the other-axis direction, if the three spatial dimensional axes are x, y, and z. It is preferable that we think of the anomalous fluctuation around the charged particle in the three dimensional liquid as the hedgehog-like fluctuation (soliton), taking into account the curvature. We adopt the parameter $\rho(r,u) = \rho^a (a = 1,2,3,4)$, which is similar to that in the Sachdev and Nelson model [25]. The SO(4) quadruplet fields, A^a_μ, are spontaneously broken through the Higgs mechanism similar to the way in which the fluid host is broken around a charged particle. When the hedgehog-like fluctuation (soliton) around a charged particle is created, we set the symmetry breaking of the quadruplet fields, $\langle 0|\rho|0\rangle$, equal to $(0,0,0,\nu_4)$. Now, we can introduce the approximate Lagrangian as follows,

$$\begin{aligned}
L = \;& \psi^{+}(i\partial_0 - g_2 T^a A^a_0)\psi - \frac{1}{2m}\psi^+(i\nabla - g_2 T^a A^a_{\mu\neq 0})^2\psi \\
- \;& \frac{1}{4}(\partial_\nu A^a_\mu - \partial_\mu A^a_\nu + g_1 e_{abc} A^b_\mu A^c_\nu)^2 \\
+ \;& \frac{1}{2}(\partial_\mu \rho^\beta - e_{\beta\alpha\gamma} A^\alpha_\mu \rho^\gamma)^2 + \frac{m_1^2}{2}[(A^1_\mu)^2 + (A^2_\mu)^2 + (A^3_\mu)^2] \\
+ \;& m_1(A^1_\mu \partial_\mu \rho^2 - A^2_\mu \partial_\mu \rho^1) + m_1(A^2_\mu \partial_\mu \rho^3 - A^3_\mu \partial_\mu \rho^2) + m_1(A^3_\mu \partial_\mu \rho^1 - A^1_\mu \partial_\mu \rho^3) \\
+ \;& g_1 m_1\{\rho^4[(A^1_\mu)^2 + (A^2_\mu)^2 + (A^3_\mu)^2] - A^4_\mu(\rho^1 A^1_\mu + \rho^2 A^2_\mu + \rho^3 A^3_\mu)\} \\
- \;& \frac{m_2^2}{2}(\rho^4)^2 - \frac{m_2 g}{2m_1}\rho^4(\rho^a)^2 - \frac{m_2^2 g^2}{8m_1^2}(\rho^a\rho^a)^2. \qquad (1)
\end{aligned}$$

where m_1 is $\nu_4 \cdot g$ and m_2 is $2\sqrt{2}\lambda_3 \cdot \nu_4$. The effective Lagrange density, L_{eff}, represents three massive vector fields A^1_μ, A^2_μ, A^3_μ, and one massless vector field A^4_μ. The generation function $Z[J]$ for Green's functions is shown as follows;

$$\begin{aligned}
Z[J] \;=\; & \int \mathcal{D}A\mathcal{D}B\mathcal{D}\rho\mathcal{D}c\mathcal{D}\bar{c}\mathcal{D}\psi^+\mathcal{D}\psi \\
& \cdot \exp\left[\int_0^\beta d\tau \int d^3x\,(\mathcal{L}_{eff} + \mathcal{L}_{\mathrm{GF+FP}} + J\cdot\Phi)\right], \qquad (2)
\end{aligned}$$

$$\mathcal{L}_{\mathrm{DF+FP}} = B^a\partial^\mu A^a_\mu + \frac{1}{2}\alpha B^a B^a + i\bar{c}^a\partial^\mu \mathcal{D}_\mu c^a, \qquad (3)$$

where B^a and c^a are the Nakanishi-Lautrup(NL) fields and Faddeev-Popov fictitious fields, respectively, and $\beta = 1/T$, whose T is Temperature.

$$J\cdot\Phi \;\equiv\; J^{a\mu}A^a_\mu + J^a_B B^a + J_\rho\cdot\rho^a + \bar{\eta}\psi + \eta\psi^+ + \bar{J}^a_c c^a + J^a_c \bar{c}^a \qquad (4)$$

BRS-quartet [26,27] in the present theoretical formula are$(\phi_1, B^1, C^1, \bar{C^1})$, $(\phi_2, B^2, C^2, \bar{C^2})$, $(\phi_3, B^3, C^3, \bar{C^3})$, $(A^4_{L,\mu}, B^4, C^4, \bar{C}^4)$ where $A^4_{L,\mu}$ is the longitudinal component of A^4_μ. So we need these fields are unobservable and fictious ones. Since these masses are created through the Higgs mechanism by introducing a charged particle in the fluid host, the massive gauge fields A^1_μ,A^2_μ,A^3_μ are localized around the charged particle. In metallic liquids, A^4_μ also becomes massive due to screaning effect of conduction electrons.

Kanazawa [21,22] has proposed the explanation for the increase of the positron diffusion length with temperature in liquid metals. That is, as the temperature increases, the effective mass of the positron decreases due to the restoration of the spontaneously broken density around the positrons in the liquid phase. The diffusivity D_+ of positron is introduced approximately from the previous theoretical formula[21,22],

$$D_+ = \frac{2}{3}\frac{k_B T}{M_{eff}}\tau_r \propto \frac{k_B T \cdot \tau_r}{\left[4\,c_3^2 - \left(\frac{4}{3}\lambda_3 + \frac{g^2}{2}\right)T^2\right]^{\frac{1}{2}}} \tag{5}$$

Where τ_r is the relaxation time, for example, the phonon relaxation time is $\tau_{ph} \sim T^{\frac{3}{2}}$. Restoration of the spontaneously broken density reduces the effective mass M_{eff} as temperature increases as shown in Figure 1. From eq.(5), it is seen that the diffusivity D_+ increases with temperature. From Fig.2, it is seen that eq.(5) is consistent with experimental results [15].

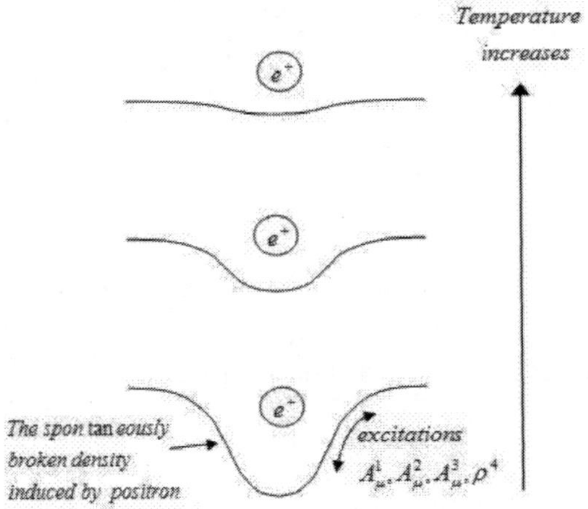

Figure 1. Intuitive picture of the reduction of the positron effective mass due to restoration of the spontaneously broken density around positron, with increase in temperature.

Now, we recall that the electro-hydrodynamic stress tenser S_{el}. is given in the Hubbard-Onsager theory [3,4] as $8\,\pi\,S_{el} = DE+ED-E\cdot DI$, where I is the unit tensor and $D = E+4\,\pi\,P$. In the presence of a uniform periodic external field E_{ext}, the induction D, and the total electric field are given by

$$D = \epsilon(\omega)\,E_{ext} + \epsilon(\omega = 0)\,E_0$$

$$E = E_{ext} + E_0\ ,$$

where ϵ is the dielectric constant, E_0 is the field due to an ion field in a dielectric continuum. In the present theoretical formula, $E_0 \propto -\partial_t\,A^4_{\mu(\mu\neq 0)} - \partial_{\mu(\mu\neq 0)}\,A^4_0$. A^4_μ is the massless gauge field. The divergence of electical stress S_{el} defines a local force density f which we write as

$$f = \frac{1}{2}\left\{\frac{[\epsilon(\omega) - E(\omega = 0)]}{4\,\pi}\right\}[D \times (E_{exp} \times E_0) + \nabla(E_0 \cdot E_{ext})]\ . \tag{6}$$

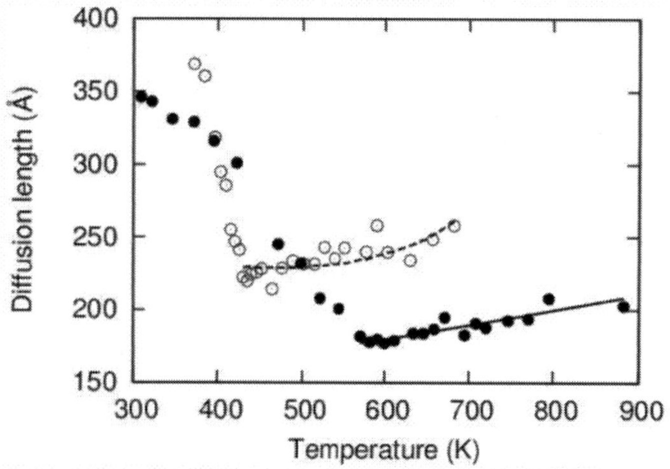

Figure 2. Positron diffusion length versus temperature in the solid and liquid In and Pb metals. Open circles and solid circles correspond to experimental data points of In and Pb metals [15]. A dotted line is fitted one [23]. A solid line (the present work) is fitted one by using eq. (5).

3. Conclusion

We have discussed positron and ion diffusions in liquids, by using the gauge-invariant effective Lagrange density with the spontaneously broken density and internal Yang-Mills gauge fields. In addition, the relation to the Hubbard-Onsager theory has been discussed.

References

[1] Hubbard J, Onsager L 1977 *J. Chem. Phys.* **67** 4850
[2] Hubbard J 1978 *J. Chem. Phys.* **68** 1649
[3] Wolynes P G 1978 *J. Chem. Phys.* **68** 473
[4] Colonomos P, Wolynes P G 1979 *J. Chem. Phys.* **71** 2644
[5] Iaknbov I T, Khrapak A G 1982 *Rep. Prog. Phys.* **45** 697
[6] Hernandez J P 1991 *Rev. Mod. Phys.* **63** 675
[7] Miller B N, Resse T L 1989 *Phys. Rev.* A **39** 4735
[8] Miller B N, Resse T L 1990 *Phys. Rev.* A **42** 6068
[9] West R N, Borland R E, Cooper J R A, Cusak N E 1967 *Proc. R. Soc. London Ser.* A **92** 195
[10] Gustafson D R, Mackintosh A R 1963 *Phys. Lett.* A **45** 234
[11] Gustafson D R, Mackintosh A R, Zaffarano D J 1963 *Phys. Rev.* **130** 1455
[12] Brandt W, Wanng H F 1968 *Phys. Lett.* A **27** 700
[13] Crisp V H C, Lock D G, West R N 1974 *J. Phys.* F **4** 830
[14] Gramsch E, Lynn K G, Throwe J, Kanazawa I 1991 *Phys. Rev. Lett.* **67** 1282
[15] Gramsch E, Lynn K G, Throwe J, Kanazawa I 1999 *Phys. Rev.* B **59** 14282
[16] Kanazawa I 1993 *Phys. Lett.* A **176** 246
[17] Kanazawa I 1997 *Prog. Theor. Phys. Suppl.* **126** 397
[18] Kanazawa I 2001 *J. Non-Crst. Solids* **293-295** 615
[19] Kanazawa I 2002 *J. Non-Crst. Solids* **312-314** 608
[20] Suzuki H, Kanazawa I 2010 *Intermatallics* **18** 451
[21] Kanazawa I 1996 *J. Radioanal. Nucl. Chem.* **210** 451
[22] Kanazawa I 2000 *Rad. Phys. Chem.* **58** 457
[23] Matsushita Y, Kitahata H, Kanazawa I 2007 *Phys. Stat. Sol.* (C) **4** 3546
[24] Kitahata H, Matsushita Y, Kanazawa I 2008 *Appl. Surf. Sci.* **244** 167
[25] Sachdev S, Nelson D R 1984 *Phys. Rev. Lett.* **53** 1947
[26] Becchi C, Rouet A, Stora R 1975 *commun. Math. Phys.* **42** 127
[27] Kugo T, Ojima I 1979 *Prog. Theor. Phys. Suppl.* **66** 1
[28] Kanazawa I, Suzuki H, Kitahata J 2010 *J. Phys.* CS **225** 012024
[29] Yamada K, Saito M, Suzuki H, Kanazawa I 2013 *Mater. Sci. Forum* **733** 132

Journal of Physics: Conference Series **502** (2014) 012013 doi:10.1088/1742-6596/502/1/012013

Spectromicroscopic analysis of lithium intercalation in spinel $LiMn_2O_4$ for lithium-ion battery by 3D nano-ESCA

Naoka Nagamura[1,2,4], Shota Ito[1], Koji Horiba[1,2], Toshihiro Shinohara[1], Masaharu Oshima[1,2], Shin-ichi Nishimura[3], Atsuo Yamada[3] and Noritaka Mizuno[1]

[1]Department of Applied Chemistry, Graduate School of Engineering, The University of Tokyo, 7-3-1 Hongo, Bunkyo-ku, Tokyo 113-8656, Japan
[2]Synchrotron Radiation Research Organization, The University of Tokyo, 1-1-1 Kouto, Sayo-cho, Sayo-gun, Hyogo 679-5198, Japan
[3]Department of Chemical System Engineering, Graduate School of Engineering, The University of Tokyo, 7-3-1 Hongo, Bunkyo-ku, Tokyo 113-8656, Japan

E-mail: nagamura@tagen.tohoku.ac.jp

Abstract. The process of lithium-intercalation in spinel-structured lithium manganese oxide ($LiMn_2O_4$) particles was probed by a highly spatially resolved X-ray photoelectron microscopy technique using our "3D nano-ESCA" (three-dimensional nanoscale electron spectroscopy for chemical analysis) equipment. We found the change of the line shape in Li 1*s*-Mn 3*p* core-level spectra with lithiation/delithiation of the sample. Moreover, we have succeeded in detecting the photoelectron intensity mapping of chemically lithiated/delithiated $Li_{1\pm\delta}Mn_2O_4$ crystals. It suggests the homogeneous intercalation/deintercalation from the crystal surfaces instead of the crystal edges, and the particle size dependence of the intercalation tendency.

1. Introduction

Spinel-structured lithium manganese oxide ($LiMn_2O_4$)[1-3] has been extensively studied as a promising cathode material for lithium-ion batteries by virtue of its high voltage, low cost, and non-toxicity. However, high rate capability and long cycle life are needed for the application to high-power batteries for commercial use. In order to improve the specific power density, nanostructured electrodes including nanoparticles[4], nanowires[5], and nanofilms[6] have been widely investigated taking advantage of the large surface-to-volume ratio and the short lithium ion diffusion length.

Understanding the lithiation/delithiation process which is spatially inhomogeneous in a realistic battery[7] helps to realize stable cycle characteristics. However, lithium intercalation in $LiMn_2O_4$ nanostructures still remains unclear compared to other cathode materials such as $LiCoO_2$ and $LiFePO_4$.

Spectromicroscopic analysis with high spatial resolution is a powerful tool for elucidating the lithium intercalation mechanism in $LiMn_2O_4$ particles. Here we have performed nondestructive soft X-ray scanning photoelectron microscopy measurements using "3D nano-ESCA", equipped with a

[4] Present address: Institute of Multidisciplinary Research for Advanced Materials, Tohoku University, 2-1-1 Katahira, Aoba-ku, Sendai 980-8577, Japan.

Journal of Physics: Conference Series **502** (2014) 012013 doi:10.1088/1742-6596/502/1/012013

focused X-ray probe, a piezo-driven sample stage, and an angle-resolved photoelectron spectrometer[8].

2. Experimental

$LiMn_2O_4$ powders were synthesized by conventional solid-state reaction using Li_2CO_3 and MnO_2. Stoichiometric amounts of precursors were thoroughly mixed and then pre-sintered at 850 °C for 24 h in air. Then optional 10 wt% lithium chloride (LiCl) was added into the pre-sintered powder in order to accelerate crystal growth. The self-assembled spinel $LiMn_2O_4$ composite was obtained by mixing them and sintering at 850 °C for 24 h in air again. The $Li_{1-\delta}Mn_2O_4$ particles are prepared by chemical oxidation reaction of $LiMn_2O_4$ with ammonium persulfate ($(NH_4)_2S_2O_8$) in pure water. They were stirred for 72 h, then filtered and dried in air. To obtain a partially lithiated sample $Li_{1+\delta}Mn_2O_4$, chemical treatment was carried out by reacting $LiMn_2O_4$ with lithium iodide (LiI) in acetonitrile at 60 °C for 48 h in Ar atmosphere. The product was filtered and washed, and then dried in vacuum. The $Li_{1\pm\delta}Mn_2O_4$ crystals were characterized by X-ray diffraction measurements[9] and scanning electron microscopy (SEM) images.

Spectromicroscopic measurements were performed by the 3D nano-ESCA system which has an X-ray focusing system with a Fresnel zone plate and a modified VG Scienta R3000 analyzer. The spatial resolution is better than 100 nm. This system has been installed at the University-of-Tokyo outstation beamline, BL07LSU at Spring-8. The following results were all taken at excitation photon energy of 1000 eV.

3. Results and Discussion

Figure 1 shows a SEM image of $LiMn_2O_4$ particles at a magnification of 3000×. Spinel $LiMn_2O_4$ octahedral crystals with different size ranging from 1 μm to 5μm are observed.

Figure 1. SEM image of as-prepared single crystalline spinel $LiMn_2O_4$ particles.

Figure 2 displays the X-ray photoemission spectroscopy (XPS) spectra of Li 1*s* and Mn 3*p* core-levels for $LiMn_2O_4$ and chemically lithiated/delithiated $Li_{1\pm\delta}Mn_2O_4$ particles. In order to compensate charging effects, the C 1*s* peak at 284.5 eV was used as a reference for the determination of the core-level binding energies (BEs). The Li 1*s* core level has a BE of 55.0 eV. This energy shift to higher BE from Li metal (54.7 eV) indicates that lithium exists as Li^+ ions[10]. The peaks which have larger intensity at around 49.3 eV are attributed to the Mn 3*p* components. As Li concentration increases, the Li 1*s* components clearly increase. The Mn 3*p* peaks are deconvoluted into four peaks corresponding to the Mn^{3+} and Mn^{4+} contributions and their multiplets.

We have also measured valence band spectra for these samples. As shown in Figure 3, the valence band becomes sharper with increasing Li content, which is consistent with the previous report[4]. Furthermore, the Mn 3*d* e_g shoulder peak appears in the $Li_{1+\delta}Mn_2O_4$ sample, suggesting the increase of Mn^{3+} component.

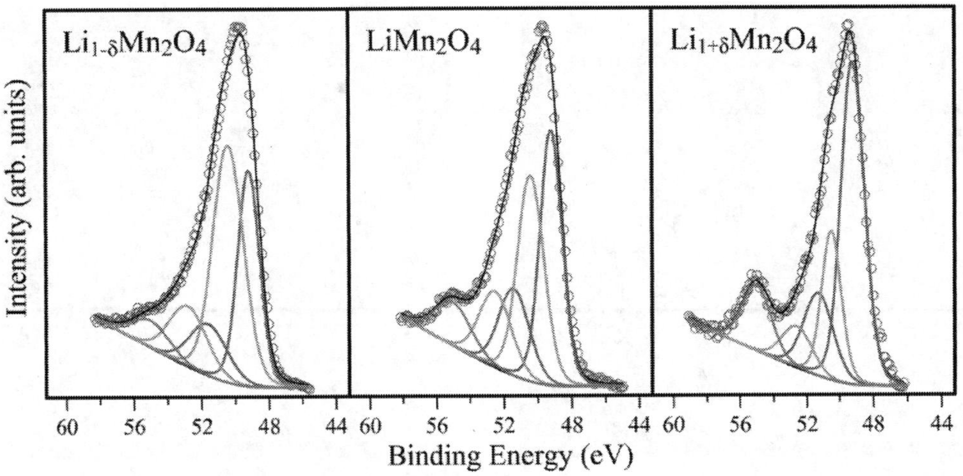

Figure 2. Photoelectron spectra of Mn 3*p* and Li 1*s* and their deconvolution for $LiMn_2O_4$ and $Li_{1\pm\delta}Mn_2O_4$ samples. Red curves indicate Li 1*s*. Green (Mn^{4+}) and blue (Mn^{3+}) curves indicate Mn 3*p* components.

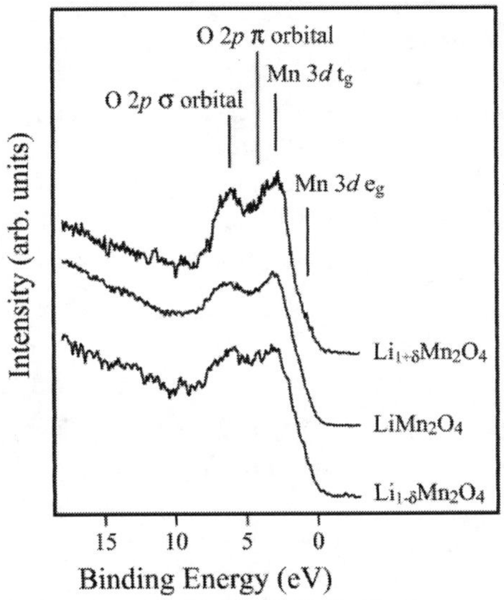

Figure 3. Photoelectron spectra of valence bands for $LiMn_2O_4$ and $Li_{1\pm\delta}Mn_2O_4$ samples.

For $Li_{1\pm\delta}Mn_2O_4$ particles, imaging analysis of XPS spectra has been performed. Figure 4 presents a Li 1*s* and Mn 3*p* photoelectron intesity map with binding energy ranging from 40 eV to 64 eV for the $Li_{1-\delta}Mn_2O_4$ particle. Pin-point measurements of XPS spectra exhibit the line shape shown in the left-hand side of Fig. 2 at any point on the $Li_{1-\delta}Mn_2O_4$ crystal in Fig. 4. This homogeneity suggests a possibility that lithium ions might be intercalated or deintercalated from the crystal surface instead of the particle edges.

On the other hand, Figure 5 shows a photoelectron intesity map covering the Li 1*s* and Mn 3*p* core-level region for the $Li_{1+\delta}Mn_2O_4$ particles. The spectrum of each point in the map is decomposed into two peaks, Li 1*s* components (blue contrast) and unresolved Mn 3*p* components consisting of Mn^{4+} and Mn^{3+} (grayscale). Fully charged particles (enclosed by white circles in Fig. 5) and fully discharged particles are dominant although a state-of-charge particle (enclosed by white dashed circles in Fig. 5) is also detected, which conflicts with the core-shell model[11]. Futhermore, the averaged size of lithium-rich particles seems to be smaller than that of lithium-poor particles. This difference in the particle size suggests that smaller particles are lithiated preferentially.

Figure 4. X-ray photoelectron intensity map of a chemical oxidized $Li_{1-\delta}Mn_2O_4$ particle.

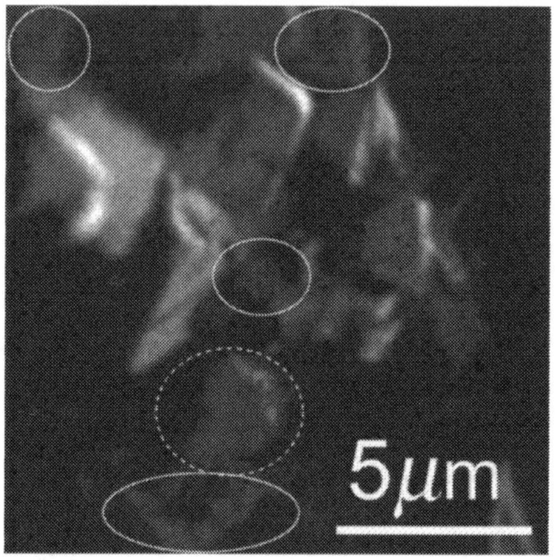

Figure 5. X-ray photoelectron intensity map of a chemical reduced $Li_{1+\delta}Mn_2O_4$ particle. The grayscale corresponds to the Mn 3*p* core-level intensity and the blue contrast to the Li 1*s* core-level intensity, respectively.

In conclusion, we have investigated the lithiation/delithiation process of spinel $LiMn_2O_4$ crystal particles by using the 3D nano-ESCA, a spectromicroscopic technique with a high spatial resolution. We have confirmed that the line shape of Li 1*s* core level spectra change as the sample are lithiated/delithiated. The photoelectron intensity mapping of the Li 1*s* and Mn 3*p* core-level region for $Li_{1\pm\delta}Mn_2O_4$ particles suggests the homogeneous intercalation from crystal surfaces and the particle size dependence of the lithiation process.

Acknowledgments
This research is supported by the Japan Society for the Promotion of Science (JSPS) through its "Funding Program for World-Leading Innovative R&D on Science and Technology (FIRST Program)."

References
[1] Gummow R J *et al.* 1994 *Solid State Ionics* **69** 59.
[2] Amatucci G and Tarascon J M, 2002 *J. Electrochem. Soc.* **149** K31.
[3] Hirayama M *et al.* 2007 *J. Electrochem. Soc.* **154** A1065.
[4] Talik E, Załóg A *et al.* 2012 *Cryst. Res. Technol.* **47** 351.
[5] Hosono E *et al.* 2009 *Nano Lett.* **9** 1045.
[6] Tang S B *et al.* 2006 *Electrochimica Acta* **52** 1161.
[7] Chueh W C *et al.* 2013 *Nano Lett.* **13** 866.
[8] Horiba K *et al.* 2011 *Rev. Sci. Instrum.* **82** 113701.
[9] Thackeray M M *et al.* 1984 *Mater. Res. Bull.* **19** 179.
[10] Ramana C V *et al.* 2005 *Surf. Interface Anal.* **37** 412.
[11] Padhi A K *et al.* 1997 *J. Electrochem. Soc.* **144** 1188.

Journal of Physics: Conference Series **502** (2014) 012014 doi:10.1088/1742-6596/502/1/012014

Anomalous x-ray scattering studies of functional disordered materials

S Kohara[1], H Tajiri[1], C H Song[1], K Ohara[1], L Temleitner[1,2], K Sugimito[1], A Fujiwara[1], L Pusztai[2], T Usuki[3], S Hosokawa[4], Y Benino[5], N Kitamura[6] and K Fukumi[6]

[1] JASRI/ SPring-8, 1-1-1 Kouto, Sayo-cho, Sayo-gun, Hyogo 679-5198, Japan
[2] Wigner Research Centre for Physics, Hungarian Academy of Sciences, H-1525 Budapest, P. O. Box 49, Hungary
[3] Department of Material and Biological Chemistry, Faculty of Science, Yamagata University, 1-4-12 Koshirakawa, Yamagata 990-8560, Japan
[4] Department of Physics, Graduate School of Science and Technology, Kumamoto University, 2-39-1 Kurokami, Kumamoto, 860-8555, Japan
[5] Faculty of Environmental Science and Technology, Okayama University, 1-1-1 Tsushima-naka, Kita-ku, Okayama, 700-8530, Japan
[6] Reserch Institute for Ubiquitous Energy Devices, National Institute of Advanced Industrial Science and Technology, 1-8-31 Midorigaoka, Ikeda, Osaka 563-8577, Japan

E-mail: kohara@spring8.or.jp

Abstract. We have developed anomalous x-ray scattering (AXS) spectrometers, that employ intrinsic Ge detectors and crystal analyzers, at SPring-8. The use of LiF analyzer crystal provides us with an energy resolution of ~ 12 eV. Furthermore, it has been established that the use of AXS technique is essential to reveal the relationship between the atomic structure and its function of a fast phase-change material, $Ge_2Sb_2Te_5$. We were able to address the issue of why the amorphous phase of fast phase change materials is stable at room temperature for a long time despite the fact that it can rapidly transform to the crystalline phase by using a combination of AXS and large scale density functional theory-based molecular dynamics simulations.

1. Introduction

Scattering experiments provide a powerful probe of the structure of disordered materials. X-ray and neutron diffraction have widely been used to make direct measurements of coordination numbers and bond lengths: Fourier transforming the total structure factor, $S(Q)$, yields the pair distribution function, $g(r)$, ($Q=4\pi\sin\theta/\lambda$, 2θ: scattering angle, λ: wavelength of photons or neutrons) a function in real space, by the following equation:

$$g(r) = 1 + \frac{1}{2\pi^2 \rho r} \int_{Q_{min}}^{Q_{max}} Q[S(Q)-1]\sin(Qr)dQ, \qquad (1)$$

where ρ is the atomic number density. In order to obtain detailed structural information from scattering experiments, it is necessary to measure $S(Q)$ up to high Q, since the resolution in real space is determined by Q_{max}. In the case of neutron diffraction, spallation sources make it possible to obtain high Q data with the time of flight technique, so that pulsed neutron diffraction is a powerful tool for

[1] To whom any correspondence should be addressed.

structural studies of disordered materials. On the other hand, with the advent of third generation synchrotron sources, it has become possible to perform precise x-ray diffraction experiments on disordered materials, taking accurate diffraction data up to high Q by the use of high intensity photons of high-energy (E > 50keV) [1]. Furthermore, a combination of x-ray and neutron diffraction opens up the way to study the structure of disordered materials at atomic to nanoscale [2,3].

Since the development of novel functional disordered materials, such as fast phase-change materials and bulk metallic glasses, has become very important, it is of particular interest to understand the relationship between the atomic/electronic structure and the function of materials. To study complicated functional disordered materials by diffraction/scattering techniques, atom specific probe is necessary. Although x-ray absorption fine structure (XAFS) has widely been used to probe local environment of disordered materials, atom sensitive structure information beyond nearest-neighbor correlations would be desirable for understanding the origin of 'order within disorder' [4], hidden in disordered materials.

The anomalous x-ray scattering (AXS) technique [5,6] is a powerful tool to probe atom sensitive structure in disordered materials. Although the technique of AXS was proposed in the last century, it has not been a particularly popular technique due to difficulties during experiment and data analysis. In this article, we report recent instrumental developments of dedicated AXS spectrometers at SPring-8 and review recent activities on fast phase-change materials, by combining AXS and large scale density functional theory (DFT)-based molecular dynamics (MD) simulations.

2. Anomalous x-ray scattering (AXS)[5,6]

The AXS technique utilizes the anomalous change of the atomic form factor of a specific element if the incident x-ray energy is near an absorption edge of the respective element. The complex atomic form factor of an element is given as

$$f(Q,E) = f_0(Q) + f'(E) + if''(E), \tag{2}$$

where $f_0(Q)$ is the usual energy-independent term, and $f'(E)$ and $f''(E)$ are the real and imaginary parts of the anomalous term, respectively. In general, $f(Q,E)$ is governed by the Q-dependent $f_0(Q)$ in a normal x-ray scattering process, and the anomalous term is negligible. When the incident x-ray energy approaches an absorption edge of a constituent element, however, $f'(E)$ has a large negative minimum and $f''(E)$ shows an abrupt jump near the corresponding absorption edge energy of each element. One can utilize the difference between two scattering spectra near an absorption edge of the ith element $\Delta_i I$, where one is typically measured at about 10 eV below, while the other at about 100 eV below the absorption edge (E_{near} and E_{far}, respectively). This differential intensity is expressed as

$$\alpha_i \Delta_i I(Q, E_{far}, E_{near}) = \Delta_i\left[\langle f^2\rangle - \langle f\rangle^2\right] + \Delta_i\left[\langle f\rangle^2\right]\Delta_i S(Q), \tag{3}$$

where α_i is a normalization constant, and $\Delta_i[\ ]$ indicates the difference of values in the brackets at the energies of E_{far} and E_{near}, close to the absorption edge of the ith element. The $\Delta_i S(Q)$ functions are given by a linear combination of $S_{ij}(Q)$ as

$$\Delta_i S(Q) = \sum_{i=1}^{N}\sum_{j=1}^{N} w_{ij}(Q, E_{far}, E_{near}) S_{ij}(Q), \tag{4}$$

where the weighting factors are given by

$$w_{ij}(Q, E_{far}, E_{near}) = x_i x_j \frac{\Delta_i\left[f_i f_j\right]}{\Delta_i\left[\langle f\rangle^2\right]}. \tag{5}$$

It should be noted that compared to $S(Q)$, $\Delta_i S(Q)$ highly enhances the contribution of the ith element related partial structure factors and suppresses other partials.

3. Instrumentation and typical AXS data

Two AXS spectrometers were developed at the single crystal structure analysis beamline BL02B1 and at the surface and interface structures beamline BL13XU [7] Figure 1 (a) shows a photograph of the AXS spectrometer developed at BL02B1. The standard setup of the 4-circle diffractometer with a Ge detector is used. The energy resolution of the spectrometer is ΔE=500eV at 12keV. The other AXS spectrometer built at beamline BL13XU, which is the dedicated high-resolution spectrometer employing a LiF analyzer crystal, is shown in Fig. 1(b). The spectrometer consists mainly of the standard six-circle diffractometer, slits, a beam stop, an analyzer crystal, and a NaI(Tl) scintillation detector.

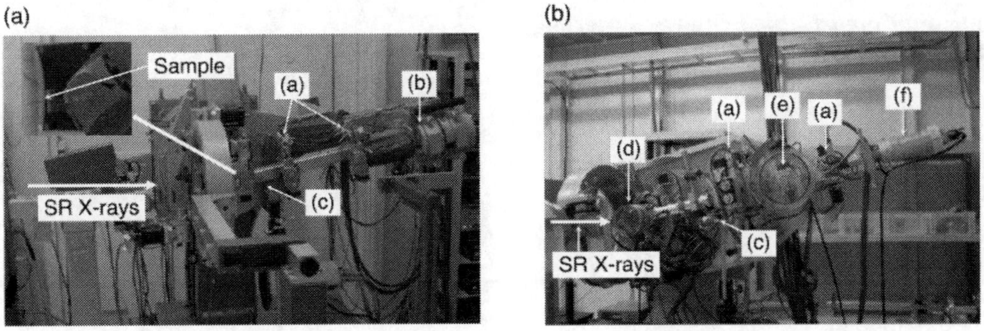

Figure 1. AXS spectrometer developped at (a) BL02B1 and at (b) BL13XU beamlines. (a) slits, (b) Ge detector, (c) beam stop, (d) vacuum chamber for sample, (e) LiF analyzer crystal, (f) NaI(Tl) scintillation detector

The ideal energy resolution ($\Delta E/E$) of an analyzer crystal is given by

$$\Delta E/E = \Delta\theta/\tan\theta_B, \quad (6)$$

where $\Delta\theta$ and θ_B are the rocking-curve width and the Bragg angle, respectively. By choosing an analyzer crystal with an adequate rocking-curve width, both a sufficient counting statistics and a sufficient energy resolution, to discriminate the contribution of fluorescence and of Compton scattering, can be achieved. In the case of 12 keV X-rays, LiF(200) reflection is appropriate choice for this purpose. In addition, it is notable that a LiF crystal is a low-cost product. A vacuum chamber is installed to suppress air scattering around the sample, which is very important to measure accurately when dealing with disordered materials. The vacuum chamber is useful for hygroscopic samples, too. High resolution in terms of energy is an important factor for precise AXS measurements, because fluorescence is induced by incident x-rays while measuring near the absorption edge. Figure 2(a) shows a typical rocking curve of the LiF(200) crystal for x-rays near 12 keV. The energy resolution is estimated to be 12 eV in FWHM, which allows us to discriminate the contributions from fluorescence and Compton scattering. Furthermore, the energy resolution of the LiF(200) crystal is approximately seven times better than that obtained by sagittal focusing of a cylindrical mosaic graphite crystal [6]. The total structure factor $S(Q)$ of amorphous $Ge_2Sb_2Te_5$ (a-GST) was measured at the Sb near edge. As it can be seen (Figure 2(b)), it is confirmed that raw scattering data can properly be normalized to give $S(Q)$, demonstrating that the high-energy resolution spectrometer succeeded in discriminating the contribution from fluorescence of Sb and from Compton scattering.

Figure 3(a) shows the differential structure factors $\Delta S(Q)$ for Nb of $25BaO$-$50Nb_2O_5$-$25P_2O_5$ glass, together with the total structure factor $S(Q)$ obtained by high-energy x-ray diffraction (high-energy x-ray diffraction beamline BL04B2). Corresponding real space data are shown in Fig. 3(b). The contrast between normal x-ray diffraction data (black) and Nb K edge AXS data is excellent. Although the resolution in real space is different between two data sets, due to different Q_{max}, it is noted that the Ba-O correlation can be reasonably eliminated by AXS data. Furthermore, significant oscillations can be observed in 6 - 9 Å of AXS data, implying that niobium atoms exhibit specific ordering beyond the first coordination distance.

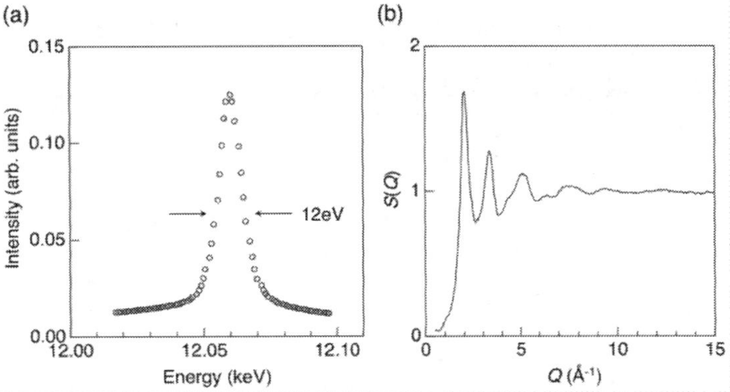

Figure 2. (a) Rocking curve of the LiF(200) crystal for x-rays near 12 keV and (b) total structure factor $S(Q)$ for a-GST measured at Sb near edge (E=30.462 keV).

4. Application to fast-phase change materials [8]

Although data storage on a DVD/Blu-ray is part of our digital world, the physical basis of the storage mechanism has not been fully understood in detail. Information is stored on a DVD in the form of microscopic bits in a thin layer of a polycrystalline alloy containing several elements. The bits can have a disordered, amorphous or an ordered, crystalline structure. The transition between the two phases lasts only tens of nanoseconds and can be triggered by a laser pulse [9]. The combination of AXS, HEXRD and Reverse Monte Carlo [10-12] / DFT-MD simulations has enabled the structures of both phases to be determined for the first time and allowed the development of a model that explains the rapid phase change and high durability. Figure 4(a) shows total scattering structure factors, $S(Q)$, for crystalline (c-) GST and a-GST. The diffraction pattern of c-GST exhibits sharp Bragg peaks and a diffuse scattering pattern, while only the latter appears for a-GST. It is demonstrated that $S(Q)$ of the RMC model (solid line) agrees well with experimental data (open circles). Differential structure factors, $\Delta S(Q)$, for Sb and Te, obtained from AXS measurements performed at BL02B1 using a germanium detector are shown in Fig. 4(b). As it can be seen in the figure, there are substantial differences between Sb- and Te-related data at $Q < 6$ Å^{-1}, demonstrating that AXS measurements can distinguish Sb (atomic number is 51) related correlation and Te (atomic number is 52) related correlation beyond the nearest neighbor distance.

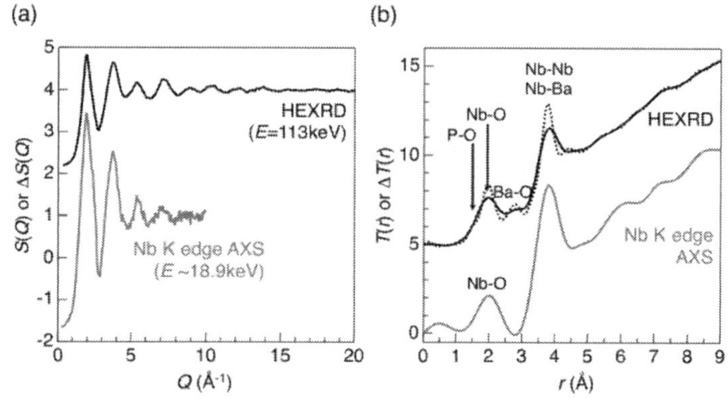

Figure 3. (a) Differential structure factors $\Delta S(Q)$ for Nb of $25BaO$-$50Nb_2O_5$-$25P_2O_5$ glass together with the total structure factor $S(Q)$ obtained by high-energy x-ray diffraction measurement and (b) corresponding total correlation functions. The $S(Q)$ and $T(r)$ of HEXRD are displaced upward by 3 and 5, respectively, for clarity. Solid curves and dashed curve of $T(r)$ are obtained by Fourier transformation with Q_{max}=9.9 Å^{-1} and 25 Å^{-1}, respectively.

Coordination numbers in a-GST are calculated up to 3.2 Å based on the RMC/DFT-MD model. In c-GST, both Ge and Sb are sixfold coordinated while the total coordination number of Te is 4.8, due to a vacancy on the Ge/Sb mixing site. These results are in line with the average structure of rocksalt, suggesting that disorder in the crystalline phase does not have any influence on average coordination numbers. Baker et al. reported on the basis of EXAFS measurements [13] that both Ge and Sb obey the so-called '8-N rule' and that Te is overcoordinated in a-GST. Akola et al. reported overcoordinated Sb and Te on the basis of atomic configurations obtained by RMC/DFT-MD simulation using HEXRD data only [14]. In our new RMC model, refined by employing AXS data, total coordination numbers for Ge, Sb, and Te are 3.83, 3.06, and 2.45, respectively. Antimony has a lower coordination number than in the previous RMC/DFT-MD model (N_{Sb}=3.31 in ref. 14). As a result, now both Ge and Sb obey the '8-N rule' and only Te is overcoordinated. These results agree well with results of EXAFS measurements [13] and with that in

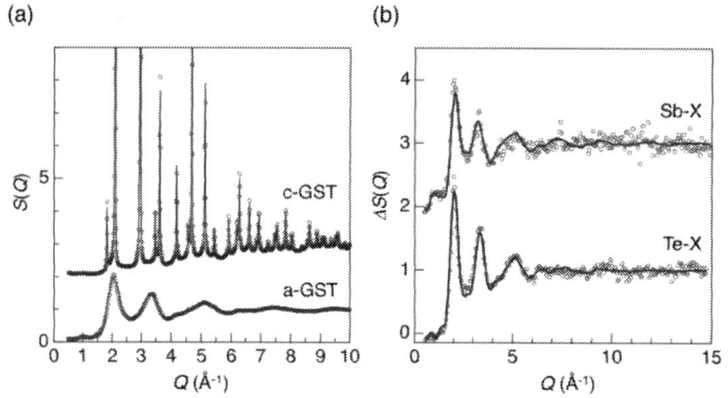

Figure 4. Comparison between the experimental structure factors (open circle) and results of RMC modelling (solid line). (a) Total scattering data for c-GST and a-GST obtained by HEXRD measurements. (b) $\Delta S(Q)$ for a-GST obtained by AXS measurements at Sb and Te K absorption edges. The $S(Q)$ of c-GST and the $\Delta S(Q)$ for Sb are displaced upward by 2 for clarity.

Journal of Physics: Conference Series **502** (2014) 012014 doi:10.1088/1742-6596/502/1/012014

Olson et al.'s argument, in which they mentioned the possibility of overcoordinated Te if there are no homopolar (Ge-Ge, Sb-Sb, Te-Te) bonds in a-GST [15].

For obtaining statistical structural features beyond the nearest coordination distance, ring statistics were calculated using a shortest path analysis [16]. As can be seen in Fig. 5(a), fourfold and sixfold rings consisting of '–Ge(Sb)–Te–' units are dominant in c-GST. The existence of sixfold rings is due to the 20% vacancies on Ge/Sb mixing sites. As discussed in ref. 14, small '–Ge(Sb)–Te–' rings such as fourfold, fivefold, and sixfold rings are dominant in a-GST. The numbers of rings consisting of '–Ge–Te–' units and of '–Sb–Te–' units in a-GST are shown in Fig. 5(b). Our new RMC model obtained by the introduction of AXS data suggests that large fractions of fourfold and sixfold rings are mainly formed by Ge-Te bonds. Therefore it is suggested that the core network constructed by Ge-Te covalent bonds in a-GST is similar to that in c-GST, and that this core network plays an important role in stabilizing the amorphous phase at room temperature for long time.

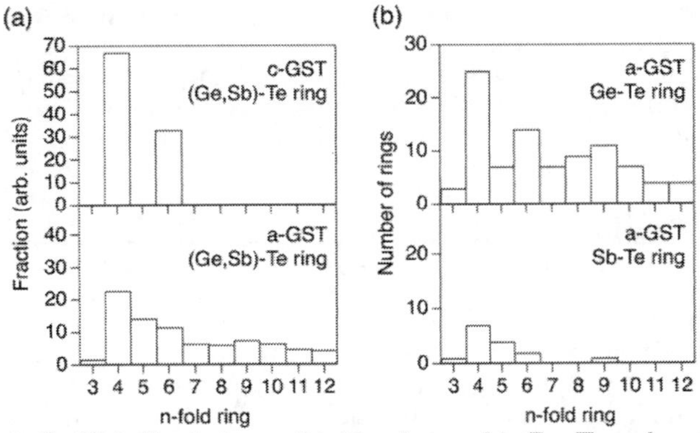

'–Ge(Sb)–Te–' rings. (b) Numbers of '–Ge–Te– rings' and '–Sb–Te– rings' in a-GST.

To understand the atomic ordering by Ge-Te and Sb-Te bonds in detail, connectivities of atoms were calculated, with varying the maximum distance within which atomic pairs were considered to be connected. As can be seen in Fig. 6(a), about 60% of Ge-Te bonds up to 3.2 Å form a continuous core network whereas Sb-Te bonds up to 3.2 Å do not form such a network. This distance corresponds to the covalent bond distance determined by DFT-MD simulation [17] and both Ge and Sb satisfy the '8-N rule' at 3.2 Å in our new RMC model (N_{Ge}=3.83 and N_{Sb}=3.06). Atomic configurations of Ge-Te and Sb-Te with bonds considered up to 3.2 Å are shown in Fig. 6(b). It can be clearly distinguished that Ge-Te bonds form a core network, which stabilizes the amorphous phase, whereas Sb-Te bonds do not. Furthermore it is evident that Ge-Te bonds form large fractions of fourfold (highlighted by blue) and sixfold (highlighted by light blue) rings. On the other hand, about 70 % of the Sb-Te pairs form a pseudo network, as can be seen in Figs. 6(a) and 6(c), when bonds are considered up to 3.5 Å (this way, the coordination numbers of Te around Ge ($N_{Ge\text{-}Te}$) and Sb ($N_{Sb\text{-}Te}$) are 4.08 and 3.08, respectively). In other words, the pseudo Sb-Te network becomes visible when we extend the correlation distance up to 3.5 Å, while such a feature cannot be observed for Ge-Te connectivities.

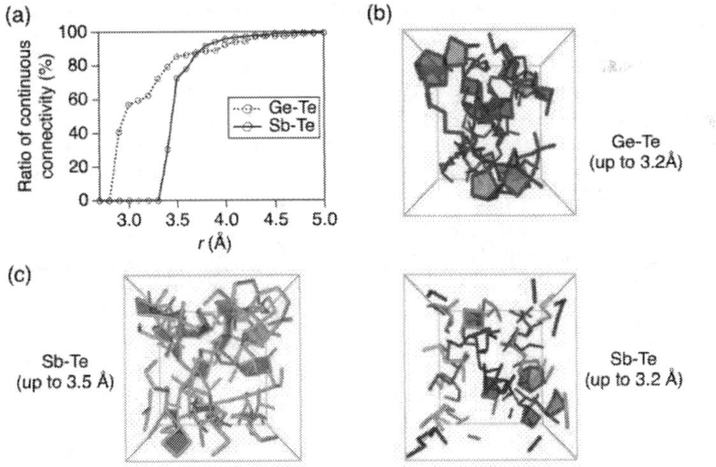

Figure 6. Atomic configuration and connectivity of Ge-Te and Sb-Te in a-GST obtained by the RMC model. (a) Connectivity of Ge-Te and Sb-Te with different r_{max}. (b) Atomic configuration of Ge-Te and Sb-Te (connectivity of Ge-Te and Sb-Te considered up to 3.2 Å). (c) Atomic configuration of Sb-Te (connectivity of Sb-Te considered up to 3.5 Å). The theeefold, fourfold, fivefold, and sixfold rings are highlighted by red, blue, green, and light blue, respectively.

On the basis of structural features in a-GST displayed above, we discuss the role of each elements in the phase-change process. As highlighted by red in Fig. 6(b), Ge-Te bonds significantly contribute to the network containing by fourfold rings. Therefore Ge and Te can be recognized as network forming elements that stabilize the amorphous phase at room temperature for long time. The core Ge-Te network may remain in the crystalline phase, which feature explains rapid crystallization. As highlighted by yellow in Fig. 6(c), Sb-Te correlations beyond the nearest coordination distance form a pseudo Sb-Te network. This unusual atomic ordering in terms of Sb-Te correlations can be ascribed to the combination of two positively charged atoms (Ge, -0.22 electrons; Sb, 0.32 electrons; Te, 0.22 electrons [19]) and allows the amorphous phase to form critical nuclei via forming Sb-Te bonds by small atomic displacements of antimony and tellurium atoms. Thus our finding can reasonably explain the reason why the crystallization of amorphous GeTe (a-GT) is very fast in PC-RAM [18] and the reason why the crystallization of a-GST is faster than that of a-GT in DVD [19].

5. Conclusion

We have succeeded in developing AXS spectrometers at SPring-8. The high-energy resolution spectrometer provides us with sufficient energy resolution to discriminate the contribution of fluorescence and of Compton scattering. In addition, the photon flux of x-rays from a SPring-8 undulator is high enough to perform AXS measurement of disordered materials. Therefore a combination of HEXRD and AXS is a powerful tool to understand the atomic arrangement of complicated functional disordered materials. Furthermore a combination of advanced synchrotron radiation measurements and massive super computing is essential to uncover the relationship between the structure of functional disordered materials at atomic/ electronic level [20] and their function.

Acknowledgements

The synchrotron radiation experiments were approved by the Japan Synchrotron Radiation Research Institute (Proposal Nos. 2007A1180, 2008A1998, 2009A1058, 2010A1217, 2010B1182, 2010B1183, 2011A0030, 2012A1618, 2012A1216, 2012B1185, 2013A1141, 2013A1349). Financial support for LP and LT was provided by the National Development Agency of Hungary (NFÜ), via grant no. TÉT_10-1-2011-0004 and by the National Scientific Research Fund of Hungary (OTKA), grant no. 083529.

References

[1] Poulsen H F, Neuefeind J, Neumann H-B, Schneider J R and Zeidler M D 1995 *J. Non-Cryst. Solids* **188** 63
[2] Kohara S, Itou M, Suzuya K, Inamura Y, Sakurai Y, Ohishi Y and Takata M 2007 *J. Phys.: Condens. Matter* **19** 506101
[3] Benmore C J, 2012 *ISRN Mater. Sci.* doi:10.5402/2012/852905
[4] Salmon P S 2002 *Nature Mater.* **1** 87
[5] Waseda Y 2002 *Anomalous x-ray scattering for materials characterization* (Heidelberg, Springer)
[6] Hosokawa S, Wang Y, Bérar J-F, Greif J, Pilgrim W-C and Murase K 2002 *Z. Phys. Chem.* **216** 1219
[7] Sakata O, Furukawa Y, Goto S, Mochizuki T, Uruga T, Takeshita K, Ohashi H, Ohata T, Matsushita T, Takahashi S, Tajiri H, Ishikawa T, Nakamura N, Ito M, Sumitani K, Takahashi T, Shimura T, Saito A and Takahasi M 2003 *Surf. Rev. Lett.* **10** 543
[8] Ohara K, Temleitner L, Sugimoto K, Kohara S, Matsunaga T, Pusztai L, Itou M, Ohsumi H, Kojima R, Yamada N, Usuki T, Fujiwara A and Takata M 2012 *Adv. Func. Mater.* **22** 2251
[9] Wuttig M and Yamada N 2007 *Nature Mater.* **6** 824
[10] McGreevy R L, Pusztai L 1988 *Molec. Simul.* **1** 359
[11] Mellergård A, McGreevy R L 1999 *Acta. Cryst.* **55** 783
[12] Gereben O, Jóvári P, Temleitner L and Pusztai L 2007 *J. Optoelectron. Adv. Mater.* **9** 3021
[13] Baker D A, Paesler M A, Lucovsky G, Agarwal S C and Taylor P C 2006 *Phys. Rev. Lett.* **96** 255501
[14] Akola J, Jones R O, Kohara S, Kimura S, Kobayashi K, Takata M, Matsunaga T, Kojima R and Yamada N. 2009 *Phys. Rev. B* **80** 020201
[15] Olson J K, Li H and Taylor P C 2005 *J. Ovonic Res.* **1** 1
[16] Guttman L 1990 *J. Non-Cryst. Solids* **116** 145
[17] Akola J and Jones R O 2008 *J. Phys.: Condens. Matter* **20** 465103
[18] Bruns G, Merkelbach P, Schlockermann C, Salinga M, Wuttig M, Happ T D, Philipp J B and Kund M, 2009 *Appl. Phys. Lett.* **95** 043108
[19] Coombs J H, Jongenelis A P J M, Es-Spiekman W v and Jacobs B A J 1995 *J. Appl. Phys.* **78** 4918
[20] Akola J, Kohara S, Ohara K, Fujiwara A, Watanabe Y, Masuno A, Usuki T, Kubo T, Nakahira A, Nitta K, Uruga T, Weber J K R and Benmore C J 2013 *Proc. Natl. Acad. Sci. USA* **110** 10129

Development of an in-vacuum diffractometer for resonant soft X-ray scattering

H Nakao[1,2], Y Yamasaki[1], J Okamoto[1], T Sudayama[1], Y Takahashi[1,2], K Kobayashi[1], R Kumai[1,2], and Y Murakami[1]

[1]Condensed Matter Research Center and Photon Factory, Institute of Materials Structure Science, High Energy Accelerator Research Organization (KEK), Tsukuba 305-0801, Japan
[2]CREST, Japan Science and Technology Agency (JST), Tokyo 102-0076, Japan

E-mail: hironori.nakao@kek.jp

Abstract. Resonant soft X-ray scattering measurements become one of powerful techniques to probe charge, orbital and spin orderings in strongly correlated electron systems. We have developed new in-vacuum diffractometer with a large vacuum chamber to solve experimental limitations of in-vacuum measurement in soft X-ray energy region, and for future instrumental developments. In this paper, the features of the diffractometer are presented.

1. Introduction

Various interesting physical properties, such as colossal magnetoresistance e ect and magneto-electric e ect, were discovered in strongly correlated electron systems (SCES). There the strong correlation among orbital, charge, and spin degrees of freedom play important roles. The study of these electronic ordering states is essential to understand the phenomena microscopically. Resonant X-ray scattering (RXS) at the K-edge becomes a powerful tool for observing the spatial ordering of charge and orbital degrees of freedom in $3d$ transition metal oxides. [1, 2] The RXS signal at K-edge ($1s \rightarrow 4p$ transition energy) reflects the $4p$ electronic state. On the other hand, the RXS at $L_{2,3}$-edge ($2p \rightarrow 3d$ transition energy) can probes the $3d$ electronic state directly, and the signal of the resonant magnetic scattering is strongly observed at the $L_{2,3}$-edge of d electron system. Moreover, K-edges of O, S, and P ions, which play an key role for the itinerancy through the hybridization with the metal ion, becomes also observable by using soft X-ray. Hence the resonant soft X-ray scattering (RSXS) measurements become to be performed globally. [1, 2]

We have developed a new in-vacuum di ractometer to perform the RSXS experiment. The RXS signal reflecting the change of physical properties is weak in general. Therefore, the improvement of signal-to-noise ratio is important to detect the signal. Here we built the di ractometer with a large vacuum chamber, and slit and detector are installed in the vacuum chamber to reduce the background, such as fluorescence. The speci cations of the di ractometer and future plans are presented.

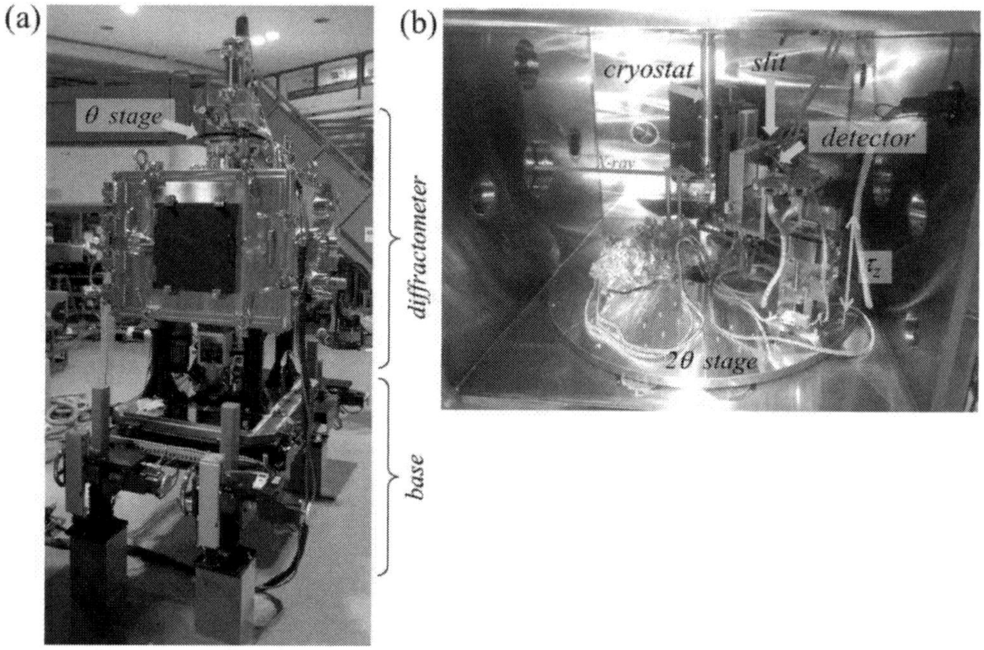

Figure 1. (a) Appearance of the di ractometer and the alignment base. (b) The di ractometer of the inside of the vacuum chamber. The stages, cryostat, and detector system are shown.

2. Instrument

2.1. Diffractometer

We developed a general in-vacuum di ractometer dedicated to the resonant soft X-ray scattering measurement. This di ractometer with a large vacuum chamber is shown in Fig. 1. Utilizing the large vacuum chamber, we can install several devices, the slits, detectors, and so on, on the di ractometer. Moreover, it have a possibility to install new devices in future. Vacuum pumping is performed by a turbo molecular pump (TG800FCAB) and a dry pump (NeoDry30E). The base pressure of $1 \times 10^{\ 5}$ Pa has been achieved up to now.

The di ractometer mainly consists of 2θ, θ, and τ_z stages as shown in Fig. 2(a), and the speci cations are listed in Table 1. The horizontal plane is the scattering plane of this di ractometer. In soft x-ray region, the wavelength is one order longer than that of hard x-ray. Hence one order lower angle resolution is enough for the experiments. In general two-circle di ractometer, the tilt of sample is aligned to set properly a scattering vector with respect to the scattering plane. By changing the height of the detector using the τ_z stage, on the other hand, this di ractometer can detect the signal out of the scattering plane without an alignment of sample tilt. This experimental con guration gives us new option of the measurements.

The sample position can be aligned by three orthogonal translation, x_s, y_s, and z_s, stages. In temperature dependence experiments, the sample position moves owing to the thermal expansion of cold head of cryostat. In organic materials, radiation damage often becomes a problem. The beam position on the sample should be changed depending on the irradiation time. Moreover, the stroke of z_s stage is large enough to set several samples on the cryostat, and then the sample for measurement can be changed automatically using the z_s stage.

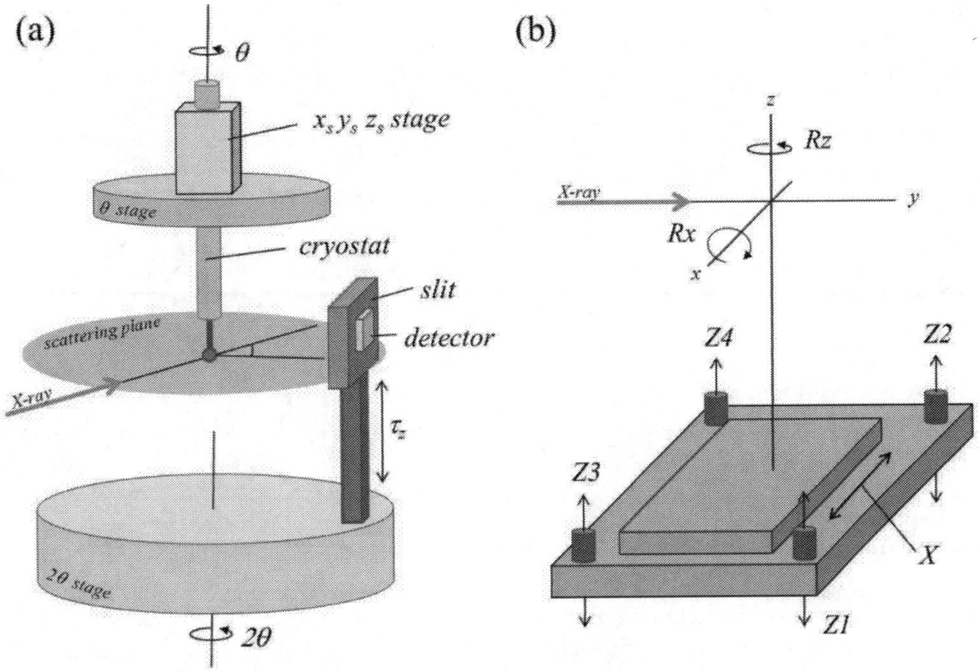

Figure 2. Schematic view of (a) di ractometer and (b) alignment base. The movements of rotation and linear stages are illustrated.

Table 1. Speci cation lists of di ractometer and alignment base.

Speci cations	2θ	θ	τ_z	x_s	y_s	z_s
Travel range	360°	360°	80 mm	25 mm	25 mm	200 mm
Resolution	0.0003°	0.0003°	1 μm	0.1 μm	0.1 μm	0.2 μm
Repeatability	$<$ 0.0005°	$<$ 0.0005°	$<$ 2 μm	$<$ 1 μm	$<$ 1 μm	$<$ 1 μm

Speci cations	X	$Z1, Z2, Z3, Z4$
Travel range	50 mm	200 mm
Resolution	0.4 μm	0.1 μm
Repeatability	$<$ 0.5 μm	$<$ 0.5 μm

2.2. X-ray detecting system

In the RXS experiment, the fluorescence from the sample usually becomes main background. Hence the reduction of background is important to detect the weak RXS signal reflecting the change of physical property. First, the cross slit screen with four blades is placed between the sample and the detector. [3] The intensity of the fluorescence is proportional to the solid angle of detecting area of detector, while the Bragg spot has a sharp pro le. The slit is used to tune the solid angle in order to improve the signal-to-noise ratio. Next a Silicon drift detector (SDD) with good energy resolution (about 100 eV) is installed. The fluorescence with energy far from the incident X-ray energy can be completely eliminated by using the SDD. By using this system, we succeeded in observing a weak RXS signal in several experiments. Even in x-ray absorption spectroscopy (XAS) measurement, the signal-to-noise ratio is remarkably improved, since the fluorescence from other elements can be excluded by the SDD.

2.3. Cryostat

He-flow type cryostat is installed for the low temperature experiment, and achieves the cold head temperature from 300 K to <10 K. However, radiation shield, which is generally used in low temperature experiments, cannot be utilized in soft X-ray region. Because the incident and scattered X-rays are also absorbed by the radiation shield. Hence the sample surface is heated up by the radiation, and it is di cult to estimate the real sample temperature. By using this cryostat and the di ractometer, the transition temperature less than 20 K could be detected up to now.

2.4. Alignment base

In order to align the di ractometer within the synchrotron beam, the alignment base are used as shown in Figs. 1(a) and 2(b) Generally, the alignment can be performed by two linear (X, Z) and two rotational (R_x and R_z) dimensions. The base has the X and Z stages, and the R_x rotation is virtually performed using $Z1$, $Z2$, $Z3$, and $Z4$ axes. The R_z rotation is substituted by the 2θ rotation.

3. Future prospects

In RXS experiments, polarization and azimuthal angle dependences become a key information especially to determine the symmetry of ordered orbital and the direction of magnetic moment. Moreover, the charge, orbital, and spin ordered states are coupled each other in SCES. The RXS signal generally contains multiple components reflecting charge, orbital, and spin states and local crystal structures. The RXS components can be separated utilizing the polarization and azimuthal angle dependences. Therefore, we plan to install a sample rotating stage and a polarizer. The analyzer stage can be mounted, taking advantage of the large vacuum chamber.

4. Conclusion

In order to perform the RXS measurement in soft X-ray region, we have developed new in-vacuum di ractometer and the related systems. The di ractometer was installed in the beamlines, BL-16A (300~1500 eV) and BL-11B (1.7~5.0 keV), at the Photon Factory, and was commissioned. Utilizing the developed di ractometer, not only the RXS but also the XAS measurements were carried out to elucidate the electronic states, and the ordering of the charge, spin, and orbital states in SCES. [4, 5, 6]

Acknowledgments

The authors thank Y. Honda for special supports to design and construction of the di ractometer. This work was supported by a Grants-in-Aid for Scienti c Research (S) (No. 21224008) and (B) (No. 25286090) from the Ministry of Education, Culture, Sports, Science, and Technology, Japan, by JSPS, FIRST program "Quantum Science on Strong Correlation", and by CREST, Japan Science and Technology Agency. This study has been carried out under the approval of the Photon Factory Program Advisory Committee (Proposal Nos. 2009S2-008 and 2012S2-005).

[1] Matsumura T, Nakao H, and Murakami Y 2013 *J. Phys. Soc. Jpn.* **82** 021007
[2] Beale T A W, et al. 2012 *Eur. Phys. J. Special Topics* **208** 89
[3] Okamoto J, Horigane K, Nakao H, Amemiya K, Kubota M, Murakami Y and Yamada K 2013 *J. Phys.: Conf. Ser.* **425** 202003
[4] Okamoto J, Horigane K, Nakao H, Amemiya K, Kubota M, Murakami Y and Yamada K 2011 *J. Electron Spectrosc. Prlat. Phenom.* **184** 224
[5] Takahashi R, Okazaki R, Yasui Y, Terasaki I, Sudayama T, Nakao H, Yamasaki Y, Okamoto J, Murakami Y, and Kitajima Y 2012 *J. Appl. Phys.* **112** 073714
[6] Y. Takahashi, H. Nakao, R. Kumai, S. Ishibashi, S. Horiuchi, M. Kohyama, K. Kobayashi, Y. Yamasaki, J. Okamoto, T. Sudayama, Y. Murakami and Y. Tokura, to be submitted to J. Phys.: Conf. Ser..

Construction of a soft X-ray diffractometer with a 7.5-T superconducting magnet

Jun Okamoto[1], Hironori Nakao[1,2], Yuichi Yamasaki[1], Takaaki Sudayama[1], Kensuke Kobayashi[1], Yukari Takahashi[1,2], Hiroyuki Yamada[3], Akihito Sawa[3], Masato Kubota[4], Reiji Kumai[1,2], and Youichi Murakami[1]

[1]Condensed Matter Research Center and Photon Factory, IMSS, KEK, Tsukuba, Ibaraki 305-0801, Japan
[2]CREST, Japan Science and Technology Agency, Tokyo 102-0076, Japan
[3]National Institute of Advanced Industrial Science and Technology (AIST), Tsukuba, Ibaraki 305-8562, Japan
[4]Quantum Beam Science Directorate, JAEA, Tokai-mura, Ibaraki 319-1195, Japan

E-mail: jun.okamoto@kek.jp

Abstract. We have constructed a soft X-ray diffractometer with a 7.5-T superconducting magnet in order to study ordered electronic structures under high magnetic field. In this paper, we present the features of this system and an example of resonant soft X-ray scattering measurement under the magnetic field of $(LaMnO_3)_3(SrMnO_3)_3$ superlattice with a remarkable negative magnetoresistance effect.

1. Introduction

Resonant soft X-ray scattering (RSXS) is a powerful method for investigating ordered electronic structures in strongly correlated electron systems (SCES). For example, RSXS at transition-metal (TM) $L_{2,3}$ edges of *3d* TM compounds is much more sensitive to the magnetic structures formed by TM *3d* electrons than the resonant X-ray scattering (RXS) at the TM *K* edge, since the TM $L_{2,3}$-edge RSXS directly measures the TM *3d* states through the *2p* to *3d* transition, whereas the TM *K*-edge RXS measures the TM *4p* states through the *1s* to *4p* transition [1]. Recently, the external-field effect has become an important issue concerning ordered TM *3d* electronic structures, since novel physical properties, such as multiferroics [2] and colossal magnetoresistance [3], are observed at the boundaries of various electric and magnetic phases, and are very sensitive to the external field. Therefore, it is required to measure RSXS under various external conditions besides temperature. Magnetic field is an important parameter since multiferroics and colossal magnetoresistance are closely related with magnetic structure and some of them show metamagnetic transition under high magnetic field [4]. Most previous RSXS measurement systems, however, can apply only a low magnetic field below 1 T by using a permanent magnet and/or electromagnet; no RSXS measuring systems with a superconducting magnet (SCM) have been used to apply a high magnetic field. Since light in the soft X-ray (SX) region is strongly absorbed, these RSXS systems need to be set in a high vacuum (HV) chamber.

Journal of Physics: Conference Series **502** (2014) 012016 doi:10.1088/1742-6596/502/1/012016

2. Instruments

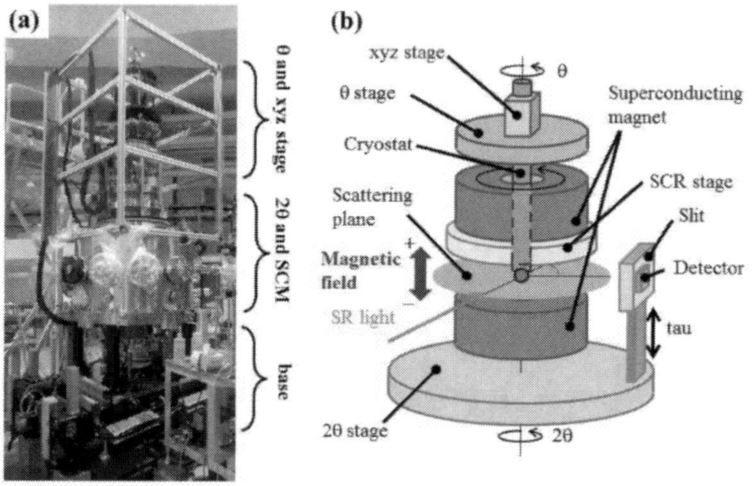

Figure 1. (a) Appearance of the SCM-RSXS system. (b) Schematic view of the diffractometer and the SCM. The magnetic field is applied orthogonal to the scattering plane.

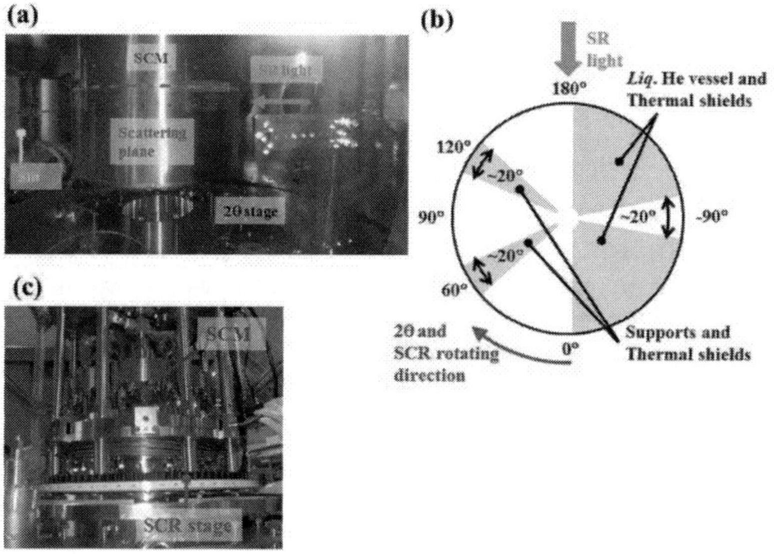

Figure 2. (a) Inside the vacuum chamber. The red broken curve shows the scattering plane. (b) A cross-section of the SCM at the scattering plane. Shaded areas show the regions screened by the thermal shields, supports, and the *liq*. He vessel. (c) Appearance of the SCR stage.

We have constructed a SX diffractometer with a 7.5-T superconducting magnet in HV (SCM-RSXS), in order to measure the RSXS signal reflecting ordered electronic structures under high magnetic field. Figure 1(a) shows the appearance of the SCM-RSXS system. This system has a large vacuum chamber in order to install the SCM and other devices such as slits and detectors in HV. The base pressure of the chamber is 1×10^{-5} Pa. Figure 1(b) shows the schematic view of the diffractometer and the SCM.

Samples are cooled down by the He-flow cryostat and the lowest temperature is 8 K at the sample holder. Scattered light from the sample is detected by a silicon drift detector (SDD) with good energy resolution of about 100 eV. In front of the SDD detector, a four-blade slit is placed in order to measure weak RSXS signals of superstructures by subtracting backgrounds such as fluorescence [5]. This diffractometer has a quasi χ axis called tau in order to detect the signal out of the scattering plane without changing the sample tilt. The effective range of χ is ±2.5°. The scattering plane is set between the two coils and the SCM applies the magnetic field orthogonal to the scattering plane up to 7.5 T. Magnetic field applied upward is defined as positive. The SCM has a large 13-liter *liq*. He vessel to cool the coils, and the SCM can be used for about 30 hours without adding *liq*. He, even at the magnetic field of 7.5 T.

Figure 2(a) shows the arrangement of devices inside the vacuum chamber. The red broken curve shows the scattering plane. The SCM has windows to transmit the incident and scattered light toward the sample and the detector, respectively, but there are blind areas in the scattering plane. Figure 2(b) shows the cross-section of the SCM at the scattering plane. Shaded areas show the regions screened by the thermal shields, supports between the two coils, and the *liq*. He vessel. In order to compensate the 2θ regions, we rotate the SCM itself around the same rotation axis of θ and 2θ, called SCR, as shown in Fig. 2(c). Finally we cover the 2θ range from 0° to 165° in the SCM-RSXS system; the 2θ region above 165° is limited by the slit.

3. Experiment and Results

Superlattices formed by the heterostructures of strongly-correlated insulating oxides have been fascinating materials since the discovery of superconductivity at the interface between $SrTiO_3$ and $LaAlO_3$ [6]. Among them, superlattices composed of antiferromagnetic insulators $LaMnO_3$ and $SrMnO_3$ layers are of particular interest, since a wide variety of electronic and magnetic phases depending on the stacking structures have been reported [7]. These $LaMnO_3$-$SrMnO_3$ superlattices are fabricated on LSAT (001) substrate by pulsed laser deposition, since lattice matching between the layers and the substrate is also crucial to the physical properties of the superlattices [7,8]. The Mott insulator $(LaMnO_3)_3(SrMnO_3)_3$ superlattice (denoted as L3S3) is located at the boundary of the antiferromagnetic-insulator and ferromagnetic-insulator phases and shows a remarkable negative magnetoresistance effect under the magnetic field of 7 T at 10 K [7]. Namely, ferromagnetic-metal (FM) phase is expected to exist at 7 T. In order to study the Mn *3d* electronic structures which form the magnetically ordered structures in the expected FM phase, we measured the RSXS for L3S3 at the Mn $L_{2,3}$ edge by using the SCM-RSXS system.

The RSXS measurement was done at the SX undulator beamline BL-16A in the Photon Factory, KEK. We applied a magnetic field of ±7 T orthogonal to the scattering plane; the magnetic moment of L3S3 is expected to be nearly orthogonal to the scattering plane. In order to measure the resonant magnetic scattering component effectively [1], we set the electric-field vector of the incident linearly polarized light in the scattering plane (π polarization), as shown in the inset of Fig. 3(b).

For the ferromagnetic structures in L3S3, Mn *3d* magnetic moments in $LaMnO_3$ layers are considered to mainly contribute to the ferromagnetism, as shown in Fig. 3(a). In this model, magnetic structures are observed at the 001 reflection. The reflection was indexed using the unit cell of the superlattice whose lattice constant is about 22.8 Å.

Figure 3(b) shows the Mn $L_{2,3}$-edge RSXS spectra of the 001 reflection for L3S3 under the ±7 T magnetic field at 10 K. Comparing with the XAS spectrum, there are strong resonant structures at the Mn $L_{2,3}$ edges in the RSXS spectra of the 001 reflection. In addition, RSXS spectra of the 001 reflection take different up-and-down line shapes at the Mn $L_{2,3}$ edge under the ±7 T magnetic field as shown in Fig. 3(b). We consider that this difference originates from the reversal of magnetic moment under the positive and negative magnetic field. The scattering intensity of the 001 reflection contains Thomson scattering, reflecting the superlattice structure of L3S3 in addition to the resonant magnetic scattering. When the magnetic moment of L3S3 is reversed by reversing the external magnetic field, the sign of the resonant magnetic scattering of the 001 reflection also reverses [1], although the sign of

Thomson scattering of the 001 reflection does not change. Then, the difference under the positive and negative magnetic field appears at the 001 reflection via the interference between Thomson scattering and resonant magnetic scattering.

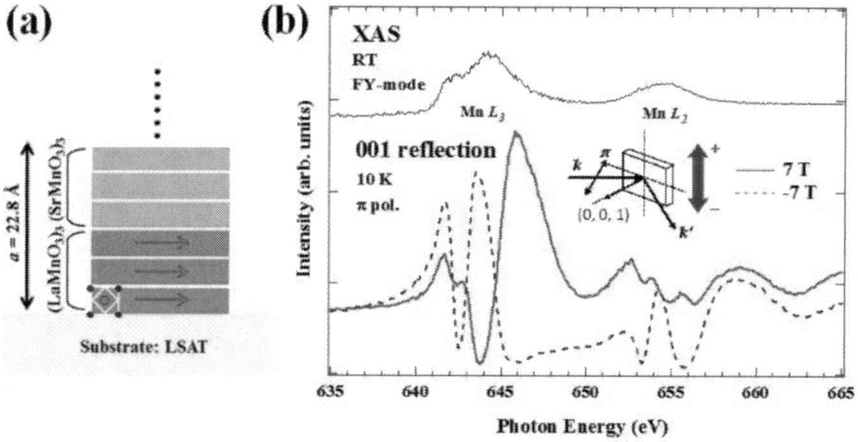

Figure 3. (a) Schematic view of the $(LaMnO_3)_3(SrMnO_3)_3$ superlattice. Red arrows show the magnetic moments in $LaMnO_3$ layers. (b) Mn $L_{2,3}$-edge RSXS spectra of the 001 reflection for $(LaMnO_3)_3(SrMnO_3)_3$ superlattice measured under the ±7 T magnetic field at 10 K with π polarization. XAS spectrum of the superlattice measured with FY-mode at room temperature. Inset shows the experimental setting.

4. Conclusions

We have constructed a soft X-ray diffractometer with a 7.5-T superconducting magnet in order to perform resonant soft X-ray measurements under high magnetic field. We measured the Mn $L_{2,3}$-edge RSXS spectra of the 001 reflection for the $(LaMnO_3)_3(SrMnO_3)_3$ superlattice in the ferromagnetic-metal phase under a ±7 T magnetic field, and observed the difference of RSXS spectra corresponding to the reversal of the Mn *3d* magnetic moment.

Acknowledgements

The authors would like to thank Y. Honda and K. Nishi for their special support in the design and construction of the diffractometer and the superconducting magnet. This work was supported by a Grant-in-Aid for Scientific Research (S) (No. 21224008) from the Ministry of Education, Culture, Sports, Science, and Technology, Japan and by the Funding Program for World-leading Innovative R&D in Science and Technology (FIRST). This study was performed with the approval of the Photon Factory Program Advisory Committee (Proposal No. 2009S2-008 and 2012S2-005).

References

[1] Hannon J P, Trammell G T, Blume M, and Gibbs D 1988 *Phys. Rev. Lett.* **61** 1245
[2] Kimura T, Goto T, Shintani H, Ishizaka K, Arima T, and Tokura Y 2003 *Nature* **426** 55
[3] Urushibara A, Moritomo Y, Arima T, Asamitsu A, Kido G, and Tokura Y 1995 *Phys. Rev. B* **51** 14103
[4] Kimura T, Lashley J C, and Ramirez A P 2006 *Phys. Rev. B* **73** 220401(R)
[5] Okamoto J, Horigane K, Nakao H, Amemiya K, Kubota M, Murakami Y, and Yamada K 2013 *J. Phys.: Conf. Ser.* **425** 202003
[6] Ohtomo A and Hwang H Y 2004 *Nature (London)* **427** 423
[7] Yamada H, Xiang P-H, and Sawa A 2010 *Phys. Rev. B* **81** 014410
[8] Yamada H, Ogawa Y, Ishii Y, Sato H, Kawasaki M, Akoh H, and Tokura Y 2004 *Science* **305**,646

Journal of Physics: Conference Series **502** (2014) 012017 doi:10.1088/1742-6596/502/1/012017

In-situ observation of synthesizing process of $Mm_xCo_4Sb_{12}$ utilizing x-ray di raction under high temperatures and high pressures

Chihiro Sekine[1], Hirotaka Kato[1], Masatoshi Kanazawa[1], Yukihiro Kawamura[1], Keiki Takeda[1], Mizushi Matsuda[1], Kunihiro Kihou[2], Chul-Ho Lee[2], and Hirotada Gotou[3]

[1]Muroran Institute of Technology, Muroran, Hokkaido 050-8585, Japan
[2]National Institute of Advanced Industrial Science and Technology, Tsukuba, Ibaraki 305-8568, Japan
[3]Institute for Solid State Physics, University of Tokyo, Kashiwa, Chiba 277-8581, Japan

E-mail: sekine@mmm.muroran-it.ac.jp

Abstract. The antimony-based partially-filled skutterudite compound $Mm_xCo_4Sb_{12}$ (Mm=mischmetal) are expected for n-type thermoelectric materials with high-perfomance and low-cost. In order to prepare high-quality samples of the compound, synthesizing processes at high pressures and high temperatures have been studied by means of in-situ x-ray di raction measurements using synchrotron radiation and a cubic-anvil high pressure apparatus. In series of the experiments, we could obtain the optimum condition for synthesizing the compound under high pressure.

1. Introduction

The skutterudite compounds have attracted much attention owing to their potential as improved, next generation thermoelectric materials [1, 2, 3]. The binary compound $CoSb_3$ crystallizes with the $CoAs_3$-type structure of space group Im$\bar{3}$ ($T^5{}_h$, No. 204) [4]. This structure has eight formula units, including two voids, per cubic unit cell. The material shows excellent thermoelectric properties (large Seebeck coefficients and high hole mobilities)[5]. The dimensionless figure-of-merit, ZT, of thermoelectric materials is given by a combination of electrical and thermal transport coefficients, $ZT{=}S^2\sigma T/\kappa$, where S is the Seebeck coefficient, T is the absolute temperature, σ is the electrical conductivity, and κ is the thermal conductivity, which has both electronic and lattice components, κ_{el} and κ_{lat}, respectively. Although the power factor, $S^2\sigma$, for $CoSb_3$ is quite large, the lattice thermal conductivity, κ_{lat} is too high. However, the filling of the voids partially in the binary skutterudite structure with rare-earth (R) ions can reduce κ_{lat} greatly[3]. The R ions are located inside the oversized cages, formed by twelve Sb ions, and the undersized ions are believed to show random motion (rattling) around the equilibrium positions. Thus, it produces a large phonon scattering, reducing the lattice thermal conductivity κ_{lat}. Owing to the reduced thermal conductivity, partially-filled skutterudite compounds show excellent thermoelectric performance. Recently, however, the use of rare earth metals has serious problems about cost and supply amount. Therefore, we focus to mischmetal (Mm) as the void-filling ions. Mischmetal is an alloy of rare earth elements in various naturally occurring proportions. A typical composition includes approximately 50% Ce and 25% La, with small amounts of Nd and Pr. Mischmetal is low-cost compared with individual rare earths because the extraction processes is unnecessary.

A preparation of high-quality samples for the antimony-based skutterudite compounds is quite difficult at ambient pressure because the high vapor pressure of Sb and impurity phases tend to be formed easily. High-pressure synthesis technique is one of the useful methods to prepare high quality samples of skutterudites. In this study, we have tried to observe synthesizing processes of $Mm_xCo_4Sb_{12}$ (Mm=mischmetal) in-situ at high temperature and high pressure to obtain the optimum condition for synthesis.

2. Experimental

In-situ x-ray diffraction patterns were taken by an energy-dispersive method using synchrotron radiation and a solid-state detecter. High pressure was applied using a multi-anvil assembly 6-6 (MA6-6) with a cubic-anvil high-pressure apparatus, the MAX80 system, installed at the beam line AR-NE5C, at Photon Factory (PF) in High Energy Accelerator Research Organization (KEK) (Tsukuba, Japan) (Fig. 1). The MA6-6 consists of six small second-stage anvils with an anvil guide, and can be compressed by a cubic-anvil apparatus (Fig. 2) [6]. The truncated edge length (TEL) of the second-stage anvil made of tungsten carbide is 6 mm. The anvil guide is made of tool steel (SUS304), with an outer edge length of 28 mm. The anvil guide has holes along one of the diagonal direction for access to the incident and exiting x-rays. The TEL of the first-stage anvil is 27 mm. The sample container made of boron-epoxy is formed a cube of 9 mm on an edge. The starting materials, which are mixture of Mm, Co and Sb, are put into a BN crucible. The crucible with a graphite heater is inserted in a cube-shaped boron-epoxy solid pressure medium. Pressure was determined by the lattice constant of NaCl internal pressure marker. The details of the in-situ observation method were described in Ref. 7.

$Mm_xCo_4Sb_{12}$ (x=0.2, 0.4, and 0.6) samples were prepared at high temperatures and high pressures using a wedge-type cubic anvil high-pressure apparatus. The sample cell assembly is similar to that used for the high-pressure synthesis of P-based filled skutterudite compounds [8].The prepared samples were characterized by powder x-ray diffraction using Co $K\alpha_1$ radiation and silicon as a standard. Resistivity was measured by a standard dc four-probe method. Seebeck coefficient and thermal conductivity were measured with a thermal transport option of Quantum Design PPMS.

Figure 1. A photograph of MA6-6 in first stage anvils of the MAX80 system. 1, first stage anvil; 2, hole for x-ray.

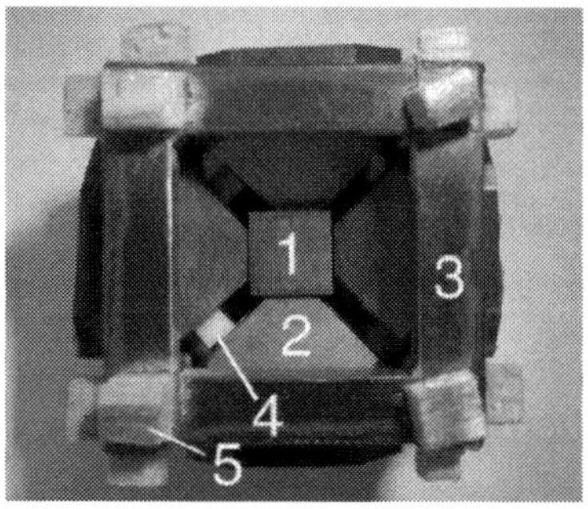

Figure 2. A photograph of MA6-6. 1, pressure medium; 2, second stage anvil; 3, anvil guide; 4, preformed gasket; 5, wood spacer.

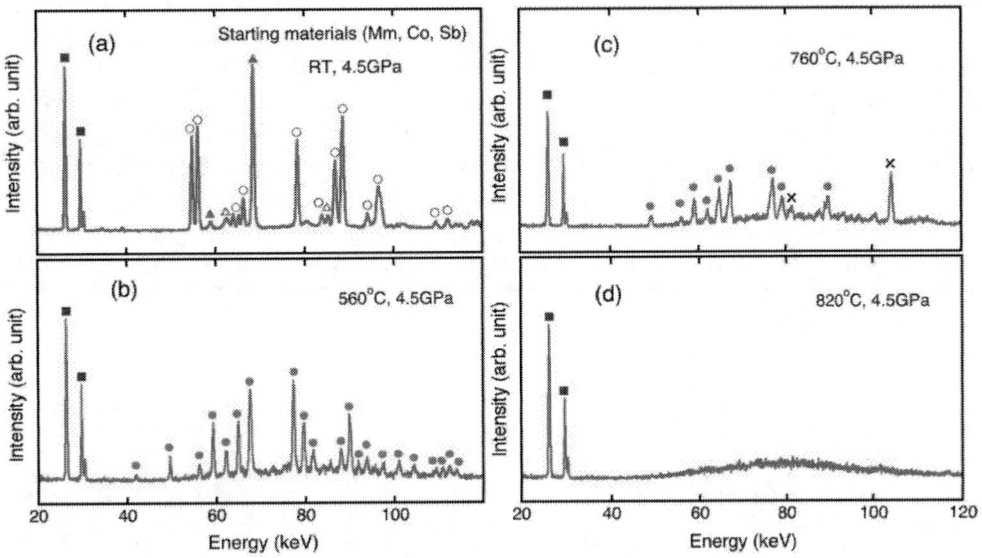

Figure 3. X-ray diffraction patterns of synthesizing process of $Mm_{0.4}Co_4Sb_{12}$ at 4.5 GPa. Solid triangles, open triangles and open circles designate the Bragg peaks of Co, Ce and Sb, respectively. Solid squares indicate the characteristic x-ray for Sb. Solid circles designate the Bragg peaks of $Mm_{0.4}Co_4Sb_{12}$. Crosses indicate the peaks of an impurity phase. (a) The starting materials at room temperature, (b) 560 C, (c) 760 C and (d) 820 C.

3. Results and discussion

Figure 3(a) shows x-ray diffraction patterns of the starting materials, which are mixture of each metal and antimony powder in the atomic ratio of Mm : Co : Sb = 0.4 : 4 : 12, at room temperature and 4.5 GPa. Open triangles, solid triangles, and open circles designate the Bragg peaks of Ce, Co and Sb, respectively. Solid squares indicate the characteristic x-ray for Sb. With increasing temperature, the Bragg peaks of the starting materials faded out, and then the peaks of the skutterudite structure were observed above 450 C. At 560 C, all diffraction lines were indexable using the skutterudite structure (closed circles in Fig. 3(b)). Then, the peaks of an impurity phase, which are indicated by crosses, appeared at 760 C (Fig. 3(c)). This indicates that decomposition products of $Mm_{0.4}Co_4Sb_{12}$ (such as $CoSb_2$) were generated above this temperature. Elevating temperature up to 820 C, the Bragg peaks disappeared and only the peaks of the characteristic x-ray for Sb were observed (Fig. 3(d)). This indicates that the sample is fully liquid at this temperature. In series of the experiments, we found an appropriate temperature range of 560-760 C for synthesizing $Mm_xCo_4Sb_{12}$ at 4.5 GPa.

We have actually synthesized $Mm_{0.4}Co_4Sb_{12}$ at 4.5 GPa and 600 C using a wedge-type cubic anvil high-pressure apparatus based on the synthesis condition obtained by the in-situ experiments. Figure 4 shows a powder x-ray diffraction pattern of $Mm_{0.4}Co_4Sb_{12}$ prepared under high pressure. While a small amount of impurity phase (metallic Sb) was detected, most of the observed diffraction lines were indexable using the skutterudite structure. The lattice constant determined by a least-squares fit to the data was 9.0714 Å. We have also synthesized $Mm_{0.2}Co_4Sb_{12}$ and $Mm_{0.6}Co_4Sb_{12}$ under the same condition. The x-ray diffraction patterns of the compounds $Mm_{0.2}Co_4Sb_{12}$ and $Mm_{0.6}Co_4Sb_{12}$ indicate that the main phase consists of the filled skutterudite. The lattice constants of $Mm_{0.2}Co_4Sb_{12}$ and $Mm_{0.6}Co_4Sb_{12}$ were determined in the same manner and found to be 9.0531 and 9.0853 Å, respectively. The inset of Fig. 4 plots the lattice constants of $Mm_xCo_4Sb_{12}$ along with the value of $CoSb_3$ reported earlier [9]. The lattice constant of $Mm_xCo_4Sb_{12}$ becomes large linearly with increasing x. This suggests that the partially-filling of Mm to $CoSb_3$ was succeeded up to $x = 0.6$ at least.

Furthermore, we have evaluated the thermoelectric properties of $Mm_{0.6}Co_4Sb_{12}$ and $CoSb_3$, prepared under high pressure, as a reference. The seebeck coefficient of $Mm_{0.6}Co_4Sb_{12}$ shows negative value, which means n-type conductivity, while $CoSb_3$ is a p-type conductor. The temperature dependence of the dimensionless figure of merit ZT for $CoSb_3$ and $Mm_{0.6}Co_4Sb_{12}$ is shown in Fig. 5. $Mm_{0.6}Co_4Sb_{12}$ shows a large ZT value of 0.07 compared with $CoSb_3$ at 300K. Therefore, the partially-filled skutterudite compound $Mm_xCo_4Sb_{12}$ are expected for n-type thermoelectric materials with high-perfomance and low-cost.

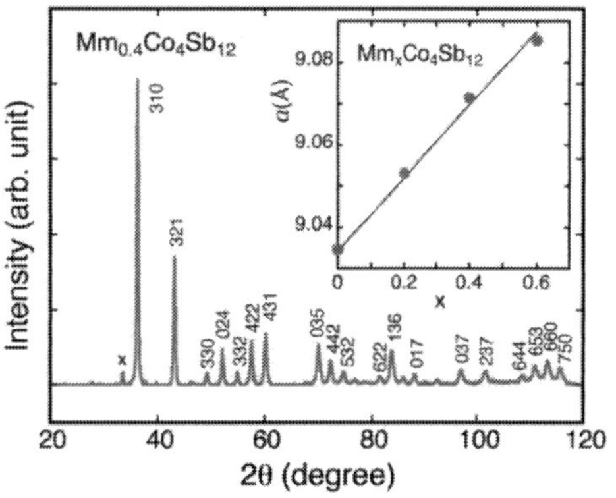

Figure 4. X-ray diffraction pattern of $Mm_{0.4}Co_4Sb_{12}$ prepared under high pressure. Cross indicates the peak of an impurity phase (metallic Sb). The inset shows lattice constants a of $Mm_xCo_4Sb_{12}$ along with the value reported earlier for $CoSb_3$ [9].

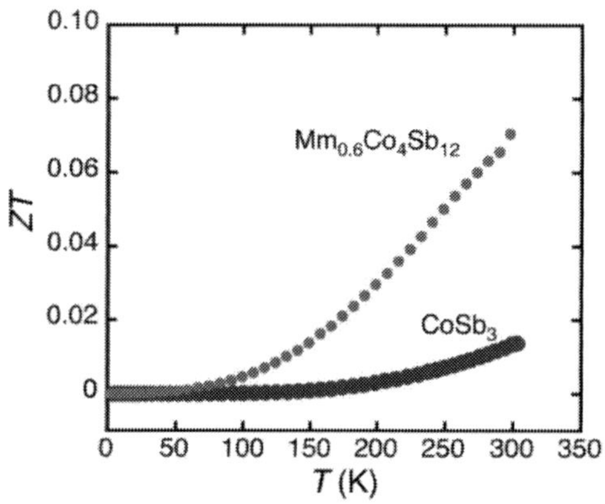

Figure 5. Temperature dependence of the dimensionless figure of merit ZT for $CoSb_3$ and $Mm_{0.6}Co_4Sb_{12}$.

4. Summary

We could obtain the optimum condition for synthesizing $Mm_xCo_4Sb_{12}$ under high pressure. In-situ x-ray experiments are excellent method for determining the condition for obtaining only target material without impurity phases for solid-phase reaction synthesis under high pressure.

Acknowledgments

This work was supported by a Grant-in Aid for Scientific Research (B) (No. 23340092) of the Japan Society for the Promotion of Science.

References

[1] Morelli D T and Meisner G P 1995 *J. Appl. Phys.* **77** 3777
[2] Sales B C, Mandrus D, Chakoumakos B C, Keppens V and Thompson J R 1997 *Phys. Rev.* B **56** 15081
[3] Nolas G S, Cohn J L and Slack G A 1998 *Phys. Rev.* B **58** 164
[4] Zhuravlev N N and Zhdanov G S 1956 *Sov. Phys. Crystallogr.* **1** 404
[5] Slack G A and Tsoukala V G 1994 *J. Appl. Phys.* **76** 1662
[6] Nishiyama N, Wang Y, Sanehira T, Irifune T and Rivers M L 2008 *High Pressure Research* **28** 307
[7] Sekine C, Kachi T, Yoshida T, Abe R, Namiki T, Akahira K and Ito K 2010 J. Phys.: Conf. Ser. **215** 012141
[8] Kihou K, Shirotani I, Shimaya Y, Sekine C, and Yagi T 2004 Mater. Res. Bull. **39** 317
[9] Matsui K, Hayashi J, Akahira K, Ito K, Takeda K and Sekine C 2010 J. Phys.: Conf. Ser. **215** 012005

A study of magnetic moments of $CeRh_3B_2$ by X-ray magnetic diffraction experiments

M. Ito[1], K. Suzuki[1], T. Tadenuma[2], R. Nagayasu[2], Y. Sakurai[3], Y. Onuki[4], E. Nishibori[3] and M. Sakata[3]
[1]Faculty of Science and Technology, Gunma University, Kiryu, Gunma 376-8515, Japan
[2]Graduate School of Engineering, Gunma University, Kiryu, Gunma 376-8515, Japan
[3]JASRI/SPring-8, Sayo, Hyogo 679-5198, Japan
[4]Graduate School of Science, Toyonaka, Osaka, Japan

E-mail: itom_phys@gunm-u.ac.jp

Abstract. X-ray magnetic diffraction experiments of a ferromagnetic rare-earth compound $CeRh_3B_2$ have been performed, and the spin and orbital magnetic form factors have been measured. Density distributions of the spin and orbital magnetic moments in real space have been obtained by using Maximum Entropy Method. Low peaks at Rh sites as well as high peaks at Ce sites are observed in these distribution maps. We have estimated the spin and orbital magnetic moments at the Ce and Rh sites in the distribution maps, and have obtained small but nonnegligible spin and orbital magnetic moments of Rh. This is probably the first experimental evidence showing existence of the spin and orbital moments of Rh in this compound.

1. Introduction

A rare-earth compound $CeRh_3B_2$ has a hexagonal $CeCo_3B_2$-type crystal structure as shown in Fig. 1 [1], and is ferromagnetic below 115K [2]. This Curie temperature is the highest among the compounds of Ce with nonmagnetic elements. On the other hand, the magnetic moment per Ce atom, $0.4\mu_B$ at 4.2K [3], is almost the smallest among them. This moment value is much smaller than the calculated value, $2.14\mu_B$, for Ce^{3+} free ion under Hund's rule. These magnetic properties of this compound have been attracting much attention of many researchers, and a large number of theoretical and experimental studies have been made to reveal the origin of the magnetic properties. After arguments between theoretical models of itinerant electrons and those of localized ones, currently accepted models are those of localized Ce 4f electrons hybridized with Ce 5d [4-6], Rh 4d [7], or Ce 4f electrons of neighbouring atoms [8]. Among them, the Monte Carlo study [7] presented a possibility

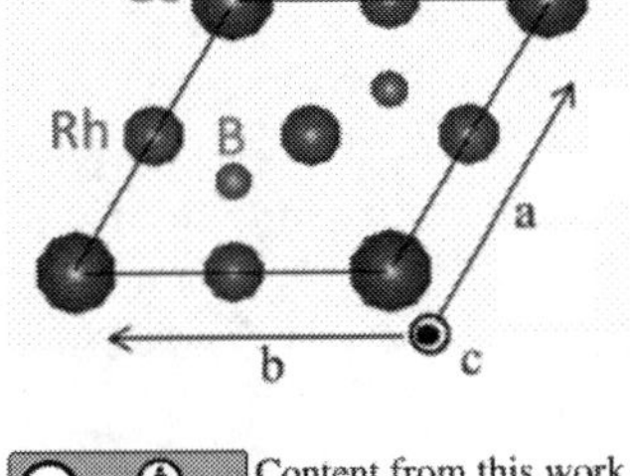

Fig. 1. (Color online.) Crystal structure of $CeRh_3B_2$. Large, medium and small balls represent Ce, Rh and B atoms, respectively. Ce and B atoms are located on the plane of z=0, and Rh atoms are on the plane of z=1/2.

of polarization of Rh 4d electrons which provided magnetic moments of $0.15\mu_B$.

The polarized neutron diffraction (PND) experiments showed that the magnetic moments were contributed mainly by localized Ce 4f electrons ($0.56\pm0.02\mu_B$), slightly by delocalized electrons distributing along the c-axis between Ce atoms ($-0.18\pm0.0\mu_B$), but negligibly by Rh and B [9, 10]. The magnetic Compton scattering (MCS) experiments showed that the spin moments were contributed considerably by Ce 4f ($-0.69\pm0.08\mu_B$) and Ce 5d electrons ($-0.41\pm0.04\mu_B$) but negligibly by Rh 4d electrons ($-0.03\pm0.04\mu_B$), which suggested hybridization of Ce 4f with Ce 5d electrons [11, 12].

In this study, we have applied a new experimental method of X-ray magnetic diffraction (XMD) to $CeRh_3B_2$. XMD is nonresonant magnetic X-ray Bragg scattering, which utilizes elliptically polarized X-rays. A specific feature of the XMD is that we can measure spin and orbital magnetic form factors of ferromagnets separately [13]. In this study, we have measured these form factors of $CeRh_3B_2$, and have estimated the spin and orbital magnetic moments of the constituent atoms. We have paid attention on these moments of Rh. The preliminary result of the XMD data analysis was presented elsewhere [14].

2. Experimental

The XMD experiments have been performed on the beamline 3C of Photon Factory (PF) in High Energy Accelerator Research Organization (KEK). On this beamline the dedicated XMD experimental system is installed, which consists of a four-circle diffractometer, an electromagnet, a refrigerator and a pure-Ge solid-state detector (SSD) with related electronic circuits to count X-ray photons. Details of this system were presented elsewhere [15]. White X-rays of elliptically polarized synchrotron radiation from the bending magnet were incident on the specimen, and X-ray intensities diffracted at the specimen with 90° scattering angle were measured with the SSD. The specimen was magnetized with the electromagnet along a direction perpendicular to the c-axis of $CeRh_3B_2$. We have measured relative change R in the diffraction intensity accompanied by reversal of the magnetic field direction. Magnetic field strength was 21.5kOe that was enough to saturate the magnetization of this compound along any direction in the plane perpendicular to the c-axis. As the geometrical relation between the crystal axis and the magnetic field direction in the XMD is different from that of the PND [9], the XMD could provide complimentary information to the PND. The specimen temperature was kept at 15 K that was far below the Curie temperature of this compound.

The R is expressed as, $R = \gamma f_P \mu_\alpha(\mathbf{k})/F(\mathbf{k})$ [16]. Here, γ is the energy term expressed as $\gamma=E/E_0$ where E is diffracted X-ray energy and E_0 is electron rest mass energy. The term f_P is the polarization factor expressed as $f_P=P_C/(1-P_L)$ where P_L and P_C are the linear and circular polarization of the incident X-rays. The term $\mu_\alpha(\mathbf{k})$ is a combined magnetic form factor represented as $\mu_\alpha(\mathbf{k})=\mu_L(\mathbf{k})(\cos\alpha+\sin\alpha)+\mu_S(\mathbf{k})\sin\alpha$ where $\mu_S(\mathbf{k})$ and $\mu_L(\mathbf{k})$ are the spin and orbital magnetic form factors and α is the angle between the directions of the incident X-rays and the specimen magnetization. $F(\mathbf{k})$ is the crystal structure factor, and $\mathbf{k}$ denotes a reciprocal lattice point. When the angle α is 135°and 0°, R is expressed as, $R_{135} = \gamma f_P \mu_S(\mathbf{k})/\sqrt{2}F(\mathbf{k})$ and $R_0 = \gamma f_P \mu_L(\mathbf{k})/F(\mathbf{k})$, respectively. By setting the angle α to 135° and 0°, we can measure the spin magnetic form factor $\mu_S(\mathbf{k})$ and the orbital magnetic form factor $\mu_S(\mathbf{k})$, respectively.

3. Results and discussion

We have measured the spin and orbital magnetic form factors for twenty reciprocal lattice points of *hk*0 series. In this XMD data analysis, we have taken into account anomalous dispersion terms of atomic scattering factors, $\Delta f'$ and $\Delta f''$, which are the real and imaginary part, and these values have been quoted from the literature 17. The obtained spin and orbital magnetic form factors, $\mu_S(s)$ and $\mu_L(s)$, are shown as circles and squares, respectively, in Fig. 2. Here, $s=|\mathbf{k}|=\sin\theta/\lambda$, θ is Bragg angle and λ is X-ray wavelength. In Fig. 2 solid lines are theoretical curves of $\mu_S(s)$ and $\mu_L(s)$ for Ce 4f electrons of a Ce^{3+} free ion under the dipole approximation [18], which are represented as $\mu_{Sdipole}(s)=\mu_{S0}\langle j_0(s)\rangle$ and $\mu_{Ldipole}(s)=\mu_{L0}\{\langle j_0(s)\rangle+\langle j_2(s)\rangle\}$. Here μ_{S0} and μ_{L0} are the spin and orbital magnetic moments of Ce 4f, and <$j_n(s)$> is the radial integral of the wavefunction of Ce 4f electrons multiplied by the n-th order spherical Vessel function. We use the values of $\mu_{S0} = -1.16\mu_B$ and $\mu_{L0} = 1.61\mu_B$ from the literature 12 as is described below.

In Fig. 2 the followings are noted. (i) The observed values of $\mu_S(s)$ are negative whereas those of $\mu_L(s)$ are positive. This indicates that the major part of the magnetic moments is the orbital moment and the spin moment is oppositely coupled with the orbital moment. (ii) The observed $\mu_L(s)$ is fairly reproduced by the theoretical dipole-approximation curve. (iii) The observed $\mu_S(s)$ decreases more slowly than $\mu_L(s)$ and is far from the theoretical curve. This behavior of $\mu_S(s)$ is discussed later. (iv) The observed $\mu_S(s)$ and $\mu_L(s)$ data are composed of two different series. One is the *hk*0 series where both *h* and *k* are even (even series), and the other is the one where either *h* or *k* is odd (odd series). Absolute values of the odd series are slightly smaller than those of the even series. Here let us remind the crystal structure factor of $CeRh_3B_2$, $f_{Ce}+3f_{Rh}$ for the even series and $f_{Ce}-f_{Rh}$ for the odd series where f_{Ce} and f_{Rh} are the atomic scattering factor of Ce and Rh (boron contribution is ignored here). Then the characteristic behavior of the observed $\mu_S(s)$ and $\mu_L(s)$ for the even and odd series suggests that both $\mu_S(s)$ and $\mu_L(s)$ are contributed by Rh as well as by Ce.

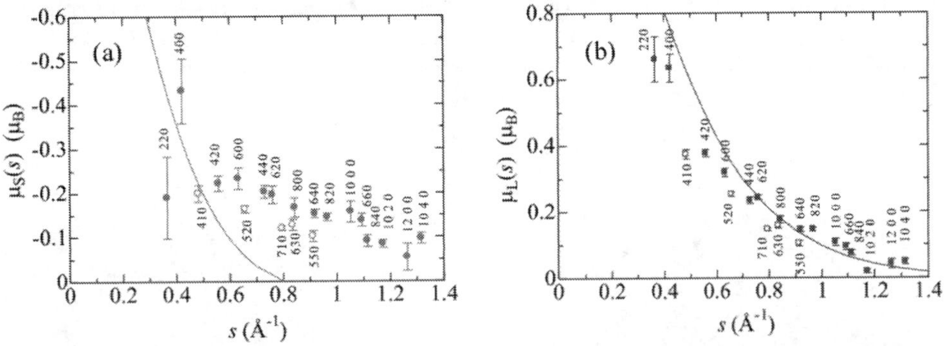

Fig. 2. (Color online.) Observed (a) spin magnetic form factor $\mu_S(s)$ and (b) orbital magnetic form factor $\mu_L(s)$ for twenty reciprocal lattice points of *hk*0. Solid circles and squares are for the even series and open circles and squares are for the odd series. It is noted that the direction of the ordinate of $\mu_S(s)$ is reversed. Solid lines denote theoretical dipole-approximation curves of Ce 4f electrons of a Ce^{3+} free ion.

We have estimated the spin and orbital magnetic moments of Ce and Rh. We have adopted the maximum entropy method (MEM), and have obtained the density distributions of the spin and orbital magnetic moments in real space, $M_S(\mathbf{r})$ and $M_L(\mathbf{r})$. The obtained distribution maps projected along the c-axis are presented in Fig. 3 by using VESTA [19]. In Fig. 3, we notice high peaks at Ce sites and low peaks at Rh sites for both $M_S(\mathbf{r})$ and $M_L(\mathbf{r})$. Peaks of $M_S(\mathbf{r})$ are sharper than those of $M_L(\mathbf{r})$, and the peaks of $M_L(\mathbf{r})$ at the Rh sites seem to extend toward Ce sites. Though the reason for latter is not clear at this moment, we discuss the reason for the former. The PND experiments [9, 10] have shown the existence of the magnetic (probably spin) moment of delocalized electrons which are distributed along the c-axis. The anisotropic distribution of the spin moment along the c-axis may relate to the sharp peaks of $M_S(\mathbf{r})$ in the map projected along the c-axis, which could be concerned with the slow decrease of $\mu_S(s)$ in the reciprocal space as shown in Fig. 2 (a).

We have estimated the spin and orbital magnetic moments of Ce and Rh by integrating the density values of the maps in Fig. 3 as follows,

$$K^{S(L)}\left(A_{Ce}^{S(L)} + 3A_{Rh}^{S(L)}\right) = \mu_0^{S(L)}, \qquad (1)$$

$$\mu_{\rm Ce}^{\rm S(L)} = K^{\rm S(L)} A_{\rm Ce}^{\rm S(L)}, \quad (2)$$

$$\mu_{\rm Rh}^{\rm S(L)} = K^{\rm S(L)} A_{\rm Rh}^{\rm S(L)}. \quad (3)$$

Here, $K^{\rm S(L)}$ is a conversion factor for the spin (orbital) magnetic moment, $A_{\rm Ce}^{\rm S(L)}$ and $A_{\rm Rh}^{\rm S(L)}$ are integrated density values at the Ce and Rh sites for the spin (orbital) magnetic moment, and $\mu_0^{\rm S(L)}$ is the spin (orbital) magnetic moment per formula unit of $CeRh_3B_2$. We have used $\mu_0^{\rm S} = -1.16\mu_{\rm B}$, which was deduced by the MCS experiment [12]. The value of $\mu_0^{\rm L}$ has been derived from the equation $\mu_0^{\rm L} = \mu_0 - \mu_0^{\rm S}$, where μ_0 is the magnetic moment per formula unit obtained by the magnetization measurement. Here, μ_0=0.42$\mu_{\rm B}$ and $\mu_0^{\rm L}$=1.61$\mu_{\rm B}$. Then we have obtained the following values: $\mu_{\rm Ce}^{\rm S} = -0.98\pm0.12\mu_{\rm B}$, $\mu_{\rm Ce}^{\rm L} = 1.52\pm0.12\mu_{\rm B}$, $\mu_{\rm Rh}^{\rm S} = -0.06\pm0.02\mu_{\rm B}$, and $\mu_{\rm Rh}^{\rm L} = 0.03\pm0.02\mu_{\rm B}$. The values of $\mu_{\rm Rh}^{\rm S(L)}$ may be associated with 4d electrons. The values of $\mu_{\rm Ce}^{\rm S(L)}$ may be contributed by both 4f and 5d electrons but each contribution cannot be distinguished in this study. The $\mu_{\rm Ce}^{\rm S}$ value is comparable to the sum of Ce 4f spin and Ce 5d spin, $-1.10\pm0.09\mu_{\rm B}$, by the MCS experiment [12]. The magnetic moment per Ce atom, $\mu_{\rm Ce}^{\rm S} + \mu_{\rm Ce}^{\rm L} = 0.54\pm0.14\mu_{\rm B}$, is also comparable to the corresponding value, $0.56\pm0.02\mu_{\rm B}$ by the PND experiments [9, 10]. In this study, we have observed small but nonnegligible spin and orbital magnetic moments of Rh. Future MCD experiment for Rh L edge will be helpful to check the existence of the magnetic moments of Rh in this compound.

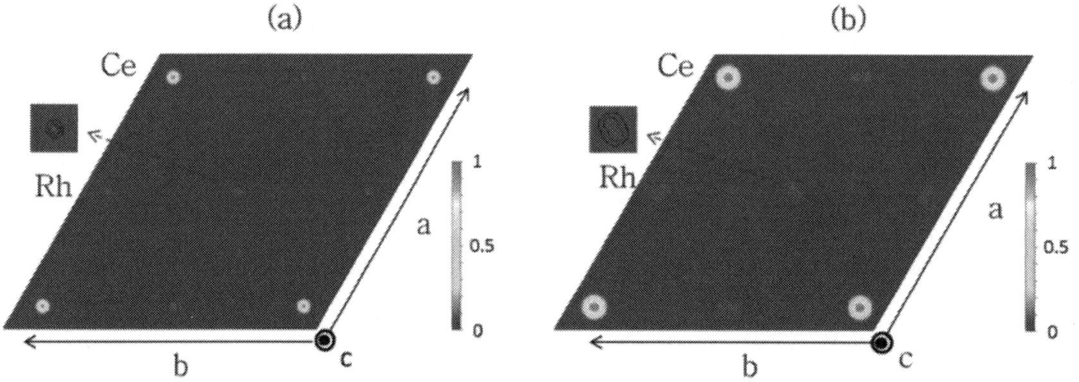

Fig. 3. (Color online.) Density distribution of (a) the spin magnetic moment and (b) the orbital magnetic moment projected along the c-axis. Upper left figures of (a) and (b) show the distributions of the central Rh sites with contours lines the interval of which is 4% of their maximum values.

4. Conclusion

X-ray magnetic diffraction experiments of a ferromagnetic compound $CeRh_3B_2$ have been performed, and the spin and orbital magnetic form factors have been measured separately. Both form factors have been shown to be contributed by Rh as well as Ce. Density distributions of the spin and orbital magnetic moments in real space have been obtained by using the Maximum Entropy Method. By integrating density values at the Ce and Rh sites, we have obtained the spin and orbital magnetic moments of Ce and Rh. As a result, we have observed small but nonnegligible spin and orbital magnetic moments of Rh. This is probably the first experimental evidence showing existence of the spin and orbital moments of Rh in $CeRh_3B_2$.

Acknowledgments
The authors thank to H. Adachi, K. Hirano and X. Zhang for their support to the XMD experiments at KEK-PF. This research was partly supported by a Grant-in-Aid for Scientific Research from Ministry of Education, Science, Sports and Culture (20540331) and the XMD measurements were performed in the KEK-PAC proposal of 2006G046 and 2008G190.

References
[1] Ku H C, Meisner G P , Acker F and Johnston D C 1980 *Solid State Commun.* **35** 91
[2] Dhar S K, Malik S K and Vijayaraghavan R 1981 *J. Phys.* C **14** L321
[3] Malik S K, Vijayaraghavan R, Wallace W E and Dhar S K 1983 *J. Mag. Mag. Mater.* **37** 303
[4] Takegahara K, Harima H and Kasuya T 1985 *J. Phys. Soc. Japan* **54** 4743
[5] Kasuya T, Kasaya M, Takegahara K, Iga F, Liu B and Kobayashi N 1987 *J. Less-Common Met.* **127** 337
[6] Yamaguchi K, Namatame H, Fujimori A, Koide T, Shidara T, Nakamura M, Misu A, Fukutani H, Yuri M, Kasaya M, Suzuki H and Kasuya T 1995 *Phys. Rev.* B 51 13952
[7] Kono H N and Kuramoto Y 2006 *J. Phys. Soc. Jpn.* **75** 13952
[8] Yamauchi K, Yanase A and Harima H 2010 *J. Phys. Soc. Jpn.* **79** 044717
[9] Alonso J A, Boucherle J X, Givord F, Schweizer J, Gillon B and Lejay P 1998 *J. Magn. Magn. Mater.* **177–181** 1048
[10] Givord F, Boucherle J X, Lelièvre-Berna E and Lejay P 2004 *J. Phys.: Condens. Matter* **16** 1211
[11] Yaouanc A, Dalmas de Réotier P, Sanchez J P, Tschencher Th and Lejay P 1998 *Phys. Rev.* B **57** R681
[12] Sakurai Y, Itou M, Tamura J, Nanao S, Thamizhavel A, Inada Y, Galatanu A, Yamamot E and Ōnuki Y 2003 *J. Phys.: Condens. Matter* **15** S2183
[13] Blume M and Gibbs D 1988 *Phys. Rev.* B **37** 1779
[14] Ito M, Nagayasu R, Tadenuma T, Suzuki K, Sato A, Adachi H, Sakurai Y and Ōnuki Y 2012 *Key. Eng. Mater.* **497** 3
[15] Suzuki K, Ito M, Tsuji N, Adachi H and Kawata H 2009 *Jpn. J. Appl. Phys.* **48** 056506
[16] Collins S P, Laundy D and Rollason A J 1992 *Philos. Mag.* B **65** 37
[17] Sasaki S 1989 *KEK Report* **88-14** 1
[18] see for example, Balcar E and Lovesey S W 1989 *Theory of Magnetic Neutron and Photon Scattering*, Clarendon Press, Oxford
[19] Momma K and Izumi F 2011 *J. Appl. Crystallogr.* **44** 1272

Journal of Physics: Conference Series **502** (2014) 012019 doi:10.1088/1742-6596/502/1/012019

Crystal Growth of Chiral Magnetic Material in $CsCuCl_3$

Yusuke Kousaka[1], Takemi Koyama[2], Marina Miyagawa[3], Kohei Tanaka[1], Jun Akimitsu[1] and Katsuya Inoue[2,3,4]

[1]Department of Physics and Mathematics, Aoyama-Gakuin University, Sagamihara, Kanagawa 252-5258, Japan
[2]Faculty of Science, Hiroshima University, Higashi-Hiroshima, Hiroshima 739-8526, Japan
[3]Graduate school of Science, Hiroshima University, Higashi-Hiroshima, Hiroshima 739-8526, Japan
[4]Institute for Advanced Materials Research, Hiroshima University, Higashi-Hiroshima, Hiroshima 739-8526, Japan

Email: koyu@phys.aoyama.ac.jp

Abstract. We report a crystallization technique to make single crystals with homochiral crystallographic domain in $CsCuCl_3$. By adapting crystallization technique with stirring, we succeeded in obtaining the mm-ordered enantiopure single crystals. The samples obtained without the stirring crystallization technique had zigzag shape, suggesting many crystalline nuclei combined during the crystallization process. The samples done with the stirring crystallization technique formed quart-crystal-like shape, suggesting only one crystalline nucleus grew. Absolute structure analysis using X-ray diffraction showed that the samples grown only by our technique had a homochiral crystalline domain.

1. Introduction

Chirality concept (right-handedness or left-handedness) is one of the most fundamental building blocks of nature, ranging from nano and biosciences to cosmic sciences. Therefore, it is very important to understand the chirality in molecules, crystals and magnetic structures both from theoretical and experimental viewpoints. Recently, the interplay between crystallographic and magnetic chiralities have been paid attention because the properties of a helical magnetic structure (pitch, sense, etc..) strongly depends on the chiral skeleton structure that could allow an antisymmetric Dzyaloshinskii-Moriya (DM) interaction [1, 2]. In inorganic compounds, it is still a big challenge to control the crystallographic chirality. Transition-metal monosilicides with B20 type chiral crystal structure are a unique example to form homochiral crystal structure [3]. The magnetic chirality follows the crystalline chirality [4]. By use of conventional spontaneous crystallization technique, a water-soluble B20 type compound $NaClO_3$ forms racemic-twinned crystals, containing the left- and right-handed crystalline domains in a crystalline sample. With stirring the solution during the crystallization process, crystalline specimen can be a homochiral domain [5]. However, in case of most of the inorganic magnetic materials, single crystalline samples form racemic-twinned crystals.

$CsCuCl_3$ has a chiral crystal structure with chiral space groups of right-handed $P6_122$ or left-handed $P6_522$. The crystal structures are shown in Figure 1. Quasi-one-dimensional chains of Cu^{2+} ions form right- or left-handed helices with a period of six ions along the c-axis. The DM vector is along the c-

axis. Magnetization measurements show an antiferromagnetic response at T_N = 10.5 K. Cooling temperature below T_N, unpolarized neutron diffraction studies show helimagnetic ordering with the magnetic propagation vector $\boldsymbol{k}_{mag}$ = (1/3, 1/3, ±δ) [6].

Crystallographic chirality is generally determined by an absolute structure analysis. The intensities of a pair of reflections at (*h*, *k*, *l*) and (-*h*, -*k*, -*l*), termed Bijvoet pairs, are not equivalent because atomic scattering amplitude in X-ray diffraction includes imaginary part, called anomalous scattering [7]. Flack parameter x can be obtained as a fitting parameter of the structure analysis, which is expressed as $I_{obs} = (1\text{-}x)I_{cal}(h, k, l) + xI_{cal}(\text{-}h, \text{-}k, \text{-}l)$ [8, 9]. Here, I_{obs} and I_{cal} are the observed and calculated intensities, respectively. We have synthesized single crystals of $CsCuCl_3$ and evaluated the crystallographic chirality using circularly-polarized resonant X-ray diffraction [10, 11]. Our findings indicate that mm-ordered single crystalline samples grown by conventional crystallization technique form only racemic twinned. The domain sizes of the left- and right-handed chiralities are tens of micrometers. To make a homochiral domain in $CsCuCl_3$, we report a crystallization technique as spontaneous crystallization with stirring.

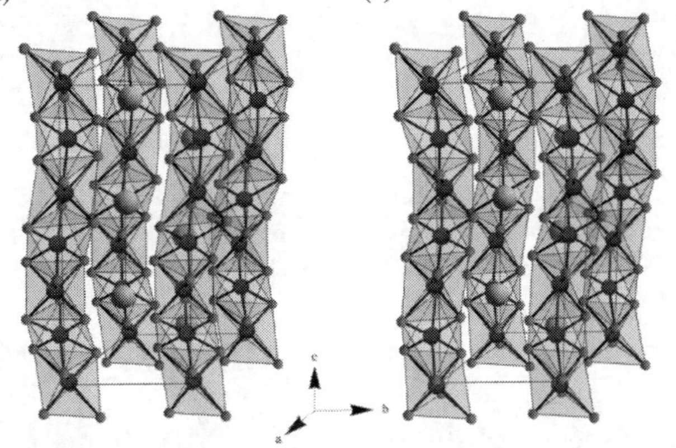

Figure 1. (Color online) Crystal structures of (a) right-handed (space group: $P6_122$) and (b) left-handed (space group: $P6_522$) $CsCuCl_3$. Large, middle and small balls represent Cs, Cu and Cl ions, respectively.

2. Experiments

Single crystals of $CsCuCl_3$ were prepared from an aqueous solution containing CsCl and $CuCl_2$ by slow evaporation of a slightly acidified solution. To avoid the undesired formation of Cs_2CuCl_4, the $CuCl_2$ content of the solution was increased by 10 % from the stoichiometric composition. Firstly 40 ml $CsCuCl_3$ solution was prepared in 100 ml beaker. During the crystallization process, we constantly stirred the solution using Teflon stirrer-bar with the size of 8 mm diameter and 35 mm length. To prevent dust particles from entering the solution, we covered the beakers with a tissue paper. To evaluate the validity of the stirring, we also grow single crystals without stirring.

By optimizing the speed of stirring and cooling temperature, we succeeded in obtaining hundreds of single crystals from a breaker with the size of 0.4 ~ 1.0 mm diameter and 1 ~ 3 mm length. To evaluate the crystallographic chirality of the obtained crystals, X-ray oscillation photographs were taken at 300 K on a Bruker APEX II area detector using Mo $K\alpha$ radiation. With applying empirical absorption correction to the observed intensity, the absolute structures were solved by direct methods and refined using the SHELXS-97 software package [12]. The chiral crystal structures were evaluated by the refined Flack parameters. Racemic twined crystals have Flack parameter = 0.5. The right-handed crystals have Flack parameter = 0 by assuming the space group of $P6_122$, and the left-handed crystals have Flack parameter = 0 by assuming the space group of $P6_522$. Due to the limitation of the X-ray beam diameter, the diameter of all the checked samples was smaller than the beam diameter. And, to evaluate the chirality of a whole sample, we obtained the data from end to end of the sample with the step of 0.7 mm along the length direction, and evaluated Flack parameters at each radiated position. In total, we checked 80 pieces of the crystalline samples with the average size of 0.7 mm diameter and 2.5 mm length.

3. Results

Figure 2 shows pictures of single crystals obtained by conventional crystallization technique without stirring and by our method with stirring. The samples obtained without stirring look zigzag shape, and the samples done with stirring look quart-crystal-like shape.

Absolute structure analysis showed that all the samples crystallized without stirring were racemic twined crystals with Flack parameter ~ 0.5. On the other hand, the samples crystallized with stirring formed single crystalline chirality domain with Flack parameter ~ 0. As an example, Table 1 shows the results of the absolute structure analysis obtained by a sample crystallized with stirring or without stirring. Flack parameters were determined by assuming the left-handed space group of $P6_522$. Both samples have the same atomic positions except for Flack parameters. While the sample crystallized without stirring forms racemic twined crystal with Flack parameter ~ 0.5, the sample done with stirring forms the right-handed homochiral structure with Flack parameter ~ 0. We note that the samples obtained with stirring can be categorized to 3 types, having only the right-handed, only the left-handed and both-handed domains. The samples having both-handed domains formed only the left-handed domain in one end, and the right-handed in the other end. Although Flack parameters at the both end of the sample were zero, the both-handed samples can be regarded as racemic twinned crystals. Table 2 shows the number of each type of the crystals obtained from a breaker. The number of left- and right-handed enantiopure crystals was nearly the same. And we can see some of the both-handed crystals. Therefore, we can conclude that the left- and right-handed crystals are energetically degenerated in a breaker.

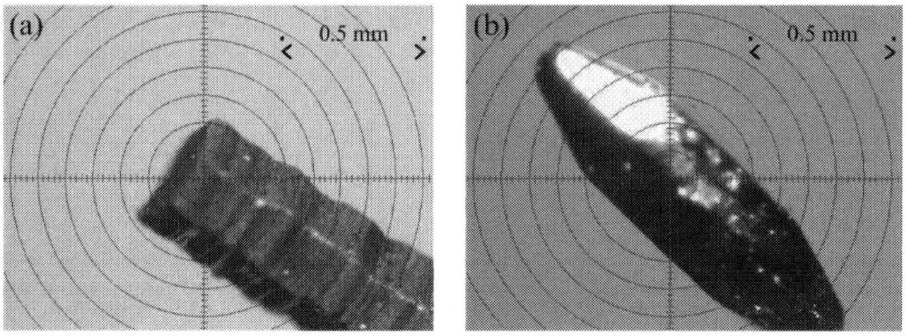

Figure 2. (Color online) Photographs of single crystalline $CsCuCl_3$, obtained by spontaneous crystallization (a) without stirring (conventional method) and (b) with stirring (our method).

Table 1. Results of absolute structure analysis obtained by single crystalline X-ray diffraction. The data was taken at one end of a sample crystallized with or without stirring the solution. Flack parameters were determined by assuming the space group of $P6_522$

Crystallization method	R	wR	Flack parameter
Without stirring	0.0330	0.0690	0.49(5)
With stirring	0.0305	0.0694	0.06(5)

Table 2. The number of each type of samples crystallized with stirring obtained from a breaker. See text for the definition of the types.

Left-handed	Both-handed	Right-handed
31	22	27

4. Discussion

To obtain enantiopure single crystals by use of conventional crystallization method without stirring the solution, we previously reported that maximum size of sample must be submillimeter-ordered [10]. Our stirring crystallization technique obviously enlarges the size of enantiopure domains, and obtains mm-ordered homochiral samples.

The circularly-polarized resonant X-ray micro diffraction indicates that the size of each homochiral domain in the racemic twined $CsCuCl_3$ is tens of micrometers [11]. In case of conventional crystallization process without stirring, many crystalline nuclei grow to tens of micrometers and then combined to one crystal. Therefore, the zigzag shape of the sample suggests the existence of several crystalline nuclei in a sample. On the other hand, the quartz-crystal-like shape samples done with our stirring technique indicates only one crystalline nucleus grew to mm-ordered single crystals. It suggests that the aqueous stream due to stirring prevents each growing nucleus from combining during the crystallization process. Obtaining some of the both-handed crystals indicates that two pieces of the mm-ordered homochiral crystals can be merged to one crystal. Therefore, the present amplitude of the water flow cannot avoid combining some of the growing nuclei completely. Kondepudi reported that most of obtained $NaClO_3$ samples crystallized with stirring in a breaker were only left- or right-handed. With several crystallizations obtained by different breakers, the number of left- and right-handed crystals finally becomes the same [5]. This behavior is different from ours because we obtained equal number of the left- and right-handed crystals in a beaker.

In summary, we succeeded in obtaining the mm-ordered enantiopure single crystals of $CsCuCl_3$ by adapting crystallization technique with stirring solution. The mm-ordered enantiopure chiral magnetic crystals enable us to perform a variety of experiments for investigating the interplay between crystalline and magnetic chirality.

Acknowlegement

This work was partially supported by a High-Tech Research Center project for private universities with a matching fund subsidy from the Ministry of Education, Culture, Sports, Science and Technology (MEXT) of the Japanese Government. Y. K. acknowledges a Grant-in-Aid for young scientist (B) (No. 24740207) and Exploratory Research (No. 24108507) from MEXT. J. A. and K. I. acknowledge a Grant-in-Aid for Scientific Research (S) (No. 25220803) from MEXT.

References

[1] Dzyaloshinskii I E 1964 *Sov. Phys. JETP* **19** 960.
[2] Moriya T 1960 *Phys. Rev.* **120** 91.
[3] Tanaka M, Takayoshi H, Ishida M and Endoh Y 1985 *J. Phys. Soc. Jpn.* **54** 2970.
[4] Ishida M, Endoh Y, Mitsuda S, Ishikawa Y and Tanaka M 1985 *J. Phys. Soc. Jpn.* **54** 2975.
[5] Kondepudi D K, Kaufman R J and Singh N 1990 *Science* 250 975.
[6] K. Adachi, N. Achiwa, M. Mekata, *J. Phys. Soc. Jpn.* 49, 545 (1980).
[7] Bijvoet J M 1954 *Nature (London)* 173 888.
[8] Flack H D 1983 *Acta Crystallogr. A* 39 876.
[9] Berardinelli G and Flack H D 1985 *Acta Crystallogr. A* 41 500.
[10] Kousaka Y, Ohsumi H, Komesu T, Arima T-H, Takata M, Sakai S, Akita M, Inoue K, Yokobori T, Nakao Y, Kaya E and Akimitsu J 2009 *J. Phys. Soc. Jpn.* **78** 123601.
[11] Ohsumi H, Tokuda A, Takeshita S, Takata M, Suzuki M, Kawamura N, Kousaka Y, Akimitsu J and Arima T 2013 *Angew. Chem. Int. Ed.* **52** 8718.
[12] Sheldrick, G. M. 1997 SHELX-97 A Programs for Crystal Structure Refinement University of Göttingen Germany.

Journal of Physics: Conference Series **502** (2014) 012020 doi:10.1088/1742-6596/502/1/012020

Development of technique for three-dimensional visualization of grain boundaries by white X-ray microbeam

K Kajiwara[1], T Shobu[2], H Toyokawa[1] and M Sato[1]

[1]Japan Synchrotron Radiation Research Institute, Sayo, Hyogo 679-5198, Japan
[2]Japan Atomic Energy Agency, Sayo, Hyogo 679-5148, Japan

E-mail: kajiwara@spring8.or.jp

Abstract. A technique for three-dimensional visualization of grain boundaries was developed at BL28B2 at SPring-8. The technique uses white X-ray microbeam diffraction and a rotating slit. Three-dimensional images of small silicon single crystals filled in a plastic tube were successfully obtained using this technique for demonstration purposes. The images were consistent with those obtained by X-ray computed tomography.

1. Introduction

The three-dimensional (3D) distribution of grains in structural materials strongly influences their mechanical properties. A technique for nondestructive observation of the 3D distribution of grains is necessary for clarifying the correlation between the mechanical properties and the grain distribution. The visualization of grains in polycrystalline material is difficult for Computed tomography (CT), which is typical 3D observation technique because absorption contrast is hardly detected at grain boundaries. X-ray diffraction technique is effective technique for observation of grain distribution detecting the deference of crystalline orientation.

A technique of grain boundary imaging by the Laue pattern (GILP) was developed at BL28B2 at SPring-8 to visualize the distribution of grain boundaries in polycrystalline materials [1]. The GILP technique uses X-ray diffraction by the transmission Laue method using white X-ray microbeams. Images of the Laue pattern were acquired using a two-dimensional (2D) detector by scanning the beam positions on the specimen. The grain boundaries were identified by using the differentials between the Laue pattern images at each position. However, this technique was not applicable to 3D observations because GILP cannot discriminate between Laue patterns from grains on a beam path in a specimen. This problem can be solved by using a rotating slit (figure 1), which was developed to restrict the observation area inside a bulky specimen in X-ray diffraction measurement using a 2D detector [2].

The purpose of this study was to apply GILP technique to 3D observation by using the rotating slit.

2. Experimental

The experiment was conducted at BL28B2 at SPring-8. The experimental setup is shown in figure 2. The white X-rays generated by a bending magnet light source were shaped in a 20 × 20 μm^2 cross section by the quadrant slit. The white X-ray microbeam was irradiated on the specimen in the transmission layout. Laue patterns from the specimen were detected by a 2D detector. A rotating slit unit was placed between the specimen and the 2D detector. This device consists of a pair of tantalum disks, as shown in figure 1. The disks were 0.5 mm thick and had diameters of 100 mm and 200 mm, respectively. Slits with the shape of an Archimedean spiral were cut into the disks. The slits were 0.2 mm wide. These slits specified the observation areas in the specimen. The disks with the slits were rotated at a rate of 120 rpm so that the aperture of the slits could cover the detection area of the 2D detector. This rotating slit made it possible for the 2D detector to discriminate the Laue patterns from the grains on the beam path. The exposure time of a Laue pattern image was 0.5 s. The position of the specimen was scanned by motorized translation stages at a step interval of 0.01 mm.

Images obtained by the new technique were compared with ones obtained by CT, for verification of the new technique. A mass of small silicon single crystals filled in a plastic tube was prepared as a sample for demonstration (figure 3) because actual distribution of grains in structural materials cannot be visualized by CT. The distribution of silicon crystals in a plastic tube was able to be visualized by both techniques. The silicon crystals were a few millimeters in diameter, and the inner diameter of the plastic tube was 7 mm. A beam path typically hit one to five silicon crystals.

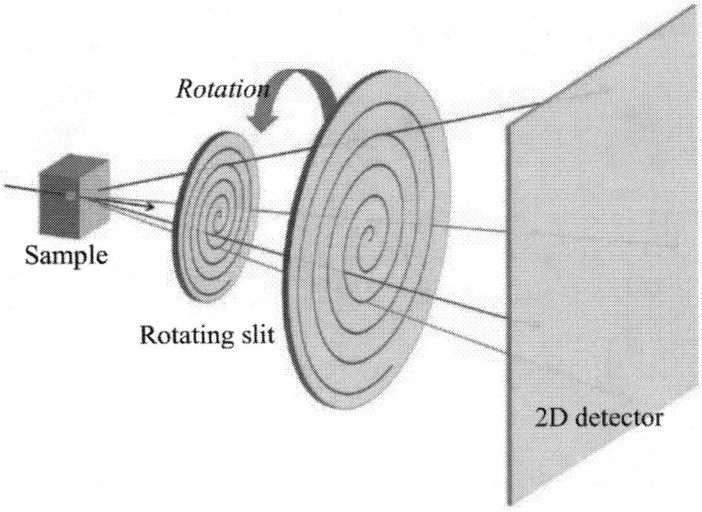

Figure 1. Schematic of a rotating slit.

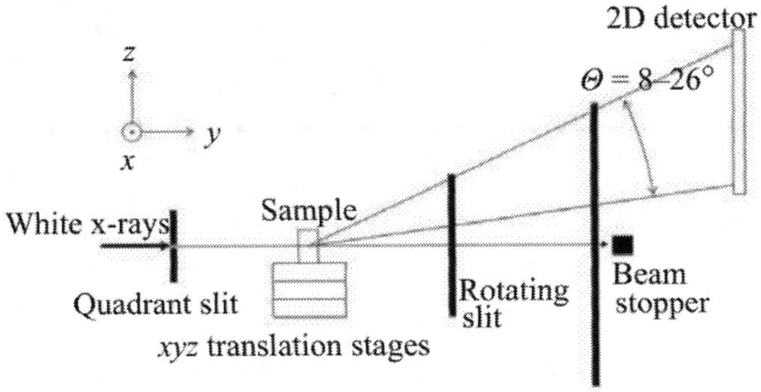

Figure 2. Schematic of the experimental setup.

Figure 3. Photograph of the sample and silicon single crystals.

3. Results and discussion

Figure 4 shows cross-sectional images of the grain boundaries measured by GILP using the rotating slit. Figure 4 (a) and (b) are images of the cross sections in the *xz* plane and *xy* plane, respectively. The black contrast indicates the grain boundaries. Figure 5 shows the edge-enhanced CT images corresponding to figure 4. The images in figure 4 consistently reproduce the morphological features shown in figure 5.

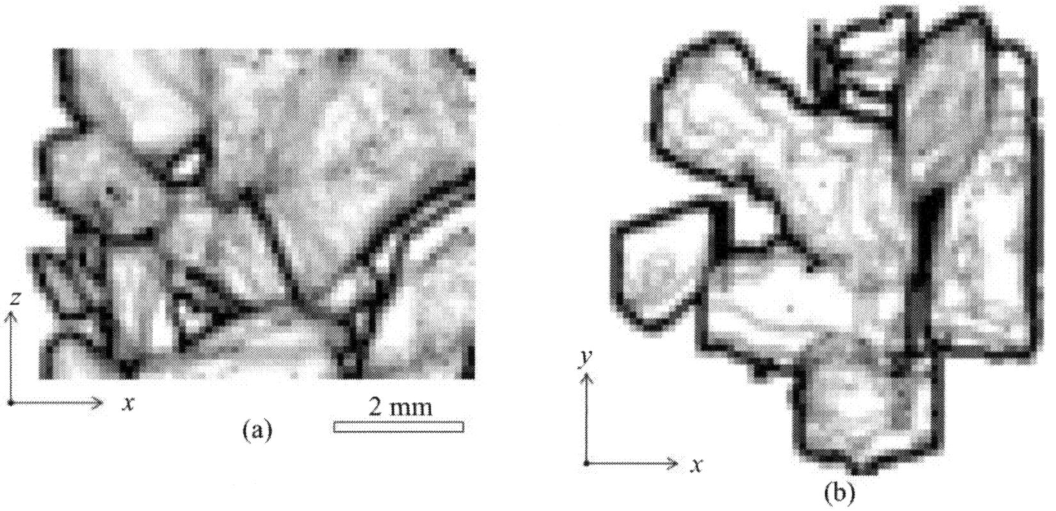

Figure 4. Reconstructed grain boundary images obtained from Laue patterns in the range of $\Theta = 17$–$26°$. Cross sections in *xz* plane (a) and *xy* plane (b).

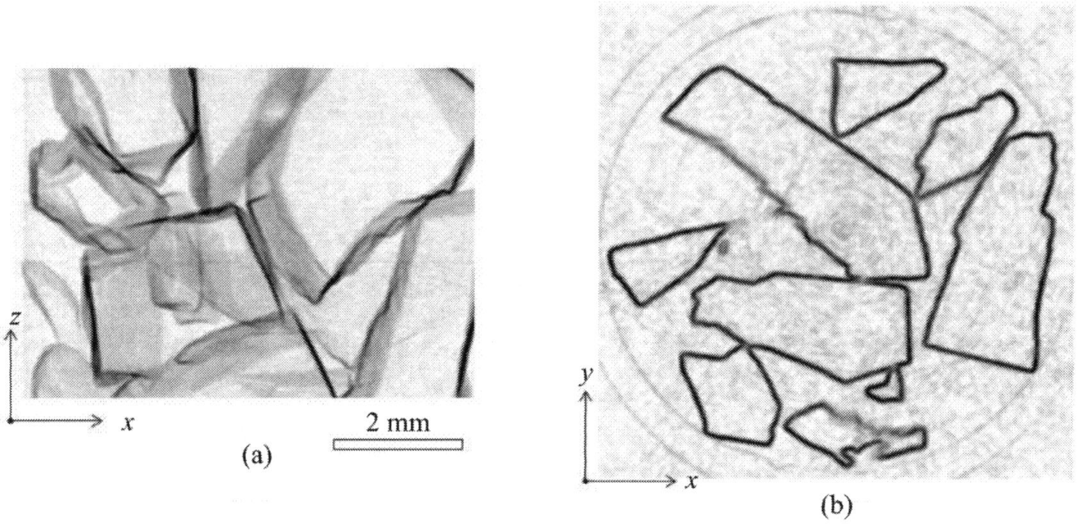

Figure 5. CT images. Cross section in *xz* plane (a) and *xy* plane (b). The images were processed by an edge enhancement filter. The X-ray energy was 30 keV. The interval of the rotation step was 0.5°. The exposure time of an image was 0.05 s.

The relationship between the diffraction angle and the spatial resolution of GILP with the rotating slit was examined. Figure 6 shows a schematic of the observation area restricted by the rotating slit. The spatial resolution of GILP with the rotating slit was expected to depend on the size of the observation area. The size of the area in the *y* direction, A_y, should increase as the diffraction angle decreases. Grain boundary images obtained from the Laue patterns in the area of low diffraction angle are shown in figure 7. The diffraction angles in figures 4 and 7 were 17–26° and 8–17°, respectively. The spatial resolution in the *y* direction in figure 7 (b) was lower than that in figure 4 (b), in agreement with the prediction. On the other hand, the difference between figures 4 (a) and 7 (a) was small. This is because the size of the observation area in the *x* and *z* directions is not affected by the diffraction angle.

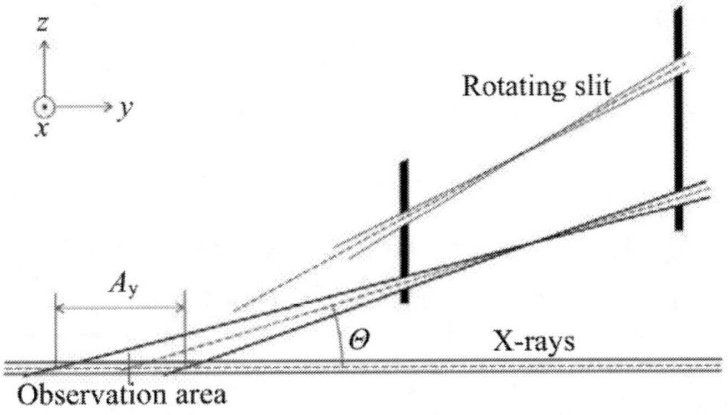

Figure 6. Schematic of the observation area restricted by the rotating slit.

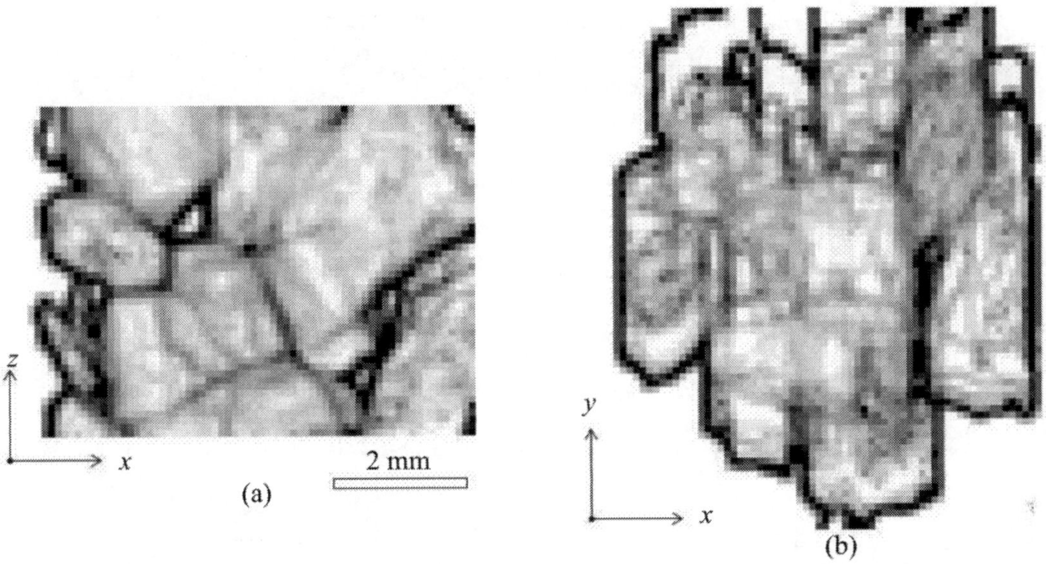

Figure 7. Reconstructed grain boundary images obtained from Laue patterns in the range of Θ = 8–17°. Cross sections in *xz* plane (a) and *xy* plane (b).

4. Conclusion

A technique for 3D observation of grain boundaries was developed by applying a rotating slit to the GILP technique. The spatial resolution of this technique depended on the diffraction angle of the Laue patterns; patterns in the area of higher diffraction angles provide a higher spatial resolution.

Acknowledgment

The synchrotron radiation experiments were performed with the approval of the Japan Synchrotron Radiation Research Institute (Proposal No. 2013A1858). This work was supported in part by Development of Systems and Technology for Advanced Measurement and Analysis (SENTAN) promoted by the Japan Science and Technology Agency.

References

[1] Kajiwara K, Sato M, Hashimoto T, Hirosawa I, Yamada T, Terachi T, Fukumura T and Arioka K 2009 *Phys. Status Solidi* A **206** 1838–1841

[2] Suzuki K, Shobu T, Shiro A and Toyokawa H 2012 *HOZENGAKU* (Maintenology) **11** 99–106 (in Japanese)

Journal of Physics: Conference Series **502** (2014) 012021 doi:10.1088/1742-6596/502/1/012021

Structural origin of ionic conductivity for $Li_7P_3S_{11}$ metastable crystal by neutron and X-ray diffraction

Y. Onodera[1], K. Mori[1], T. Otomo[2], H. Arai[3], Y. Uchimoto[4], Z. Ogumi[3], and T. Fukunaga[1]

[1]Research Reactor Institute, Kyoto University, Kumatori, Osaka 590-0494, Japan
[2]Institute of Materials Structure Science, High Energy Accelerator Research Organization, Tsukuba, Ibaraki 305-0801, Japan
[3]Office of Society-Academia Collaboration for Innovation, Kyoto University, Uji, Kyoto 611-0011, Japan
[4]Graduate School of Human and Environmental Studies, Kyoto University, Sakyo-ku, Kyoto 606-8501, Japan

E-mail: y-onodera@rri.kyoto-u.ac.jp

Abstract. Reverse Monte Carlo modeling was carried out for $Li_7P_3S_{11}$ metastable crystal and $(Li_2S)_{70}(P_2S_5)_{30}$ glass based on time-of-flight neutron and synchrotron X-ray diffraction data. Polyhedral analysis of three-dimensional structures for $Li_7P_3S_{11}$ metastable crystal and $(Li_2S)_{70}(P_2S_5)_{30}$ glass were gave us not only structural characteristics but also structural origins related to their high ionic conductivity.

1. Introduction

$Li_7P_3S_{11}$ metastable crystal, which can be synthesized by aging of $(Li_2S)_{70}(P_2S_5)_{30}$ glass at 513K, showed a high ionic conductivity in the order of 10^{-3} S/cm at room temperature (RT) [1]. In order to elucidate the relationship between the structure and the excellent electrical conduction properties, we have carried out a polyhedral analysis of three-dimensional atomic structures constructed by reverse Monte Carlo (RMC) modeling based on the time-of-flight neutron diffraction and the synchrotron X-ray diffraction data [2]. In the polyhedral analysis, it was found that a large number of vacant S_4 tetrahedra ("*ac*-[S_4] units"), which fully accept Li^+ cation, are located around a LiS_4 tetrahedron ([LiS_4] unit) in $Li_7P_3S_{11}$ metastable crystal.

In this study, we progressed the polyhedral analysis focused on "bottle-necks" and "connectivity" of [LiS_4] and *ac*-[S_4] units. These structural characteristics are strongly related to the Li^+ conductivity for $Li_7P_3S_{11}$ metastable crystal.

2. Experimental

The details of the synthesis of samples and experimental procedures are described in ref. [2]. The time-of-flight neutron diffraction (TOF-ND) measurements were carried out with the General Material Diffractometer (GEM) installed at the ISIS in the Rutherford Appleton Laboratory, UK [3]. The synchrotron X-ray diffraction (SXRD) experiments were performed with a horizontal two-axis diffractometer at the BL04B2 beam line of the SPring-8 synchrotron radiation facility [4]. In addition, the TOF-ND and SXRD data were corrected for the polarization, multiple scattering, absorption, and incoherent scattering in order to obtain the structure factors, $S(Q)$, of $Li_7P_3S_{11}$ metastable crystal and

$(Li_2S)_{70}(P_2S_5)_{30}$ glass [5-9]. To construct the three-dimensional structures of $Li_7P_3S_{11}$ metastable crystal and $(Li_2S)_{70}(P_2S_5)_{30}$ glass, the RMC modeling was carried out for $g(r)$ of $Li_7P_3S_{11}$ metastable crystal and $S(Q)$ of $(Li_2S)_{70}(P_2S_5)_{30}$ glass, using RMCA [10], where $g(r)$ is the atomic pair distribution function. The refined crystal structure of $Li_7P_3S_{11}$ metastable crystal [11] was used as the starting configuration for the RMC modeling (the detailed procedure of the RMC modelling is also described in ref. [2]).

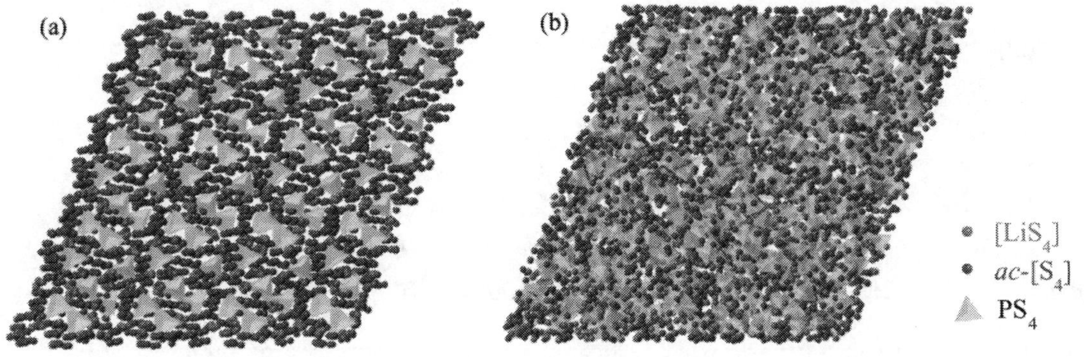

Figure 1. Spatial distribution of $[LiS_4]$ units (red spheres) and ac-$[S_4]$ units (blue spheres), and PS_4 tetrahedra (light blue tetrahedra) for (a) $Li_7P_3S_{11}$ metastable crystal and (b) $(Li_2S)_{70}(P_2S_5)_{30}$ glass.

3. Results and discussion

In our previous study [2], we found that $[LiS_4]$ units, PS_4 tetrahedra, and ac-$[S_4]$ units are distributed in the three dimensional structures of $Li_7P_3S_{11}$ metastable crystal and $(Li_2S)_{70}(P_2S_5)_{30}$ glass derived by the RMC modeling. Figure 1 shows three dimensional structures of $Li_7P_3S_{11}$ metastable crystal and $(Li_2S)_{70}(P_2S_5)_{30}$ glass, visualized with $[LiS_4]$ units, PS_4 tetrahedra and ac-$[S_4]$ units. In these three dimensional structures, it can be considered that Li^+ cations occupy $[LiS_4]$ units (red spheres) and can move into ac-$[S_4]$ units (blue spheres), in other words, Li^+ conduction pathways are adjacent to connection of $[LiS_4]$ and ac-$[S_4]$ units. Therefore, to discuss the relationship between structure and ionic conductivity more deeply, we focused on a local and a long-range connectivity of $[LiS_4]$ units and ac-$[S_4]$ units.

Figure 2 shows the local connection of $[LiS_4]$ units and ac-$[S_4]$ units. Concerning the migration of Li^+ cation from $[LiS_4]$ unit to another $[LiS_4]$ unit, it is clear that Li^+ cation needs to escaping from a $[LiS_4]$ unit and passing through ac-$[S_4]$ units. Thus, S_3 triangular planes formed a $[LiS_4]$ unit can be regarded as bottle-necks for Li^+ conduction. In order to elucidate a size of the bottle-neck, radii of all inscribed circles of S_3 triangular planes for $[LiS_4]$ units were calculated. Figure 3 shows the size distribution of radii of the inscribed circles of the S_3 triangular planes, which were calculated for the three-dimensional structures of $Li_7P_3S_{11}$ metastable crystal and $(Li_2S)_{70}(P_2S_5)_{30}$ glass. The radius of the inscribed circle for $Li_7P_3S_{11}$ metastable crystal was estimated to be 1.12 Å on average, but on the other hand, 1.08 Å for $(Li_2S)_{70}(P_2S_5)_{30}$ glass, as shown in Fig. 3. The result indicates that the larger bottle-necks in $Li_7P_3S_{11}$ metastable crystal allow Li^+ cations to move easily from $[LiS_4]$ units to ac-$[S_4]$ units in comparison with those in $(Li_2S)_{70}(P_2S_5)_{30}$ glass.

Next, we investigated the long-range connectivity of $[LiS_4]$ units and ac-$[S_4]$ units in the chain networks formed by "-$[LiS_4]$-[ac-$[S_4]$]-$[LiS_4]$-[ac-$[S_4]$]-$[LiS_4]$-[ac-$[S_4]$]-" chain. Here, $[LiS_4]$-[ac-$[S_4]$] correlations up to 2.70 Å in length were allowed as chain formers. Total number, N_{chain}, of $[LiS_4]$ and ac-$[S_4]$ units in the "-$[LiS_4]$-[ac-$[S_4]$]-$[LiS_4]$-[ac-$[S_4]$]-$[LiS_4]$-[ac-$[S_4]$]-" chain was calculated in the parallelepiped box including 5376 atoms (Li: 1792, P: 768, and S: 2816) for $Li_7P_3S_{11}$ metastable crystal and $(Li_2S)_{70}(P_2S_5)_{30}$ glass [2]. Table 1 is summarized the number of $[LiS_4]$ and ac-$[S_4]$ units in the longest "-[LiS_4]-[*ac*-[S_4]]-[LiS_4]-[*ac*-[S_4]]-[LiS_4]-[*ac*-[S_4]]-" chain and a ratio of the number of [LiS_4] and *ac*-[S_4] units in the longest chain to the total number of all [LiS_4] and *ac*-[S_4] site in the parallelepiped box for $Li_7P_3S_{11}$ metastable crystal and $(Li_2S)_{70}(P_2S_5)_{30}$ glass. As indicated in Table 1, most of [LiS_4] units and *ac*-[S_4] units connect with each other and form an exceedingly-long chain in $Li_7P_3S_{11}$ metastable crystal. In contrast, various lengths of chains exist in $(Li_2S)_{70}(P_2S_5)_{30}$ glass and the total number of [LiS_4] and *ac*-[S_4] units in the longest chain is only 515.

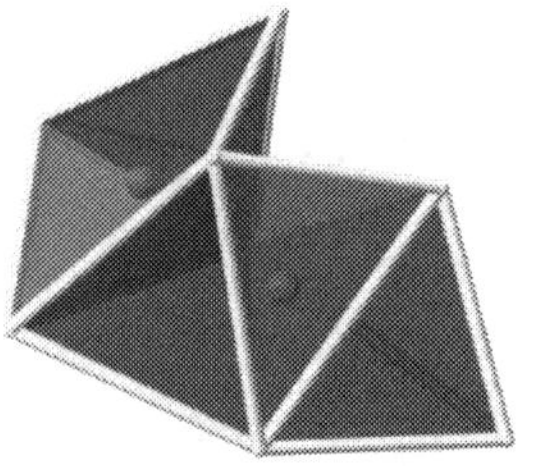

Figure 2. Local connection of [LiS_4] units (red tetrahedra) and *ac*-[S_4] units (blue tetrahedra).

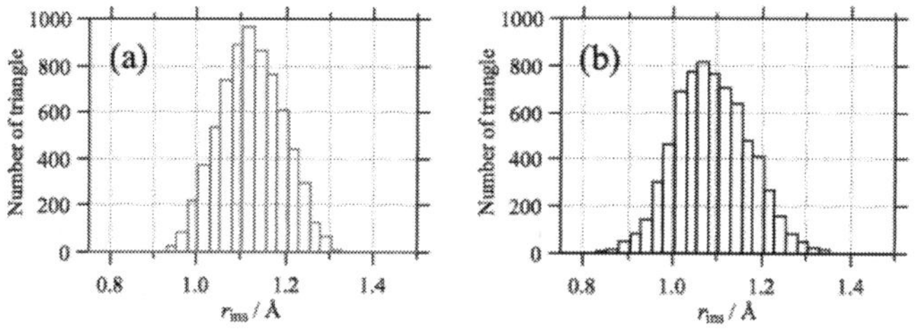

Figure 3. Distributions of the radii of inscribed circles in S_3 triangular planes for each [LiS_4] unit for (a) $Li_7P_3S_{11}$ metastable crystal and (b) $(Li_2S)_{70}(P_2S_5)_{30}$ glass.

In Fig. 4, the spatial distribution of the longest "-[LiS_4]-[*ac*-[S_4]]-[LiS_4]-[*ac*-[S_4]]-[LiS_4]-[*ac*-[S_4]]-" chain in $Li_7P_3S_{11}$ metastable crystal and $(Li_2S)_{70}(P_2S_5)_{30}$ glass are shown together with PS_4 tetrahedra. As can be seen in Fig. 4(a), continuously "-[LiS_4]-[*ac*-[S_4]]-[LiS_4]-" chain forms net-like structure throughout the ordered framework of PS_4 tetrahedra in $Li_7P_3S_{11}$ metastable crystal, whereas the chain may be shredded into narrow strips by disordered framework of PS_4 tetrahedra in $(Li_2S)_{70}(P_2S_5)_{30}$ glass. This result allows us to deeply understand that a well-developed conduction pathway of Li^+ cations exists in $Li_7P_3S_{11}$ metastable crystal.

Table 1. The analysis of the longest "-[LiS_4]-[*ac*-[S_4]]-[LiS_4]-[*ac*-[S_4]]-[LiS_4]-[*ac*-[S_4]]-" chain in $Li_7P_3S_{11}$ metastable crystal and $(Li_2S)_{70}(P_2S_5)_{30}$ glass.

	$Li_7P_3S_{11}$ metastable crystal	$(Li_2S)_{70}(P_2S_5)_{30}$ glass
Total number of [LiS_4] and *ac*-[S_4] units in the longest chain, N_{chain}	5665	515
Number of all site, N_{all} ([LiS_4] + *ac*-[S_4])	6283	4480
N_{chain} / N_{all}	0.90	0.12

4. Conclusions

Two structural origins of high ionic conductivity for $Li_7P_3S_{11}$ metastable crystal could be recognized through the polyhedral analysis of three-dimensional structures constructed by the reverse Monte Carlo modeling. Firstly, [LiS_4] units in $Li_7P_3S_{11}$ metastable crystal have larger bottle-necks of S_3 triangular planes than those in $(Li_2S)_{70}(P_2S_5)_{30}$ glass. Secondly, the exceedingly-long "-[LiS_4]-[*ac*-[S_4]]-[LiS_4]-[*ac*-[S_4]]-[LiS_4]-[*ac*-[S_4]]-" chain network corresponding to well developed Li^+ conduction pathway exists in $Li_7P_3S_{11}$ metastable crystal.

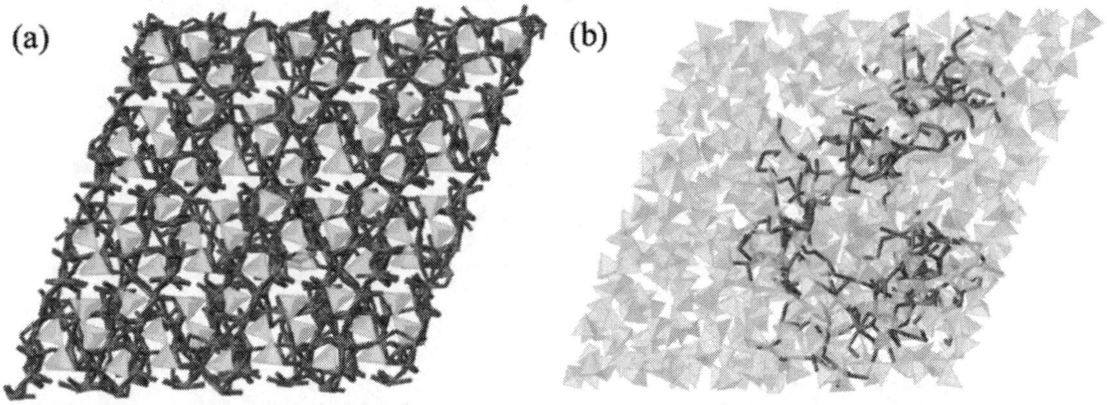

Figure 4. Spatial distribution of the longest -[LiS_4]-[*ac*-[S_4]]-[LiS_4]- chain network (red-and-blue-banded cylinders), together with PS_4 tetrahedra (light blue tetrahedra) for (a) $Li_7P_3S_{11}$ metastable crystal and (b) $(Li_2S)_{70}(P_2S_5)_{30}$ glass.

Acknowledgements

Many thanks are given to Dr. Alex Hannon of RAL-ISIS for his help in the time-of-flight neutron diffraction experiments. We also thank to Dr. Shinji Kohara of SPring-8 for his help in the synchrotron X-ray diffraction and the RMC modeling. The synchrotron X-ray diffraction experiments at BL04B2 in SPring-8 were performed with the approval of the Japan Synchrotron Radiation Research Institute (JASRI, Proposal No. 20091644). This work was supported by Grant-in-Aid for Young Scientists (B) (No. 25870371) from the Ministry of Education, Culture, Sports, Science and Technology. This work was partially supported by the "Research and Development Initiative for Scientific Innovation of New Generation Batteries (RISING project)" of the New Energy and Industrial Technology Development Organization (NEDO), Japan.

References

[1] Mizuno F, Hayashi A, Tadanaga K, Minami T, Tatsumisago M 2005 *Electrochem. Solid-State Lett.* **8** A603.
[2] Onodera Y, Mori K, Otomo T, Sugiyama M, Fukunaga T 2012 *J. Phys. Soc. Jpn.* **81** 044802.
[3] Williams W G, Ibberson R M, Day P, Enderby J E 1998 *Physica B* **241-243** 234.
[4] Kohara S, Suzuya K, Kashihara Y, Matsumoto N, Umesaki N, Sakai I 2001 *Nucl. Instrum. And Methods A* **467** 1030.
[5] Sasaki S 1991 *KEK report* **90-16**.
[6] Hubbell J H, Veigele W J, Briggs E A, Brown R T, Cromer D T, Howerton R J 1993 *J. Phys. Chem. Ref. Data* **4** 61.
[7] Paalman H H, Pings C J 1962 *J. Appl. Phys.* **33** 2635
[8] Blech I A, Averbach B L 1965 *Phys. Rev.* **137** 1113.
[9] Faber T E, Ziman J M 1965 *Philos. Mag.* **11** 153.
[10] McGreevy R L, Putztai L 1988 *Mol. Simul.* **1** 359.
[11] Onodera Y, Mori K, Otomo T, Hannon A C, Kohara S, Itoh K, Sugiyama M, Fukunaga T 2010 *J. Phys. Soc. Jpn.* Suppl. A **79** 87.

Local structure analysis of $NaNbO_3$

Y Yoneda[1], **D Fu**[2], **and S Kohara**[3]

[1]Japan Atomic Energy Agency (JAEA), Sayo-gun, Hyogo 679-5148, Japan
[2]Division of Global Research Leaders, Shizuoka University, Hamamatsu, Shizuoka 432-8651, Japan
[3]Japan Synchrotron Radiation Research Institute (JASRI), Sayo-gun, Hyogo 679-5198, Japan

E-mail: yoneda@spring8.or.jp

Abstract. $NaNbO_3$ has an antiferroelectric structure at room temperature and finds important technological applications. It exhibits an unusual complex sequence of temperature- and pressure-driven structural phase transitions. $NaNbO_3$ shows unambiguous evidence for the presence of the ferroelectric *R3c* phase coexisting with an antiferroelectric phase (*Pbcm*) over a wide range of temperatures. We have carried out atomic pair-distribution function (PDF) analysis on $NaNbO_3$ to understand the phase transitions. High-energy X-ray PDF using powder samples were carried out at the SPring-8, which is provided to become rhombohedral structure if *A*-site atoms did order and to become orthorhombic structure if *A*-site atoms did disorder.

1. Introduction

Many of antiferroelectric (AFE) crystal has orthorhombic structure. Although the AFE crystal does not show a novel characteristic, it sometimes shows a excellent characteristic by the combination with ferroelectric crystals, like $Pb(Zr,Ti)O_3$ (PZT). PZT, $Bi(Mg_{0.5}Ti_{0.5})O_3$ (BMT), and $(Li,Na)NbO_3$ (LNN) have the orthorhombic structure, but the details are not understood still well. The local structure of PZT has been revealed by the neutron pair-distribution function (PDF) analysis [1]. However, the local structure has not been reproduced by any average structure model [2, 3, 4]. $NaNbO_3$ is a well known AFE material with orthorhombic *Pbcm* structure, but there have been few reports on double hysteresis in this material [5, 6]. There have still a lot of uncertain points of these materials with AFE structure.

Among them, we are specially paying attention to $NaNbO_3$. $NaNbO_3$ shows various ferroelectric properties depending on Li substitution [7]. The structure of $NaNbO_3$ changed to a rhombohedral *R3c* with increasing the Li substitution for Na. $Li_{0.12}Na_{0.88}NbO_3$ becomes ferroelectric material and has a large electromechanical coupling [8]. This situation is very similar to $LaMnO_3$. $LaMnO_3$ is a magnetic materials and it shows various magnetic properties depending on oxygen stoichiometry. The stoichiometric $LaMnO_3$ is an antiferromagnetic (AFM) insulator with no local lattice distortion. With increased oxygen deficiency, the local atomic structure becomes distorted and deviates from the average crystallographic structure and shows ferromagnetic (FM) ordering. PDF analysis for the perovskite $LaMnO_3$ clarified the local Jahn-Teller distortion, playing a major role in the magnetic and transport transitions [9, 10].

In this study, the local structure of $NaNbO_3$ crystal was analyzed by the PDF analysis [11] obtained by high-energy synchrotron X-ray powder diffraction. The local structure analysis should be needed in such orthorhombic perovskite system.

2. Experimental

$NaNbO_3$ ceramics was prepared by a conventional powder processing technique. The starting materials were prepared by mixing high-purity Na_2CO_3 (99.99%), and Nb_2O_5 (99.99%) in appropriate ratios. The powders were mixed and calcined at 1173 K, and then sintered at 1353 - 1523 K in O_2 atmosphere. The phase purity of the ceramics was confirmed by laboratory powder X-ray diffraction, and no impurity phase appeared.

High-energy X-ray powder diffraction data were collected over a wide temperature range by using a closed-cycle helium refrigerator and an electric furnace with 60-keV X-rays at BL04B2 and BL14B1 in SPring-8. The incident X-ray beam was focused on the samples by using a bent crystal [12, 13, 14]. Samples of approximately 0.2 g were sealed between Kapton foils for the low-temperature data correction and were sealed in a quartz tube for the high-temperature data correction. Scattered radiation was collected with an intrinsic germanium detector. Both short- and long-range structural parameters were refined by the PDF method with the program PDFFIT [15] and by Rietveld refinement with the program RIETAN-FP [16]. The crystal structure was drawn using the program VESTA [17].

3. Results and discussion

Pure $NaNbO_3$ has an AFE structure at room temperature and exhibits an unusual complex sequence of temperature driven structural phase transitions. Above 913 K, $NaNbO_3$ has a paraelectric cubic structure with the space group of $Pm3m$. At 848 K, tetragonal $P4/mbm$ structure appeared, then, AFE phase with the orthorhombic $Pbcm$ structure appeared at 633 K, and finally, ferroelectric phase with rhombohedral $R3c$ structure appeared at 173 K. We obtained the X-ray diffraction profiles of the cubic, tetragonal, orthorhombic, and rhombohedral phases as shown in Fig. 1. The diffraction data of four phases can be indexed unambiguously and sufficiently satisfactory fits were obtained in all phases. The temperature dependence of the refined lattice parameters are shown in Fig. 2.

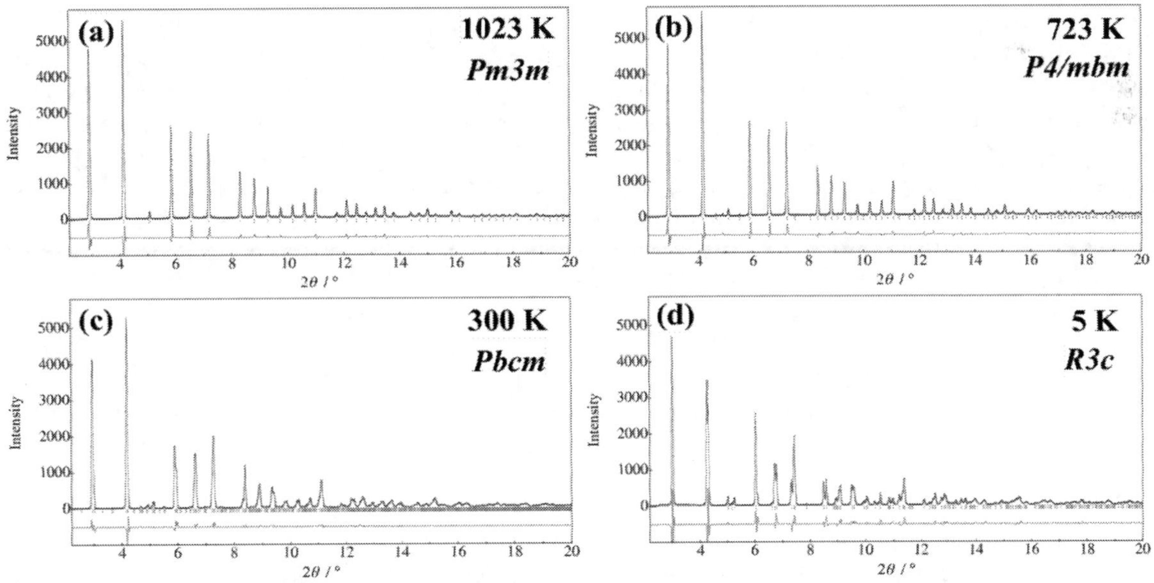

Figure 1. (Color online) Result of Rietveld refinement of $NaNbO_3$ in (a) cubic phase at 1023 K, (b) tetragonal phase at 723 K, (c) orthorhombic phase at 300 K, and (d) rhombohedral phase at 5 K.

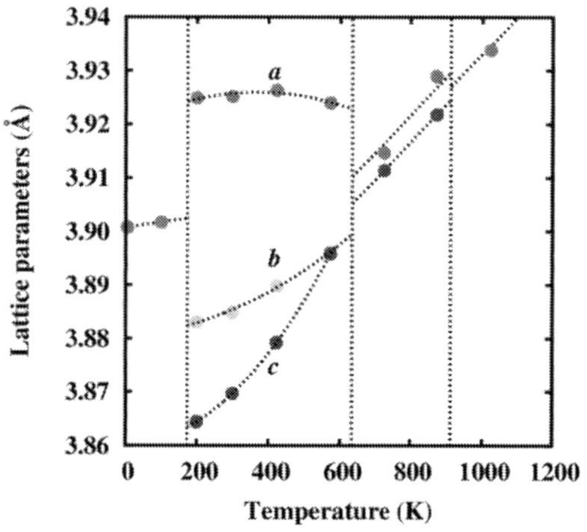

Figure 2. (Color online) Temperature dependence of lattice parameters refined by Rietveld analysis.

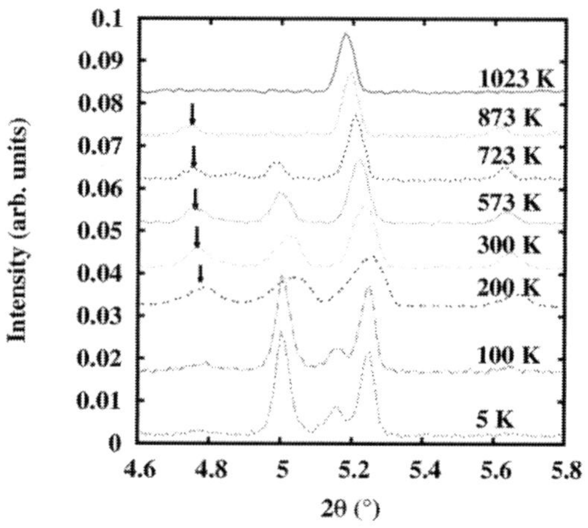

Figure 3. (Color online) Temperature evolution of Bragg reflections of $NaNbO_3$.

Figure 3 shows the evolution of powder X-ray diffraction profiles of $NaNbO_3$ with temperature in the range of 5 – 1023 K. The AFE phase consists of two or more sublattice polarizations of antiparallel nature, which give rise to superlattice reflections in the diffraction patterns. We have observed such reflection in the AFE phase, which appeared at 4.75°, as shown by arrows in Fig. 3. From the knowledge of the Miller indexes of the superlattice reflections and lattice type of the structure, as revealed by the splitting of main Bragg peaks, one can find a signature of the presence of structural phase transition. Additionally, an anomalous peak broadening was observed when the rhombohedral – orthorhombic phase transition occurred at 200 K. Below 100 K, the width of Bragg peaks were sharper than those observed above 200 K. Although all the diffraction data were collected with the same resolution, the width of Bragg peaks broadened drastically at 200 K. The crystal structure not only changed, but coherent size also changed owing to its disordered feature. In such cases, the conventional structure analysis with assuming a periodic structure is difficult to clarify the disordered structure. Unlike the conventional crystallographic analysis, lattice periodicity is not assumed in the PDF analysis. Hence it is possible to detect nonuniform atomic displacements those are not directly observed by standard crystallographic techniques. In system with which rhombohedral structure and orthorhombic structure compete, PDF analysis is effective similar to $LaMnO_3$. We performed PDF analysis on $NaNbO_3$ to clarify the phase transition mechanism.

Temperature evolution of the PDF for $NaNbO_3$ is shown in Fig. 4(a). The PDF peak lying at 2 Å seems to be singlet shape at 300 K. The doublet feature appeared at 200 K, and finally clear split peaks observed down to 5 K. A clear evidence of an order – disorder type phase transition was found in the local structure. We attempted to reproduce the experimental PDF by introducing the local displacement of Nb atoms. The calculated PDF is good agreement with the experimental PDF, as shown in Fig. 4(b). We refined the observed PDF obtained at all the temperature and could reproduce the order-disorder phase transition. Since the rattling space of NbO_6 octahedra of $NaNbO_3$ is too small to extract from the average structure analysis, the change of the atomic coordination in the rhombohedral and the orthorhombic phases is small, as shown in Fig. 5. In the low temperature ordered phase, the Nb atoms shifted from the average atomic coordination, and the local shift of Nb atoms caused the change of the rattling space, and thus, the order – disorder phase transition occurred at 200 K.

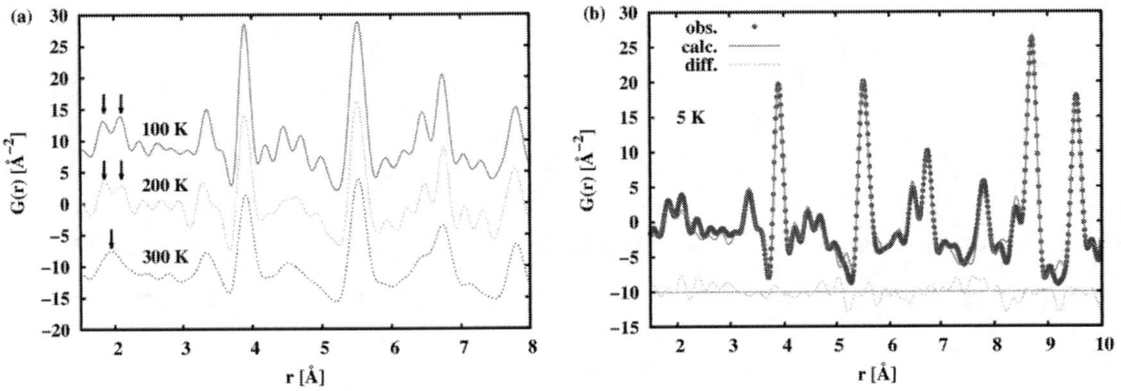

Figure 4. (Color online) (a) Temperature evolution of experimental PDF of $NaNbO_3$. (b) Result of PDF refinement of $NaNbO_3$ in rhombohedral phase at 5 K.

References

[1] Teslic S, Egami T, and Viehland D 1996 *J. Phys. Chem Solids* **57** 1537
[2] Glazer A M, Mabud A A and Clarke R 1978 *Acta Cryst.* **B34** 1060
[3] Jona F, Shirane G, Mazzi F and Pepinski R 1957 *Phys. Rev.* **105** 849
[4] Fujishita H and Hoshino S 1984 *J. Phys. Soc. Jpn.* **53** 226
[5] Fu D, Arioka T, Taniguchi H, Taniyama T, and Itoh M 2011 *Appl. Phys. Lett.* **99** 012904
[6] Mishra S K, Choudhury N, Chaplot S L, Krishna P S R, and Mittal R 2007 *Phys. Rev. B* **76** 024110
[7] Aoyagi R, Takeda A, Iwata M, Maeda M, Nishida T and Shiosaki T 2008 *Jpn. J. Appl. Phys.* **47** 7689
[8] Aoyagi R, Iwata M and Maeda M 2009 *Key Eng. Mater.* **3880** 233
[9] Louca D, Brosha E L, and Egami T 2000 *Phys. Rev. B* **61** 1351
[10] Qiu X, Proffen Th, Mitchell J F, Billinge S J L 2005 *Phys. Rev. Lett.* **94** 177203
[11] Egami T and Billinge S J L 2001 *Underneath the Bragg Peaks: Structural Analysis of Complex Materials* (Oxford: Elsevier Science).
[12] Yoneda Y, Matsumoto N, Furukawa Y and Ishikawa T 2001 *J. Synchrotron Rad.* **8** 18
[13] Yoneda Y, Matsumoto N, Furukawa Y and Ishikawa T 2001 *Nuclear Instruments and Method in Physics Research* A **467-468** 370
[14] Nisawa A, Yoneda Y, Ueno G, Murakami H, Okajima Y, Yamamoto K, Senba Y, Uesugi K, Yamamoto M, Goto S, and Ishikawa T 2013 *J. Synchrotron Rad.* **20** 219
[15] Proffen Th and Billinge S J L 1999 *J. Appl. Crystallogr.* **32** 572
[16] Izumi F and Momma K 1997 *Solid State Phenom.* **130** 15
[17] Momma K and Izumi F 2011 *J. Appl. Crystallogr.* **44** 1272

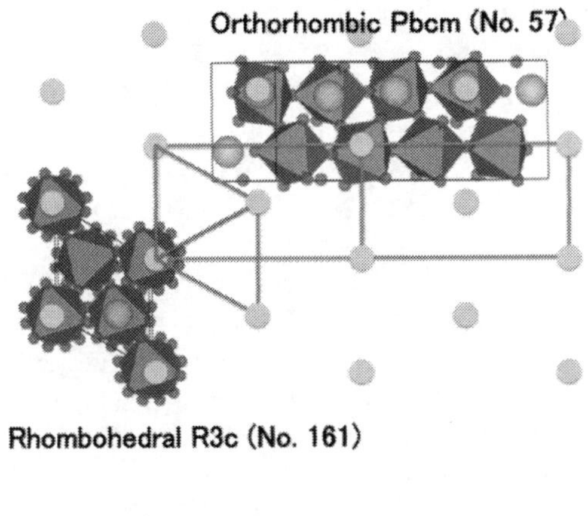

Figure 5. (Color online) Crystal structures of rhombohedral (below 200 K) and orthorhombic (above 200 K) phases in $NaNbO_3$. Rhombohedral and orthorhombic unit cells are also drawn. Atomic coordination of these phases are similar.

Journal of Physics: Conference Series **502** (2014) 012023 doi:10.1088/1742-6596/502/1/012023

A combination of anomalous x-ray scattering and neutron diffraction for structural characterizations of $Zr_{63}Ni_{25}Al_{12}$ metallic glass

S Hosokawa[1,2], J-F Bérar[3], N Boudet[3], S Kohara[4], J Stellhorn[2], W-C Pilgrim[2], A Zeidler[5], H Kato[6], Y Kawakita[7], T Otomo[8]

[1]Department of Physics, Graduate School of Science and Technology, Kumamoto University, Kumamoto 860-8555, Japan
[2]Department of Chemistry, Physical Chemistry, Philipps University of Marburg, D-35032 Marburg, Germany
[3]Institut Néel, Centre National de la Recherche Scientifique/Université Joseph Fourier (CNRS/UJF), 38042 Grenoble Cedex 9, France
[4]Research and Utilization Division, Japan Synchrotron Radiation Research Institute (JASRI/SPring-8), Sayo, Hyogo 679-5198, Japan
[5]Department of Physics, University of Bath, Bath BA2 7AY, United Kingdom
[6]Institute of Materials Research, Tohoku University, Sendai 980-8577, Japan
[7]J-PARC Center, Japan Atomic Energy Agency (JAEA), Tokai, Ibaraki 319-1195, Japan
[8]High Energy Accelerator Research Organization (KEK), Tsukuba 305-0801, Japan

E-mail: `hosokawa@sci.kumamoto-u.ac.jp`

Abstract. $Zr_{63}Ni_{25}Al_{12}$ bulk metallic glass was investigated using neutron diffraction and anomalous x-ray scattering close to the Zr and Ni K edges. The experimental results were analyzed using reverse Monte Carlo modeling (RMC) to obtain partial structure factors and pair distribution functions. The obtained partial structural results were compared with RMC results when neutron diffraction data are not included in the analysis and the report of $Zr_{60}Ni_{25}Al_{15}$ by Fukunaga et al. using high-energy x-ray and neutron total scattering data.

1. Introduction

In the last two decades, bulk metallic glasses (BMG) with distinct glass transitions have been discovered in various multi-component metallic alloys. They show extremely excellent glass-forming abilities (GFA), where even a very slow cooling rate such as $\sim$ 1 K/s can avoid crystallization. Physical and technological properties of these glasses were well investigated, including glass transition, structural change, phase stability, elastic constants, magnetic properties, etc [1]. Among these BMG's, the alloy studied in this paper, $Zr_{63}Ni_{25}Al_{12}$, has an excellent critical cooling rate of some K/s allowing to form a massive BMG with a diameter of more than 15 mm [2].

To understand the origin of such an excellent GFA in Zr-Ni-Al alloys, partial structural investigations are essential. Fukunaga et al. [3] utilized contrasts between x-ray and neutron diffraction (XD and ND) data of $Zr_{60}Ni_{25}Al_{15}$ BMG for reverse Monte Carlo (RMC) modeling to obtain partial structural information. We have recently performed anomalous x-ray scattering (AXS) on $Zr_{63}Ni_{25}Al_{12}$ BMG close to the Zr and Ni K edges, and the experimental results were analyzed using RMC modeling [4]. The comparison of our results with those obtained by Fukunaga et al. showed some discrepancies especially in the Ni-Ni pair correlations.

To investigate $Zr_{63}Ni_{25}Al_{12}$ BMG in more detail, we added ND measurements. In this paper, we report results of AXS and ND measurements and subsequent RMC modeling to obtain partial structure factors, $S_{ij}(Q)$, and partial pair distribution functions, $g_{ij}(r)$ in this BMG.

2. Experimental procedure and data analysis

A $Zr_{63}Ni_{25}Al_{12}$ alloy ingot with nominal composition was prepared by arc-melting of mixtures of Zr, Ni, and Al metals with purities of 99.5%, 99.99%, and 99.99%, respectively, in a high-purity Ar atmosphere. From this, a cylindrical rod of $\sim$ 3 mm in diameter and $\sim$ 15 mm in length was manufactured by tilt casting with a Cu mold.

AXS experiments were conducted at the beamline BM02 of the European Synchrotron Radiation Facility (ESRF) in Grenoble, France. To obtain differential structure factors, $\Delta_k S(Q)$, close to the Zr and Ni K edges, two scattering experiments were performed at two incident x-ray energies 30 eV and 200 eV below the Zr K edge (17.998 keV), and 20 eV and 200 eV below the Ni K edge (8.333 keV), using a standard $\omega - 2\theta$ diffractometer installed at the beamline. The experimental details, in particular, on a high-resolution high-intensity detecting system, is given elsewhere [5]. Following procedures given in Ref. [6], $\Delta_k S(Q)$'s were calculated from the scattering data. For the analysis, the ordinary and anomalous terms of the atomic form factors were taken from literatures [7, 8].

ND measurements were performed using the High Intensity Total Diffractometer (NOVA) installed at the beamline BL21 of Materials and Life Science Experimental Facility (MLF) in Japan Proton Accelerator Research Complex (J-PARC). The rod sample was put into a thin-walled vanadium cylindrical can with a thickness of 0.1 mm and an inner diameter of 6 mm, and set at the center of the diffractometer. A ND total structure factor, $S_{\mathrm{N}}(Q)$, was obtained from the scattered intensity after correcting background,container scattering, absorption, and multiple scattering, and normalizing to a reference sample of a vanadium rod.

RMC modeling [9] is a useful tool to construct 3D atomic configurations of disordered materials using experimental diffraction data. In the RMC calculations, atoms are moved from an initial configuration so as to minimize deviations from experimental data, e.g., in this study, two $\Delta_k S(Q)$'s, an XD total structure factor measured near the Zr K edge, $S_{\mathrm{X}}(Q)$, and $S_{\mathrm{N}}(Q)$ using a standard Metropolis Monte Carlo algorithm [10].

A starting configuration of a system containing totally 10,000 atoms with the measured number density of 46.73 nm^{-3} was generated using a hard-sphere Monte Carlo simulation (a random configuration except short bonds). To avoid unphysical atomic configurations, only a constraint of shortest atomic distances of 0.20 nm was applied to all the atomic pairs in the RMC calculations. The RMC calculations were performed using the RMC++ program package coded by Gereben et al. [11].

3. Results

Circles in Fig. 1 show experimental data of $\Delta_{\mathrm{Zr}} S(Q)$, $\Delta_{\mathrm{Ni}} S(Q)$, $S_{\mathrm{X}}(Q)$, and $S_{\mathrm{N}}(Q)$, of which the x-ray results were already published elsewhere [4]. The Q range in $\Delta_{\mathrm{Ni}} S(Q)$ is limited to about 70 nm^{-1}, owing to the low energy of incident x-rays near the Ni K edge. As shown in the figure, $\Delta_{\mathrm{Ni}} S(Q)$ has spectral features very different from $\Delta_{\mathrm{Zr}} S(Q)$, $S_{\mathrm{X}}(Q)$, and $S_{\mathrm{N}}(Q)$. In particular, out-phases are seen in the oscillations beyond 30 nm^{-1}.

Although $\Delta_{\mathrm{Zr}} S(Q)$ looks similar to $S_{\mathrm{X}}(Q)$, small differences are identified by inspecting the spectra, i.e., the amplitude of oscillations in $\Delta_{\mathrm{Zr}} S(Q)$ is larger than in $S_{\mathrm{X}}(Q)$, and a slight phase shift is realized towards smaller Q values.

Solid curves in Fig. 1 represent best fits of the RMC modeling. Although the experimental results show very different scattering features from each other, the RMC modeling reproduce

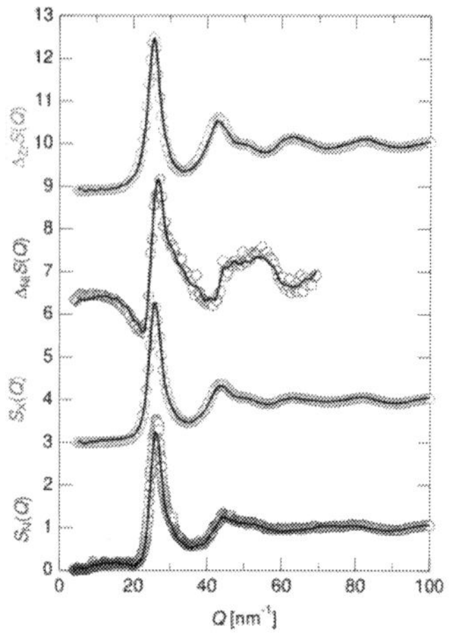

Figure 1. (Color online.) From top to bottom: $\Delta_k S(Q)$'s close to the Zr and Ni K edges, $S_{\mathrm{X}}(Q)$, and $S_{\mathrm{N}}(Q)$. Circles indicate those obtained from the experiments, and solid curves denote best fits of the RMC modeling. For clarity, the spectra are displaced upwards by 3 each.

well all of the experimental data within the experimental errors.

4. Discussion

Figures 2(a) and (b) show the $g_{ij}(r)$ and $S_{ij}(Q)$ functions, respectively, obtained from the present RMC modeling. The $g_{ij}(r)$ functions show sharp first peaks at different interatomic distances. Compared with the previous results [4], the first peaks in the Al-related partials become sharper and the unphysical peaks near the cutoff length disappear, while the other correlations remain mostly unchanged. The difference in the Al-related partials may be due to the addition of the ND data which has different weighting factors for the partial correlations as given in Table 1 of Ref. [4] to the RMC modeling.

Some interesting features in the local structures are observed: 1) The atomic distances of Zr-related partials are in good agreement with the XAFS data for $Zr_{70}Ni_{20}Al_{10}$ [12], and it is confirmed that the Zr-Ni distance is much smaller than the average of the Zr-Zr and Ni-Ni

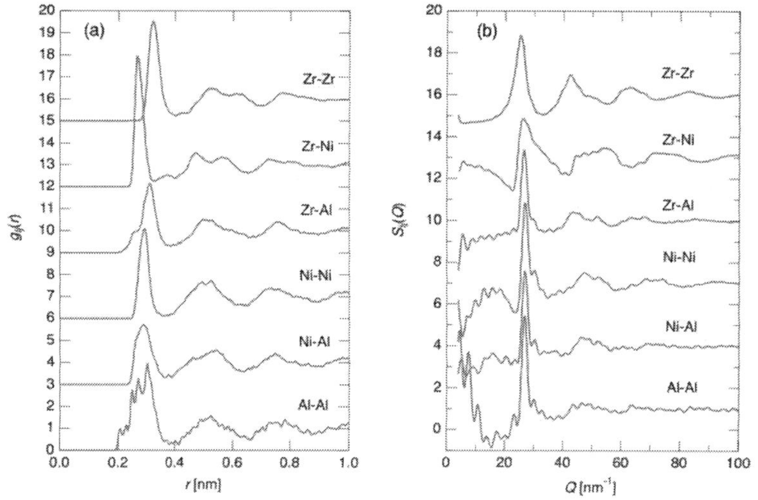

Figure 2. (Color online.) (a) $g_{ij}(r)$ and (b) $S_{ij}(Q)$ functions obtained from the RMC fits. For clarity, the spectra are displaced upwards by 3 each.

distances. 2) The difference in the atomic diameters of Zr and Ni atoms is approximately 18 %, which matches well Inoue's empirical rule [13]. 3) The width of the first peak in the Zr-Zr correlation (0.045 nm) is wider than those in the Zr-Ni (0.034 nm) and Ni-Ni (0.032 nm) ones. The $g_{\rm AlAl}(r)$ function is still noisy even in the first peak, which may be owing to the small weighting factor of 0.3 % even for the ND data.

As regards the $S_{ij}(Q)$ functions, $S_{\rm ZrZr}(Q)$ resembles that of a typical dense-packed monatomic configuration. The first peak in $S_{\rm ZrNi}(Q)$, however, has a large shoulder at the higher Q side, indicating a non-hard-sphere Zr-Ni correlation. Of special interest is that the Ni-related partials have broad prepeaks at about 15 nm^{-1}, and the Ni-Ni partial shows the most prominent one, indicating the existence of weak intermediate-range correlations around the Ni atoms, in particular, in the Ni-Ni correlations. The existence of the prepeak in $S_{\rm NiNi}(Q)$ was the large discrepancy between our previous results without ND [4] and the report by Fukunaga et al. on the $Zr_{60}Ni_{25}Al_{15}$ BMG [3], and the confirmation of the prepeak in $S_{\rm NiNi}(Q)$ is an important progress by adding the ND data. However, the prepeak in our results is much smaller in height and broader compared with the report by Fukunaga et al. [3].

$S_{\rm AlAl}(Q)$ shows a large increase in the small Q range. This feature in this region resembles well the initial spectrum of random configuration of the RMC modeling, and highly depends on the initial configuration. Thus, the reliability of the Al-Al correlations in the large r range is still doubtful even after adding the ND data in the RMC modeling.

5. Conclusion

To investigate the local structural origin of good GFA of $Zr_{63}Ni_{25}Al_{12}$ bulk metallic glass, an ND measurement was performed, and the result was included in the RMC modeling together with previous AXS data close to the Zr and Ni K edges and XD results. The qualities of the $g_{ij}(r)$ and $S_{ij}(Q)$ functions was highly improved except the enhancement of the $S_{\rm AlAl}(Q)$ in the low Q region. Although a prepeak was observed in the Ni-Ni correlation as previously reported by Fukunaga et al. [3], the prepeak width is much broader than their result. By improving the partial data we made an important step towards helping to understand the good GFA in the BMG.

Acknowledgments

AXS and ND experiments were performed at BM02/ESRF (Proposal No. HD-426) and BL21/MLF/J-PARC (Proposal No. 2013A0032), respectively.

References

[1] Cheng Y Q and Ma E 2011 *Prog. Mater. Sci.* **56** 379
[2] Inoue A, Zhang T, Nishiyama N, Ohba K and Masumoto T 1993 *Mater. Trans. JIM* **34** 1234
[3] Fukunaga T, Itoh K, Otomo T, Mori K, Sugiyama M, Kato H, Hasegawa M, Hirata A, Hirotsu Y and Hannon A C 2007 *Mater. Trans.* **48** 1698
[4] Hosokawa S, Bérar J-F, Boudet N, Kohara S, Pilgrim W-C, Joester M, Stellhorn J, Zeidler A, Maruyama K, Nasu T, Zhang W and Kato H 2012 *Int. J. Mater. Res.* **103** 1108
[5] Hosokawa S and Bérar J-F 2007 *AIP Conf. Proc.* **879** 1743
[6] Hosokawa S, Oh I, Sakurai M, Pilgrim W-C, Boudet N, Bérar J-F and Kohara S 2011 *Phys. Rev. B* **84** 014201, and references therein
[7] MacGillavry C H and Rieck G D (ed.) 1969 *International Table for X-ray Crystallography, 2nd ed.* (Kynoch, Birmningham) Vol. III
[8] Sasaki S 1989 *KEK Report 1989* (Nat. Lab. High Energy Phys., Tsukuba) p. 1
[9] McGreevy R L and Pusztai L 1988 *Mol. Simul.* **1** 359
[10] Metropolis N, Rosenbluth A W, Rosenbluth M N, Teller A. H and Teller E 1953 *J. Phys. Chem.* **21** 1087
[11] Gereben O, Jóvári P, Temleitner L and Pusztai L 2007 *J. Optelectr. Adv. Mater.* **9** 3021
[12] Sato S, Sanada T, Saida J, Imafuku M, Matsubara E and Inoue A 2005 *Mater. Trans.* **46** 2893
[13] Inoue A 1995 *Mater. Trans. JIM* **36** 866

Journal of Physics: Conference Series **502** (2014) 012024 doi:10.1088/1742-6596/502/1/012024

An x-ray fluorescence holographic study on a $Bi_2Te_3Mn_{0.1}$ topological insulator

S Hosokawa[1], N Happo[2], K Hayashi[3], A Ohnishi[4], M Kitaura[4] and M Sasaki[4]

[1]Department of Physics, Graduate School of Science and Technology, Kumamoto University, Kumamoto 860-8555, Japan
[2]Graduate School of Information Sciences, Hiroshima City University, Hiroshima 731-3194, Japan
[3]Institute of Materials Research, Tohoku University, Sendai 980-8577, Japan
[4]Department of Physics, Faculty of Science, Yamagata University, Yamagata 990-8560, Japan

E-mail: `hosokawa@sci.kumamoto-u.ac.jp`

Abstract. To search the atomic sites of Mn impurities in $Bi_2Te_3Mn_{0.1}$ topological insulator, a Mn $K\alpha$ fluorescence x-ray holography experiment was performed. The reconstructed atomic image around the central Mn atom reveals a hexagonal configuration, where the local lattice constant is slightly larger than the original Bi_2Te_3 crystal. Thus, the most plausible atomic position of Mn impurities is the substitution with Bi or Te atom. The further-distant atomic images are hardly observed unlike usual crystals, indicating a formation of flat clusters around Mn in this functional crystal.

1. Introduction

In the past, Bi_2Te_3 was well-known as a thermoelectric material having a large value of thermopower. Nowadays, this material has also achieved much attention as a topological insulator, where the so-called Dirac electrons with an almost zero mass can conduct on the surface of this material. From this reason, Bi_2Te_3 is promising as a raw material for future computer systems of very fast and energy-saving.

Bi_2Te_3 is essentially a p-type degenerate semiconductor (SC). By adding electrons, e.g., increasing Te concentration, it becomes a topological insulator (TI), and then n-type degenerate SC. However, the TI properties in $Bi_2Te_{3+\delta}$ has a very short lifetime due to surface contaminations, and thus, this material is not suitable for the applicational use.

Recently, Sasaki and coworkers [1] found that Mn- and some other impurity(M)-doped $Bi_2Te_3M_x$ alloys have a similar behavior. Up to $x = 0.05$, it shows p-type SC, and beyond $x = 0.075$ it becomes n-type SC. The TI phase is located between $x = 0.05$ and 0.075. Moreover, they found some interesting features as TI materials. Firstly, the lifetime of the TI properties is extremely long, which is very important for applications. Secondly, the Dirac electrons are not scattered by impurities (clean surface Dirac electrons), which was detected by a magnetoresistance measurement at low temperature.

Then, an important question arises: Where do the impurity atoms locate in the Bi_2Te_3 crystal for showing such excellent TI properties? Bi_2Te_3 has a layer structure with a hexagonal form [2]. Figure 1 shows one layer of the Bi_2Te_3 crystal. There are two types of Te atoms in the layer.

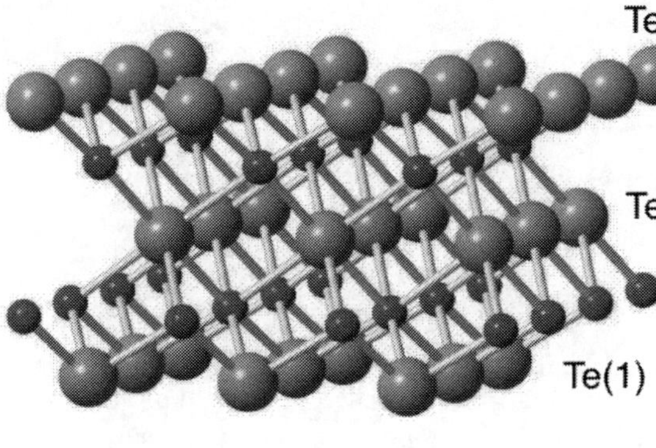

Figure 1. (Color online.) One layer in the Bi_2Te_3 hexagonal crystal [2]. Large and small balls indicate Te and Bi atoms, respectively.

One is on the surface of layers, Te(1), where a van der Waals force combines the layers, and another is at the center of layers, Te(2), which connects with Bi atoms by a covalent-ionic mixed binding. The interlayer positions would be most plausible for the impurities, but there is no reason to exclude the possibilities of substitutions with Bi or Te atoms or interstitial positions in the layer.

X-ray fluorescence holography (XFH) is a newly developed technique that enables one to draw three-dimensional (3D) atomic images around a specific element emitting fluorescent x-rays [3]. Due to the interference between the direct incident x-rays and those scattered by the surrounding atoms, the fluorescent x-ray intensity from the emitter slightly modulates with the incident x-ray angles by about 0.1%, from which 3D images can be obtained by simple Fourier transforms without any special models. We have performed the Mn $K\alpha$ XFH measurements on $Bi_2Te_3Mn_{0.1}$ single crystal at room temperature. In this paper, we report atomic images around the Mn atoms.

2. Experimental procedure and data analysis

A single crystal of $Bi_2Te_3Mn_{0.1}$ was grown by a modified Bridgman method, where Bi_2Te_3 and Mn powders were melted and crystallized in an evacuated quartz ampoule several times by slow cooling. The sample was cooled from 850°C to 550°C with a cooling rate of 10°C/h. The crystal was cut and polished so as to have a (001) flat surface larger than 5×5 mm^2. The crystalinity of the sample was examined by taking a Laue photograph.

XFH measurements were carried out at the beamlines BL6C in the Photon Factory at the High Energy Accelerator Research Organization (PF-KEK), in Tsukuba, Japan. The sample was placed on a two-axes table of a diffractometer. The measurements were performed in inverse mode by rotating the two axes, the exit angle of $0° \leq \theta \leq 75°$ in steps of 1.00°, and the azimuthal angle of $0° \leq \phi \leq 360°$ in steps of about 0.35°, of the sample stage. Incident x-rays were focused onto the (001) surface of the samples. Mn $K\alpha$ fluorescent x-rays were collected using an avalanche photodiode detector with a cylindrical graphite crystal energy-analyzer. The XFH signals were recorded at eight different incident x-ray energies from 7.0 to 10.5 keV in steps of 0.5 keV. Each scan took about 12 h. Details of the experimental setup are given elsewhere [3].

Holographic oscillation data were obtained by subtracting the background from the fluorescent x-ray intensities and normalizing them to the incident x-ray intensities. An extension of the hologram data was carried out using the crystal symmetry of the hexagonal structure [2] and the measured x-ray standing wave (XSW) lines. From the hologram patterns, 3D atomic configuration images were reconstructed using Barton's algorithm [4] by superimposing the holograms with eight different incident x-ray energies, which can highly suppress the appearance of twin images.

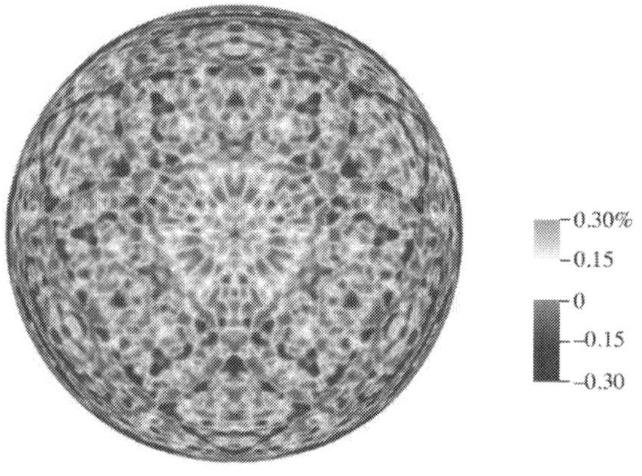

Figure 2. (Color online.) Examples of Mn $K\alpha$ hologram pattern obtained from (001) surfaces of $Bi_2Te_3Mn_{0.1}$ single crystal measured at incident x-ray energy of 8.0 keV.

3. Results and discussion

Figure 2 shows examples of the Mn $K\alpha$ hologram pattern of the $Bi_2Te_3Mn_{0.1}$ single crystal measured at an incident x-ray energy of 8.0 keV. The radial and angle directions indicate θ and ϕ, respectively, and the magnitudes are given as the color bars beside the figure. A roughly threefold symmetry including XSW signals was observed in the holographic pattern, indicating a good quality of the sample crystal.

Figure 3 shows the reconstructed atomic image of $Bi_2Te_3Mn_{0.1}$ single crystal on the (001) plane around the central Mn atom marked by the circle in the figure. The image intensities were normalized to that of the strongest atoms, and are shown as the color bar beside the figure. To avoid confusions on the noise level artifacts and to emphasize the middle intensity range, the color variation was chosen to range from 0.25 to 0.75. As a guide for eyes, cross-sections of dashed lines indicate ideal positions of the neighboring atoms obtained from x-ray diffraction (XD) [2] provided that the Mn atom is located at a Bi or Te position. As seen in Fig. 1, all of atoms on the (001) plane are the same element, Bi or Te, depending on the z value along the c axis.

As seen in Fig. 3, clear six atomic images are observed near the ideal positions of the hexagonal lattice. Thus, the Mn impurity atom does not enter an interlayer position, and would

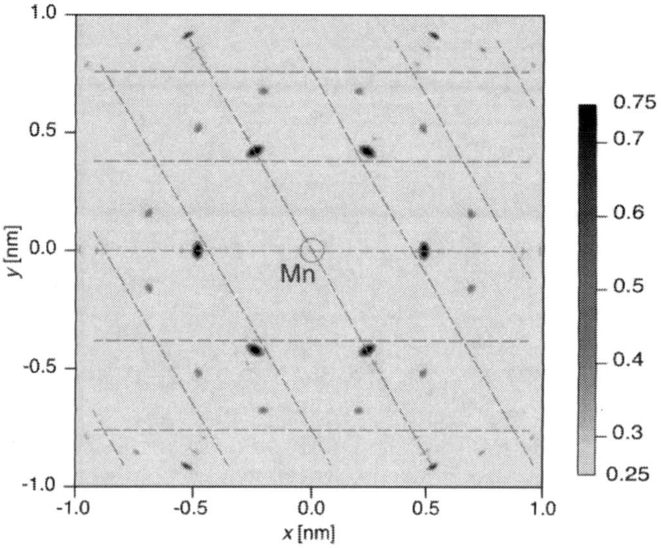

Figure 3. (Color online.) Atomic images of $Bi_2Te_3Mn_{0.1}$ single crystal on (001) plane around central Mn atoms marked by the circle. Cross-sections of dashed lines indicate ideal positions of neighboring atoms obtained from XD [2] provided that the Mn atom is located at a Bi or Te position.

be located at a substitutional position of Bi or Te atoms in the original Bi_2Te_3 crystal. Since the nearest-neighboring atoms are hardly visible (not shown in the figure) unlike usual covalent crystals, such as ZnTe [5] or GaSb [6], the Mn impurity atom would locate at a flexible site, and thus, it is most plausible that the Mn impurity atoms are substituted with an interlayer Te atom, Te(1) in Fig. 1. However, there is no reason to exclude the possibilities of substitutions with Bi or Te(2) atoms inside the layer.

Recently, an XAFS study was carried out on $(Bi_{1-x}Mn_x)_2Te_3$ epitaxial thin film crystal of impurity substitutional type, and it was concluded that the Mn impurities are located at the interlayer positions [7]. However, such positions of Mn impurity do not make the hexagonal lattice of Bi or Te atoms on the (001) plane around the central Mn atom.

The local lattice constant around the Mn impurity is slightly larger than the original Bi_2Te_3 crystal. Note that the present XFH shows only images of the second neighboring Te atoms, and the other images are hardly visible (not shown). As discussed in a previous paper on mixed crystals [5, 6, 8], an impurity induces strong fluctuations in nearest-neighboring atomic positions, which results in very weak atomic images of nearest-neighboring atoms. In this material, however, further distant neighboring atoms are not observed unlike previous mixed crystals. From these findings, it would be reasonable to conclude that a strong flat clusters are formed, and the positional fluctuations between the clusters are very large.

4. Conclusion

We have carried out a Mn $K\alpha$ fluorescence x-ray holography experiment on $Bi_2Te_3Mn_{0.1}$ topological insulator. The reconstructed atomic image around the central Mn atom reveals a hexagonal configuration, where the local lattice constant is slightly larger than the original Bi_2Te_3 crystal. The plausible atomic position of Mn impurities is the substitution with Bi or Te atom. The further-distant atomic images are hardly observed unlike usual crystals, indicating a formation of flat clusters around Mn in this functional crystal. To confirm these results, a Bi $L\alpha$ XFH experiment and a more detailed XAFS measurement are essential, which are now in progress.

Acknowledgments

The authors thank Professor S. Sasaki of Tokyo Institute of Technology for the support of the XFH experiments. NH and KH acknowledge the Japan Society for the Promotion of Science (JSPS) for Grants-in-Aid for Scientific Research (Nos. 24560016 and 22360264). This work was performed under the inter-university cooperative research program of Institute for Materials Research, Tohoku University (Project Nos. 11K0008 and 12K0034). The XFH experiments were performed at BL6C of PF-KEK (Proposal No. 2011G530).

References

[1] Kim H-J, Kim K-S, Kim M D, Lee S-J, Han J-W, Ohnishi A, Kitaura M, Sasaki M, Kondo A and Kindo K 2011 *Phys. Rev. B* **84** 125144
[2] Feutelais Y, Legendre B, Rodier N and Agafonov V 1993 *Mater. Res. Bull.* **28** 591
[3] Hayashi K, Happo N, Hosokawa S, Hu W and Matsushita T 2012 *J. Phys.: Condens. Matter* **24** 093201
[4] Barton J J 1991 *Phys. Rev. Lett.* **67** 3106
[5] Hosokawa S, Happo N and Hayashi K 2009 *Phys. Rev. B* **80** 134123
[6] Hosokawa S, Happo N, Ozaki T, Ikemoto H, Shishido T and Hayashi K 2013 *Phys. Rev. B* **87** 094104
[7] Strohm C, Bauer G, Holy V, Springholz G, Caha O, Ruzicka J and Steiner H 2013 *ESRF Report* HE-3932
[8] Happo N, Hayashi K and Hosokawa S 2009 *J. Cryst. Growth* **311** 990

Journal of Physics: Conference Series **502** (2014) 012025 doi:10.1088/1742-6596/502/1/012025

Particle size dependence in thermal disorder/order transformation of FeCo alloy

K Shinoda, S Fujieda, and S Suzuki
Institute of Multidisciplinary Research for Advanced Materials, Tohoku University, Sendai 980-8577, Japan.

E-mail: shinoda@tagen.tohoku.ac.jp

Abstract. FeCo alloy particles synthesized by liquid phase reduction are disorder distribution of the alloying elements in bcc crystalline lattice. The particle size dependence of thermally ordering behavior during heating was investigated. Superlattice lines such indices as 100, 111 etc., which appear in the x-ray diffraction pattern when the atomic arrangement is ordered, are difficult to detect due to quite small difference between Fe and Co in the atomic x-ray scattering factor. In this study, the energy near to Fe K absorption edge for incident x-ray was used in order to apply the anomalous x-ray dispersion effect. From results of the *in situ* heated x-ray diffraction experiments using the Synchrotron radiation source, different ordering-onset temperature was observed for the two samples with different particle size. This is considered due to a difference on the lattice distortion and its elimination during heating.

1. Introduction

FeCo alloys are known to be a fundamental material for many applications due to their soft magnetic properties [1]. Bulk $Fe_{50}Co_{50}$ alloy exhibits a body-centered cubic (bcc) structure below 983 °C, the temperature of transformation to a face-centered cubic (fcc) structure. And the alloy undergoes a crystalline phase transformation from disordered bcc structure (namely A2 phase) to ordered CsCl-type structure (namely B2 phase) at 730 °C [2]. Such order/disorder phase transformation has been studied using their change of physical properties such as the electrical resistivity, specific heat and the magnetization [3-6], structural change applying x-ray [3, 7], electron [8] and neutron [9, 10] diffraction techniques, or theoretical approach [11]. On the other hand, several nanometer to sub-micrometer sized particles of FeCo alloy system have been successfully synthesized by using a liquid phase reduction method namely polyol process in recent [12, 13]. In the synthesis process, poly alcohol such as ethylene glycol (EG) or tetraethyl glycol (TEG) is used as solvent. The poly alcohol (so called polyol) also act as reductant and dispersant during the precipitation of alloy particles. The particle size and the alloy composition can be controlled easily by tuning the reaction conditions such as kind of salt for metal precursor and its concentration, kind of polyol, reaction temperature, additives, etc.
Degree of ordering in FeCo alloy system is important parameter relating with the magnetic properties. To determine directly the long range order parameter *S* using observed intensity of diffraction so called superlattice line is useful to investigate the process of thermal disorder/order and order/disorder transformation. For bulk FeCo relating alloys, such investigations by determination of *S* using neutron or electron diffraction have been studied. On the other hand, there had been almost no report using x-ray diffraction due to the marginal difference between Fe and Co in the amplitude of the atomic x-ray scattering factors when ordinary x-ray source such as Cu Kα radiation. Recently, the ordering degree

in the FeCo alloy particles was investigated by using Co Kα radiation to apply the anomalous x-ray dispersion effect on Fe to enhance the difference between the atomic scattering factors of Fe and Co [14].
The atomic x-ray scattering factor f is described by the following equation,

$$f(\sin\theta/\lambda, E) = f_0(\sin\theta/\lambda) + f'(E) + i f''(E) \tag{1}$$

where, f' and f'' are the real and imaginary parts of the anomalous dispersion terms. The diffracted intensity of the hkl indices reflection I_{hkl} is related to the structure factor F_{hkl}, multiplicity factor p and Lorentz-polarization factor LP as follows:

$$I_{hkl} = |F_{hkl}|^2 \cdot p \cdot LP \tag{2}$$

F_{hkl} of (110) fundamental line is represented by the sum of f for Fe and Co and independent of the degree of ordering. On the other hand, F_{hkl} of (100) superlattice line is described by $F_{hkl} = S\,(f_{Co} - f_{Fe})$ $(0 < S < 1)$ as a function of S. The value of S can be obtained from the observed diffraction intensity ratio between (110) fundamental and (100) superlattice lines by the following equation,

$$S^2 = (I_{100} / I_{110})_{observed} / (I_{100} / I_{110})_{calc} \tag{3}$$

$(I_{100} / I_{110})_{calc}$ is the ratio calculated on the assumption of perfectly ordered, $S = 1$. The plot of $|F_{100}|^2$ variation in consideration of the anomalous dispersion terms vs. the incident x-ray energy around the Fe K (7111 eV) and Co K (7709 eV) absorption edges is shown in Fig. 1. I_{100} is almost zero even in perfectly ordered structure at the energy of Cu Kα radiation. The (100) superlattice line is detectable when Co Kα radiation is used, and it was found that the atomic distribution in the crystalline lattice of as-synthesized FeCo alloy particles is random [14]. This feature of atomic arrangement in the liquid phase synthesized particles is not found in bulk FeCo alloys. The arrangement was ordered gradually with increase of temperature at above approximately 400 °C. Such thermal "disorder/order" transition behavior is unique in the liquid phase synthesized FeCo alloy particles. Although its size-dependence can be easily predicted, the detection of weak peak intensity and accurate determination of S in the initial ordering stage are quite difficult. In this study, the XRD measurements were carried out at just below the Fe K absorption edge of 7086 eV using a synchrotron radiation to make more effective in enhancement of superlattice diffraction intensity than using Co Kα radiation.

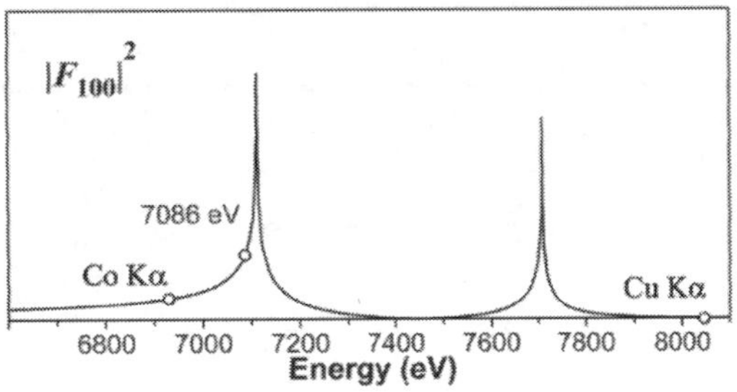

Figure 1. The square of structure factor for (100) superlattice line.

2. Sample preparation

The FeCo alloy samples with different particle size were synthesized by using the polyol process. The procedure is as follows: Iron chloride tetrahydrate ($FeCl_2 \cdot 4H_2O$), cobalt acetate tetrahydrate ($Co(CH_3COO)_2 \cdot 4H_2O$), polyvinylpyrrolidone (PVP) as dispersant and sodium hydroxide (NaOH) were dissolved in EG. The total concentration of metal salts and amount of PVP were 0.03 mol/L and

50 g/L, and 0.05 mol/L and 5 g/L for the large and the small particle samples, respectively. The concentration of NaOH was 0.2 mol/L for both. The solution was heated up to the reaction temperature of 130 °C, and refluxed for 1 hour. After the precipitation reaction the suspension was cooled to room temperature, and the formed particles were centrifuged, washed with ethanol several times. The chemical composition of Fe : Co analyzed by EPMA and average particle size are listed in Table 1. The morphology of the as-synthesized particles is shown in Fig. 2.

Table 1. Chemical composition and average particle size of as-synthesized particles.

Sample	Fe : Co	Particle size
A (large)	50.7 : 49.3	170 nm
B (small)	51.5 : 48.5	80 nm

Figure 2. SEM images of as-synthesized samples A (left) and B (right).

3. X-ray diffraction measurements

In situ x-ray diffraction (XRD) measurements to analyze the thermal variation of S were performed at BL19B2 in SPring-8 (proposal No. 2009A1777). The Debye-Scherrer camera with imaging plate mounted at the position at camera length of 286.5 mm as media for detecting the diffraction intensity profiles was used. The FeCo alloy powder samples were sealed in a glass capillary of 0.5 mm in diameter under an oxidation-free atmosphere. In the *in situ* XRD experiments, the sample was heated with a N_2 gas flow system with a conical nozzle at temperature between RT and 770 °C. At each measuring temperature, two exposures for 30 min were carried out after 30 min interval. Achieving equilibrium in the ordering degree was confirmed by comparison of the two diffraction patterns.

4. Results

Any oxide or other crystalline phases were not detected even after the high-temperature *in situ* XRD experiments, as shown in Fig. 3(a). After heating up to 770 °C, (100) and (111) superlattice lines were clearly observed. Fig. 4 shows variation of S calculated from the intensity ratio of (100) superlattice line shown in Fig. 3(b) and (110) fundamental line. The random structure in as-synthesized sample was ordered with increasing the temperature above 300 °C, and disordered again immediately after exceeding the order/disorder transition point of 730 °C. The ordering-onset temperature of sample B was higher than that of sample A.

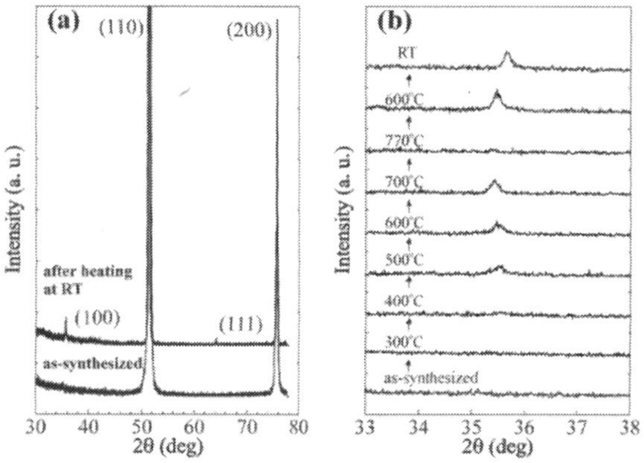

Figure 3. Observed XRD patterns of the samples as-synthesized and after *in situ* experiments (a), (100) superlattice line profiles during heating (b) for sample A.

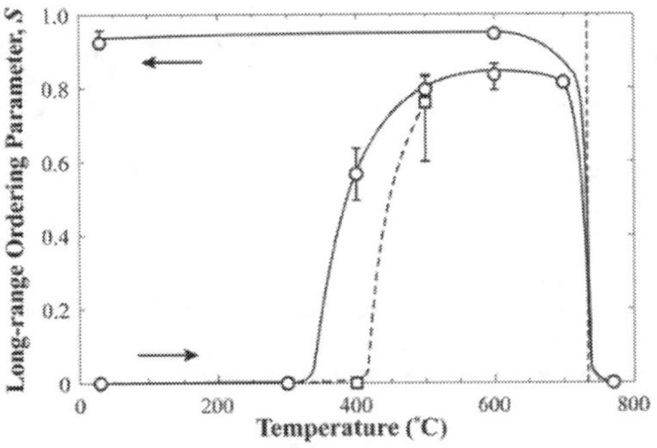

Figure 4. *S* variation during the *in situ* XRD measurements for samples A (circle) and B (square).

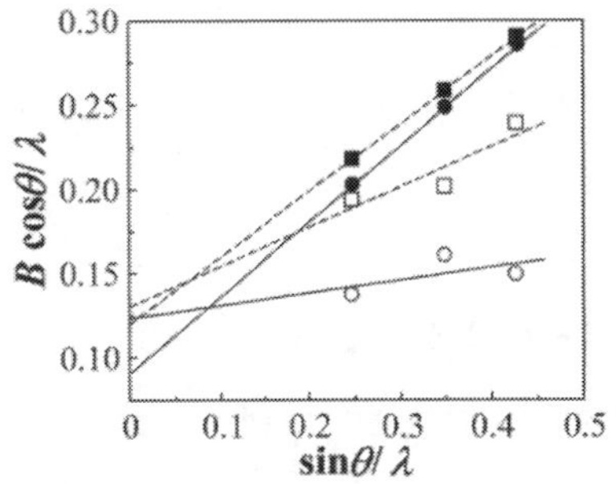

Figure 5. Williamson-Hall plot for as-synthesized (filled) and annealed at 400 °C (open) samples A (circle) and B (square).

Fig. 5 shows the Williamson-Hall plot using the three fundamental lines of (110), (200) and (211) for the samples A and B as-synthesized and after 400 °C annealing. The XRD measurements were carried out using Cu Kα at room temperature. The two as-synthesized samples were similar in both the size component from the intercept and the strain component from the slope. On the other hand, although the crystallite size of the annealed samples is not change, the strain in the smaller sample B is still large in comparison with that in another sample. This difference of strain is considered relating with the degree of ordering. The fraction of surface in sample B is larger than A. Therefore, a resistance to thermally reduction of the lattice strain in sample B is larger probably due to presence of a resist layer on the surface of particles. The surface layer will be lattice defect, surface oxidation, or deviation of the surface alloy composition from the average. The authors found an alloy compositional gradation in the surface layer on the FeCo particles synthesized by the polyol process [15].

5. Conclusion

The thermal ordering behavior was investigated by *in situ* XRD measurements applying anomalous x-ray dispersion effect. Particle size dependence in the ordering-onset temperature was indicated. Such difference in the transformation behavior is considered due to higher resistance to reduction of strain by defect, oxide layer or composition gradation in the surface of particles.

References

[1] Hall R C 1960 *J. Appl. Phys.* **31** S157
[2] Nishizawa T, Ishida K 1984 *Bul. Alloy Phase Diagrams* **5** 250
[3] Asano H, Bando Y, Nakanishi N, Kachi S 1967 *Trans. JIM* **8** 180
[4] Eguchi T, Matsuda H, Oki K 1968 *IEEE Trans. Mag.* **4** 476
[5] Orehotsky J, Schroder K 1974 *J. Phys. F: Metal Phys.* **4** 196
[6] Rossiter P L 1981 *J. Phys. F: Metal Phys.* **11** 615
[7] Ustinovshikov Y, Tresheva S 1998 *Mater. Sci. Eng.* **A248** 238
[8] Shao G 1999 *Appl. Phys. Lett.* **74** 2643
[9] Shull C G, Siegel S 1949 *Phys. Rev.* **75** 1008
[10] Smith A W, Rawlings R D 1976 *Phys. Status Solidi (a)* **34** 117
[11] Hanlumyuang Y, Ohodnicki P R, Laughlin D E, Mchenry M E 2006 *J. Appl. Phys.* **99** 08F101
[12] Kodama D, Shinoda K, Sato K, Konno Y, Joseyphus R J, Motomiya K, Takahashi H, Matsumoto T, Sato Y, Tohji K, Jeyadevan B 2006 *Adv. Mater.* **18** 3154
[13] Kodama D, Shinoda K, Sato K, Sato Y, Jeyadevan B, Tohji K 2006 *IEEE Trans. Mag.* **42** 2796
[14] Chon G B, Shinoda K, Suzuki S, Jeyadevan B 2010 *Mater. Trans.***51** 707
[15] Chon G B, Kodama D, Shinoda K, Suzuki S, Jeyadevan B 2010 *Mater. Sci. Forum* **631-632** 507

High-resolution X-ray microdiffraction from a locally strained SOI with a width of 150 nm

Y Imai[1], S Kimura[1], D Kosemura[2] and A Ogura[2]

[1]Japan Synchrotron Radiation Research Institute, Sayo, Hyogo 679-5198, Japan
[2]Meiji University, Kawasaki 214-8571, Japan

E-mail: imai@spring8.or.jp

Abstract. 150-nm-wide region in a locally strained silicon on insulator (SOI) was studied using high-resolution X-ray microdiffraction technique. Synchrotron radiation X-rays with energies of 10 keV were focused on the sample by a zone plate. The focused beam size was approximately $0.8 \times 0.4\ \mu m^2$. The sample was a model of a gate of metal-oxide semiconductor eld-effect-transistors (MOSFETs); a 30-nm-thick SOI plate with an 80-nm-thick silicon nitride capping lm with tensile inner stress. The capping lm was removed in a stripe pattern by etching with a width of 150 nm. The width of the gap where the lm is removed corresponds to the gate length of MOSFETs. Using thin SOI as a sample, we could observe only thin surface area even by X-ray diffraction that has deep penetration depth. Reciprocal space maps (RSMs) of symmetrical SOI 004 were measured around the 150-nm-gap of the lm. RSMs showed that the SOI lattice was tilted along the stripe pattern up-to 0.1 deg at both sides of the gap. The tilting spreads more than 1 μm distant from the center of the gap.

1. Introduction

Strained silicon is a promising technology to enhance carrier mobilities of metal-oxide semiconductor eld-e ect-transistors (MOSFETs). The technology improves performance of MOSFETs without further size reduction of transistors. In order to achieve the maximum performance, designed stain should be added in a speci c position. Therefore, techniques to detect strain in high-spatial-resolution are highly required. The stain and its distribution have been measured by using various methods, for example, convergent beam electron di raction (CBED), micro-Raman spectroscopy [1] and microbeam X-ray di raction [2]. Each of these methods has advantages and disadvantages. CBED has the highest-spatial-resolution. However, samples need to be thinned for transmission electron microscope investigation, that causes relaxation of the stress. Micro-Raman spectroscopy has relatively high-spatial-resolution and relatively high-sensitivity to the strain. The most important strain is the strain in a few nm thick surface that is related to carrier mobilities. For this reason, UV-Raman spectroscopy is adopted to measure strain in the surface because the penetration depth of 364 nm UV light for silicon crystals is 5 nm.[3] However, it is di cult for micro-Raman spectroscopy to obtain strains if directions of the stress are complicated. Because, strains are calculated from Raman shifts based on an assumption of directions of the stress. That is, one cannot get strain if the directions of the stress are unknown. On the other hand, X-ray di raction directly gives both angles of the lattice tilt and lattice spacing. Synchrotron radiation X-ray microdi raction is one of the most powerful tools to investigate the local strain on surfaces or interfaces of

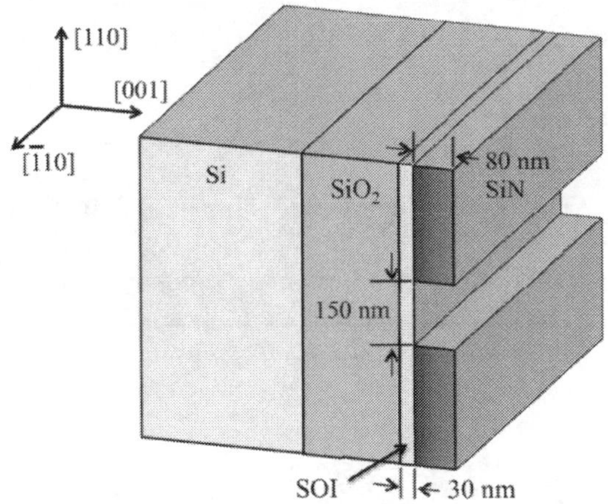

Figure 1. Schematic of the locally strained SOI structure. 80-nm-thick SiN tensile inner stress capping lm is removed in a line shape with a width of 150 nm. X-rays were incident on the sample from [110] direction.

semiconductor materials. Using SOI as a sample one can measure strain in the surface area by X-ray that has large penetration length. In addition, high intense synchrotron X-ray enables out-of-plane di raction by a few nm thick samples, whereas conventional X-ray is used with in-plane di raction that cause broadening of X-ray foot print on the sample resulting in worse spatial resolution. Furthermore, by limiting the angular divergence perpendicular to the scattering plane, both high-spatial- and high-angular-resolution have been achieved.[4]Besides, the X-ray microdi raction technique gives structural information of a region that is smaller than the beam size in some cases. This is achieved by virtue of the simultaneous measurement of the lattice tilting and the lattice spacing. For example, from a reciprocal space map measured by the X-ray microdi raction, size of domains that are one order smaller than the beam size can be estimated.[5]

2. Experiment

Figure 1 shows the schematic of the sample. The sample is a model of a gate of MOSFET. The thickness of buried-oxide is 200 nm and SOI is 30 nm. The SOI is covered by 80-nm-thick silicon nitride capping lm, that has inner stress of 2 GPa. The capping lm is removed by etching in a line shape with a width of 150 nm. The 150-nm-wide area, the gap, where the lm is removed corresponds to the gate width of MOSFET. Tensile strain along [110] is introduced to the gap by the lm.[6] The tensile strain increases carrier mobilities in n-type MOSFETs, so that performance of transistors increase. UV-Raman spectroscopy revealed that compressive strain exists at edges of the gap and tensile strain exists at the center of the gap.[6,7] We have applied the synchrotron X-ray microdi raction technique to the same sample to evaluate lattice tilting around the gap.

The microdi raction experiment was performed at BL13XU SPring-8. Synchrotron X-rays with energy of 10 keV were focused on the sample by a zone plate. The beam size at the focal point was approximately $0.8 \times 0.4\ \mu m^2$. Detail about the high-resolution microdi raction system is described in references 8 and 9.

In order to locate the center of the gap, the incident angles of the X-ray were set at lower and higher half maximum positions of the rocking curve of the substrate silicon 004, then the sample was translated in [110] direction by monitoring the di raction intensities.

There is a di erence in orientations between the silicon substrate and the SOI, so that only the di raction from the SOI can be measured.[10] Reciprocal space maps (RSMs) of symmetric SOI 004 around the 150-nm-wide gap were measured by changing the sample position in [110] direction (56 positions by 0.1 μm step). X-rays were incident on the sample from [110] direction

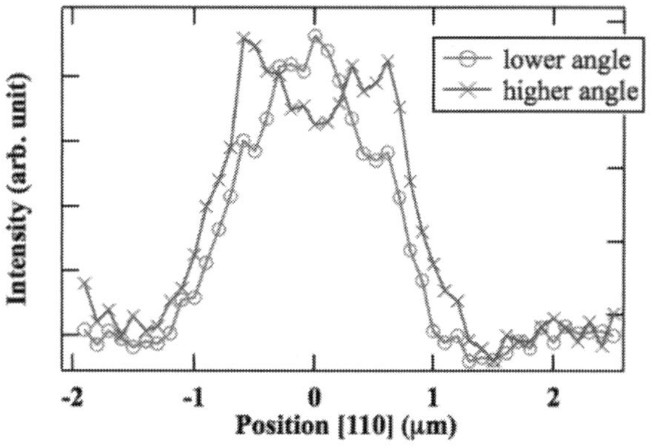

Figure 2. Intensity variation of the substrate silicon 004 di raction around the 150-nm-gap along [110]. Incident angles of the X-ray to the sample were set at lower angle (open circle) that gives half maximum and higher angle (cross mark) that gives half maximum of the 004 rocking curve.

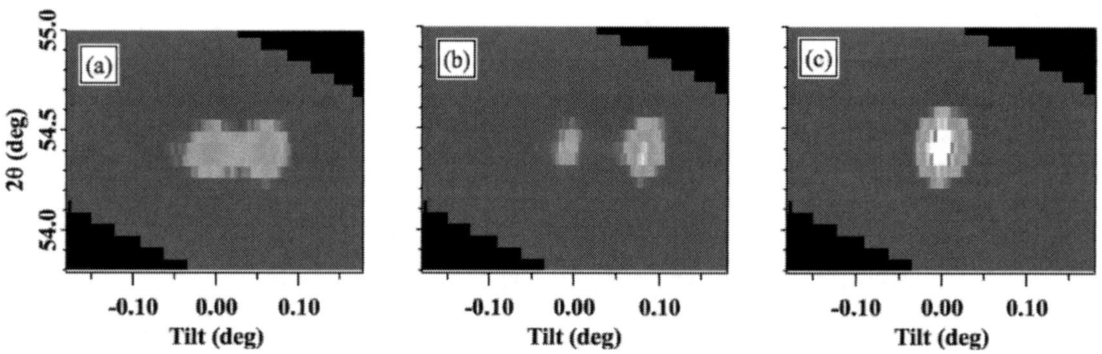

Figure 3. RSMs measured at sample positions at the center of the gap (a), 0.5 μm (b) and 2.4 μm (c) distant from the center of the gap. Spots are splitted in (a) and (b), wheres spot is single in (c).

and the sample was rotate around [110]. Therefore, lattice tilti in 2D RSMs is the rotation around [110]. A RSM was measured by rotating the sample by 0.04 deg step and observing di raction images using a CCD detector. Exposure time for a single image was 1 s. Time spent for getting a single RSM was 6.5 min.

3. Results

Figure 2 shows di raction intensities obtained with X-ray incident angles set at lower and higher sides of half maximum positions of the rocking curve. Open circles make a peak at the center of the gap. On the other hand, cross marks make double peaks beside the gap. We regard the center of these curves as the center of the gap.

Typical RSMs measured around the gap are shown in gure 3. Figure 3 (a), (b) and (c) are RSMs taken at positions: 0, 0.5 and 2.4 μm denoted in the horizontal axis of gure 2, respectively. Lengths are distances from the center of the gap in [110] direction. The RSM measured at the position of 2.4 μm (gure 3 (c)) has a single peak. In contrast, the other RSMs has two peaks (gure 3(a) and (b)). The distribution of the lattice tilt was extracted from 56 RSMs by integrating RSMs in 2θ direction. The distribution is shown in gure 4. There is 0.13 deg tilted element at the position of -0.8 μm. Most part of SOI between positions from 0.2 to 0.7 μm is tilted up-to 0.1 deg. If the distance from the center of the gap becomes larger than 2 μm, no splittings are seen. That is, the gap a ect to the lattice tilt as far as 2 μm from the center

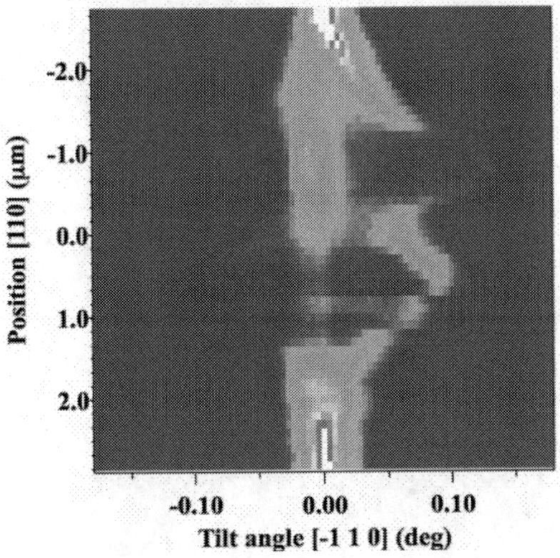

Figure 4. Lattice tilting distribution around the 150-nm-gap in the locally strained 30-nm-thick SOI measured by the high-resolution X-ray microdi raction. Vertical axis denotes the distance from the center of the gap in [110] direction. Horizontal axis is the lattice tilting along [110], that is rotation around [110].

of the gap in [110] direction. The reason of lattice tilts around [110] is still unclear because the gap is uniform in [110] direction. We need more experiments to understand these phenomena.

4. Conclusion

We have revealed the SOI lattice tilting distribution around the 150-nm-wide gap of the silicon nitride capping lm on the SOI by high-resolution X-ray microdi raction technique. The tilting was along [110] direction. The maximum tilt was 0.13 deg. The tilted region spread more than 1 μm distant from the center of the gap.

Acknowledgments

The synchrotron radiation experiments were performed at the BL13XU of SPring-8 with the approval of Japan Synchrotron Radiation Institute (JASRI) (Proposal No. 2011A1321).

References

[1] Wolf I De, Norström and Maes H E *J. Appl. Phys.* 1993 **74** 4490-500
[2] Matsui J, Tsusaka Y, Yokoyama K, Takeda S, Urakawa M, Kagoshima Y and Kimura S 2002 *J. Cryst. Growth* **237-239** 317-23
[3] Ogura A, Yamasaki K, Kosemura D, Tanaka S, Chiba I and Shimidzu R 2006 *Jpn. J. Appl. Phys.* **45** 3007-11
[4] Takeda S, Kimura S, Sakata O and Sakai A 2006 *Jpn. J. Appl. Phys.* **45** L1054-6
[5] Mochizuki S, Sakai A, Taoka N, Nakatsuka O, Takeda S, Kimura S, Ogawa M and Zaima S 2006 *Thin Solid Films* **508** 128-31
[6] Ogura A, Kosemura D, Kakemura Y, Yoshida T, Uchida H, Hattori N and Yoshimaru M 2008 *Jpn. J. Appl. Phys.* **47** 1465-8
[7] Kosemura D, Kakemura Y Yoshida T, Ogura A, Kohno M, Nishita T and Nakanishi T 2008 *Jpn. J. Appl. Phys.* **47** 2538-43
[8] Imai Y, Kimura S, Sakata O and Sakai A 2010 *AIP Conf. Proc.* **1221** 30-2
[9] Kimura S, Imai Y, Sakata O and Sakai A 2011 *Key Eng. Mater.* **470** 104-9
[10] Kimura S and Ogura A 1998 *Jpn. J. Appl. Phys.* **37** 1282-4

Structural Study of Zr-Cu-Ag Bulk Metallic Glasses Using the Anomalous X-ray Scattering Method

T Kawamata[1], K Sugiyama[1], Y Yokoyama[1] and T Fujita[2]
[1] Institute for Materials Research (IMR), Tohoku University, Sendai 980-8577, Japan
[2] Advanced Institute for Materials Research (AIMR), Tohoku University, Sendai 980-8577, Japan

E-mail: kawamata@imr.tohoku.ac.jp

Abstract. The structures of $Zr_{45}Cu_{45}Ag_{10}$ and $Zr_{40}Cu_{40}Ag_{20}$ bulk metallic glasses (BMGs) were investigated using the anomalous x-ray scattering and reverse Monte Carlo simulation (AXS-RMC) method. The fundamental structural features of $Zr_{45}Cu_{45}Ag_{10}$ and $Zr_{40}Cu_{40}Ag_{20}$ can be properly demonstrated through the common dense random packing of the hard spheres, and the addition of Ag appeared to result in no prominent formation of the particular chemical ordering units. A Voronoi analysis indicated that the fraction of the icosahedron-like coordination was the largest around the Cu in $Zr_{45}Cu_{45}Ag_{10}$ BMG, where the best glass-forming ability was realized. The improvement in the glass-forming ability in a Zr-Cu-Ag system appears to be associated with the icosahedron-like local coordination.

1. Introduction

An encouraging multi-component alloy system for the formation of BMGs at a very low cooling rate was recently found [1]. The atomic origin of such an interesting BMG formation has been studied from the viewpoint of chemical and topological ordering in the atomic arrangement. Nevertheless, the essential structural features providing superior BMG properties have yet to be elucidated. Zr-Cu-Ag is interesting for a BMG system, and its chemical short-range ordering and corresponding heterogeneity were suggested through a molecular-dynamics simulation [2]. The anomalous X-ray scattering (AXS) method utilizing the anomalous dispersion effect near the absorption edge provides environmental structural information on the atoms of interest. Additionally, an analysis using a reverse Monte Carlo (RMC) simulation coupled with AXS measurements subsequently allows us to obtain a three-dimensional (3D) atomic arrangement under the necessary conditions [3, 4]. The main purpose of this paper is to report the structural features of Zr-Cu-Ag BMGs as demonstrated using the advanced AXS-RMC method.

2. Experimental

Alloy ingots with nominal compositions of $Zr_{45}Cu_{45}Ag_{10}$ and $Zr_{40}Cu_{40}Ag_{20}$ were prepared through the conventional arc melting of the pure-metal mixtures (99.99 wt. % Zr, 99.99 wt. % Cu, and 99.99 wt. % Ag). To minimize the oxygen concentration in the sample, a high purity Zr crystal rod (less than 0.05 wt. % oxygen) was used. The tilt-casting technique was applied in a purified Ar atmosphere to avoid the formation of cold spots, which is an inherent source of unfavorable crystallization during the solidification of a melt. A platy-shaped mold with a thickness of 1.0 mm and a length of about 30 mm was employed for the casting, and the surface of each sample alloy was polished for X-ray diffraction measurements of the reflection geometry. The density of the sample was measured using the Archimedes method with *n*-tridecane (7.72 and 8.02 g/cm^3 for $Zr_{45}Cu_{45}Ag_{10}$ and $Zr_{40}Cu_{40}Ag_{20}$, respectively). Ordinary X-ray diffraction and AXS measurements of the Cu, Zr, and Ag *K* absorption edges were carried out at BL-7C and NW10A of the Photon Factory at the Institute of Material Structure Science (IMSS), High Energy Accelerator Research Organization, Tsukuba, Japan. A pair of incident energies, corresponding to 25 eV and 300 eV below each *K* absorption edge (Cu, 8.980 keV; Zr, 17.037 keV; Ag, 25.517 keV), was used. Environmental interference functions, $\Delta Qi_{Cu}(Q)$, $\Delta Qi_{Zr}(Q)$, and $\Delta Qi_{Ag}(Q)$ for each sample were obtained from the measured scattering intensities. Environmental radial distribution functions (RDFs) were obtained based on the Fourier transformation of the interference functions. In the AXS-RMC analysis, four independent interference functions were employed for modeling the structure of each sample. A model consisting of 25,000 atoms placed within a cubic hyper-cell with a classical dense random packing (DRP) of the hard spheres [5] was employed as the initial model. The number of atoms and the unit hyper-cell size were set corresponding to the chemical composition and experimental density value of the sample. It should be noted that the closest distance between two atoms were determined in order to prevent atoms coming un-physically close to one another.

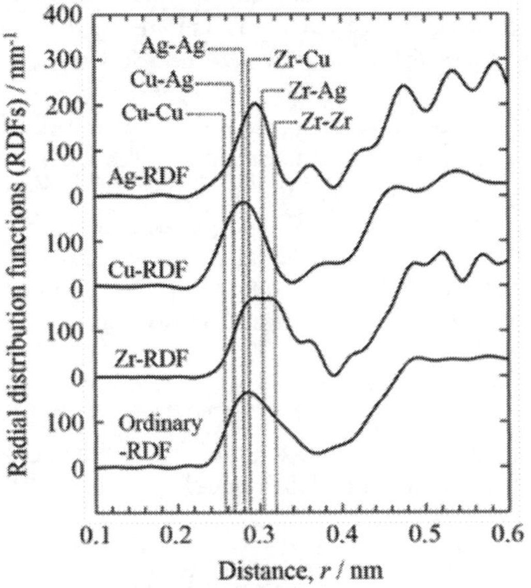

Figure 1. Ordinary and environmental RDFs around Cu, Zr, and Ag of $Zr_{45}Cu_{45}Ag_{10}$. Ordinary RDF was obtained from the intensity profile measured at 15.0 keV.

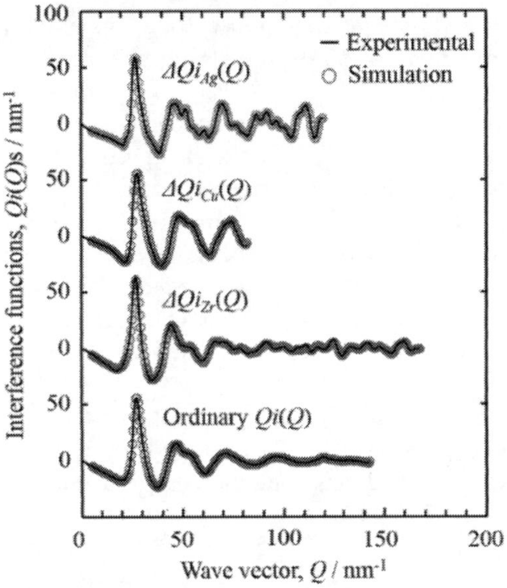

Figure 2. Ordinary and environmental interference functions. The solid and dotted lines correspond to the experimental data and the data obtained using the AXS-RMC method, respectively.

3. Results and discussion

As an example, the environmental RDFs obtained from the AXS measurements of $Zr_{45}Cu_{45}Ag_{10}$ are shown in Figure 1. The dashed lines indicate the interatomic distances estimated based on the Goldschmidt radii [6] (Zr, 0.160 nm; Cu, 0.128 nm; Ag, 0.144 nm). The correlation peak for the Zr-Zr pair was not observed in the environmental RDF for Cu and Ag, which indicates the excellent performance of the AXS analysis. A similar description is possible in the environmental RDF around Zr, where only the pair correlations associated with Zr are pronounced. Figure 2 shows four experimental interference functions, $Qi(Q)$, $\Delta Qi_{Cu}(Q)$, $\Delta Qi_{Zr}(Q)$, and $\Delta Qi_{Ag}(Q)$, together with those obtained through the AXS-RMC simulation. A pre-peak signal at the lower *Q*-region, which implies the presence of a strong chemical ordering unit, could not be observed for each function. This suggests that

Journal of Physics: Conference Series **502** (2014) 012027 doi:10.1088/1742-6596/502/1/012027

the fundamental atomic arrangement of $Zr_{45}Cu_{45}Ag_{10}$ can be approximated through the dense random packing (DRP) model. We therefore started the AXS-RMC simulation using the initial model based on the DRP, and the successive AXS-RMC simulation converged to represent the experimental interference functions, as shown in Figure 2. The atomic arrangement, such as an icosahedral structure, has been frequently suggested to contribute to the development of the glass-forming ability (GFA), and Voronoi tessellation [7] is one of the most convenient methods for discriminating the coordination topology in the local atomic arrangement. A Voronoi polyhedron can be described using a set of Voronoi index numbers, $(n_i) = (n_3, n_4, n_5, n_6, n_7...)$, where n_i is the number of *i*-edged faces of a Voronoi polyhedron. The structural parameters of the nearest-neighbor region of two BMG samples are summarized in Table 1. The averaged coordination numbers increase in the order of atomic size, suggesting that the coordination number around Cu is the smallest, corresponding to that of a icosahedron. On the other hand, the coordination number around Zr is the largest, which is closely attributable to a body-centered cubic structure. The values obtained through the AXS-RMC simulation are nearly equal those by the DRP model, implying that the fundamental structures of the two BMG samples are described well based on the classical DRP structure. Additionally, the first peak position of each partial pair distribution functions in $Zr_{45}Cu_{45}Ag_{10}$ and $Zr_{40}Cu_{40}Ag_{20}$ corresponds to the sum of the Gold Schmidt radii. These features are similar to those found in $Zr_{50}Cu_{50}$ amorphous alloy [4], and the unique local atomic arrangement based on the strong chemical ordering was not developed through the addition of Ag. Figure 3 shows the relationship between R^* and the fraction of icosahedron-like ordering units. The value of R^* is closely associated with the fraction of icosahedron-like units, and the most efficient icosahedral packing is realized at approximately R^*=0.90 of the ideal icosahedral unit. At the same time, the significant development of icosahedron-like units is readily suggested in the BMGs. These results imply that the structural features of the BMGs cannot be fully demonstrated through the common DRP model, where only the effects of the bulk chemical composition and the hard sphere sizes of the individual constituents are considered. The chemical affinity for the paired atoms, which corresponds to the chemical effect arising from the interatomic interactions, is one of the possible reasons for realizing such topological characteristics of the nearest-neighbor region of the BMGs.

Table 1. Structural parameters for the nearest-neighbor region simulated using the AXS-RMC method: r_{ij}, averaged distance; Z_{ij}, averaged coordination number; and R^*, atomic size ratio between the central atom and its ligands. Values given in parentheses are those for the initial DRP model.

	$Zr_{45}Cu_{45}Ag_{10}$			$Zr_{40}Cu_{40}Ag_{20}$		
	r_{ij} (nm)	Z_{ij}	R^*	r_{ij} (nm)	Z_{ij}	R^*
Zr-Zr	0.320 (0.324)	7.0 (7.0)	1.101 (1.100)	0.322 (0.324)	6.4 (6.2)	1.100 (1.101)
Zr-Cu	0.294 (0.298)	5.8 (5.7)		0.296 (0.298)	5.2 (5.1)	
Zr-Ag	0.313 (0.312)	1.4 (1.4)		0.315 (0.313)	3.0 (2.9)	
		14.2 (14.1)			14.5 (14.2)	
Cu-Zr	0.294 (0.298)	5.8 (5.7)	0.880 (0.878)	0.296 (0.298)	5.1 (5.1)	0.880 (0.880)
Cu-Cu	0.271 (0.266)	4.7 (4.5)		0.269 (0.267)	4.1 (4.0)	
Cu-Ag	0.279 (0.282)	1.1 (1.1)		0.278 (0.282)	2.3 (2.3)	
		11.6 (11.3)			11.5 (11.4)	
Ag-Zr	0.313 (0.312)	6.4 (6.4)	0.989 (0.990)	0.315 (0.313)	5.9 (5.7)	0.988 (0.990)
Ag-Cu	0.279 (0.282)	5.1 (5.2)		0.278 (0.282)	4.5 (4.6)	
Ag-Ag	0.301 (0.298)	1.3 (1.3)		0.298 (0.297)	2.6 (2.6)	
		12.8 (12.9)			13.0 (12.9)	

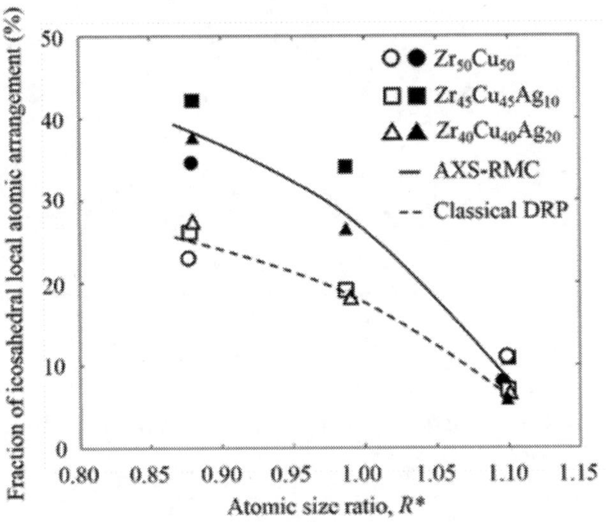

Figure 3. Relationship between R^* and the fraction of icosahedron-like ordering units. The solid and open symbols indicate the results of AXS-RMC and DRP simulations, respectively. An atom with a set of Voronoi index numbers; (0 0 12 0 0), (0 2 8 1 0 0), (0 2 8 2 0), or (0 1 10 2 0) was ranked as that with an icosahedron-like ordering. The results for $Zr_{50}Cu_{50}$ were obtained using the experimental data from a previous study [4].

4. Conclusion

The structural features of ternary $Zr_{45}Cu_{45}Ag_{10}$ and $Zr_{40}Cu_{40}Ag_{20}$ BMGs were investigated through an AXS-RMC analysis. The 3D structural models obtained reproduced the environmental interference functions well. The structural change caused by the addition of Ag to $Zr_{50}Cu_{50}$ was discussed by comparing the structural models with the classical DRP models. The fundamental structural features of the BMGs studied in this work are well represented based on the classical DRP model, and the addition of Ag indicates no formation of the particular chemical ordering units. A Voronoi analysis indicates that the fraction of icosahedron-like coordination was largest around Cu in the $Zr_{45}Cu_{45}Ag_{10}$ BMG, where the best glass-forming ability was reported in the Zr-Cu-Ag system [8]. The improvement of the glass forming ability in a Zr-Cu-Ag system appears to be associated with the icosahedron-like local coordination.

Acknowledgments

AXS experiments were performed with the approval of the Photon Factory Advisory Committee (Proposal No. 2011G171). This work was supported in part by a Grant-in-Aid for Challenging Exploratory Research (22656152) by the Japanese Society for the Promotion of Science (JSPS).

References

[1] Inoue A 2000 *Acta Mater.* **48** 279
[2] Fujita T, Guan P F, Sheng H W, Inoue A, Sakurai T and Chen M W 2010 *Phys. Rev. B* **81** 140204
[3] Saito M, Park C, Omote K, Sugiyama K and Waseda Y 1997 *J. Phys. Soc. Jpn.* **66** 633
[4] Kawamata T, Yokoyama Y, Saito M, Sugiyama K and Waseda Y 2010 *Mater. Trans.* **51** 1796
[5] Clarke A S and Wiley J D 1987 *Phys. Rev. B* **35** 7350
[6] Brandes E A and Brook G B 1992 *Smithells Metals Reference Book 7th ed.* (London: Elsevier Butterworth-Heinemann Ltd.) p. 4–41
[7] Bernal J D 1959 *Nature* **183** 141
[8] Zhang W, Jia F, Zhang Q and Inoue A 2007 *Mater. Sci. Eng. A* **459** 330

Journal of Physics: Conference Series **502** (2014) 012028 doi:10.1088/1742-6596/502/1/012028

Site Determination of Sr in $Sr_3YB_3O_9$ Crystals by Anomalous X-ray Scattering

R Simura and K Sugiyama

Institute for Materials Research, Tohoku University, 2-1-1 Katahira, Aoba-ku, Sendai, Miyagi 980-8577, Japan

E-mail:ray@imr.tohoku.ac.jp

Abstract. The distribution of Sr in $Sr_3YB_3O_9$ crystals was determined by anomalous X-ray scattering. The $Sr_3YB_3O_9$ sample was grown by the floating-zone method in dry air. The $Sr_3YB_3O_9$ crystal has three cation sites (*3a*, *3b*, and *18f*). The results of single-crystal structural analysis using anomalous X-ray scattering at the Sr *K*-edge suggested that Sr is distributed at the *3a* site as well as it is at the chemically disordered *18f* site.

1. Introduction

Crystals of the strontium–yttrium double borate ($Sr_3YB_3O_9$ or SYB; $R\bar{3}$, a = 1.2527(2) nm, c = 0.9280(2) nm) doped with photogenic rare earth elements (REEs) have potential for use in laser and neutron scintillators [1-3]. A previously reported structural analysis of such SYB crystals has suggested that the crystals contain three cation sites: two Y sites at *3a* and *3b* and one Sr site at *18f* (Fig.1). The sites at *3a* and *3b* are octahedrally coordinated by oxygen, and the interatomic distance in the case of Y_{3a}-O (0.251 nm) is greater than that for Y_{3b}-O (0.224 nm) [3]. The effective ionic radii of Y^{3+}(VI) and Sr^{2+}(VI) are 0.09 and 0.118 nm, respectively [4]. These values suggest the preferential distribution of Sr instead of Y at the *3a* site. A single-crystal structural analysis using anomalous X-ray scattering (AXS) is necessary in order to elucidate the actual distributions of Sr and Y, which have similar atomic numbers.

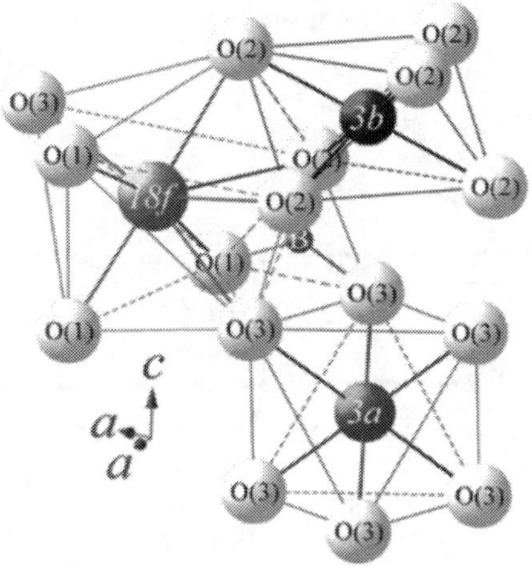

Figure 1. Schematic figure showing the three cation sites of $Sr_3YB_3O_9$. The 3*a* and 3*b* sites are both octahedrally coordinated by six oxygen atoms, while the 18*f* site is coordinated by eight oxygen atoms.

2. Experimental

The SYB sample was grown by the floating-zone (FZ) method in dry air at the growth rate of 1 mm/h. Raw $SrCO_3$, Y_2O_3, and H_3BO_3 powders were stoichiometrically mixed and heated at 1000 °C for 24 h and then sintered to form feed/seed rods 7 mm in diameter for FZ growth.

A single crystal grain with a diameter of 50 μm was cut from the grown SYB crystal and used for single-crystal X-ray diffraction analysis. The AXS-based measurements were performed at the beamline 10 A of the Photon Factory (PF), Institute of Materials Structure Science (IMSS), High Energy Acceleration Organization (KEK), Tsukuba, Japan. The PF synchrotron was operated with a stored energy of 2.5 GeV and an electron current of 450 mA. The X-ray incident energies used for the AXS-based measurements were 16.083 keV and 15.958 keV; these values corresponded to levels 25 eV and 150 eV lower than the Sr *K* absorption edge (16.108 keV). The energies were tuned by using a Si(111) monochromator. The experimental setup and the data processing procedure were almost similar to those used previously [5-7]. The intensity data for $\sin\theta/\lambda = 0.696$ were collected for each energy using a vertical-type four-circle goniometer in the ω-scan mode. The number of reflections measured for the X-ray incident energies of 16.083 and 15.958 keV were 1278 and 1280, respectively. After averaging the equivalent reflections, 790 reflection pairs were used for the AXS-based analysis.

3. Structural analysis

3.1. $R\bar{3}$ type $A_3RB_3O_9$ (A = Sr or Ba; R = rare earth element) crystal family

The crystal structures of the $R\bar{3}$-type $A_3RB_3O_9$ crystal family, which includes compounds such as $Ba_3DyB_3O_9$ (BDB) [8], $Sr_3ScB_3O_9$ (SSB) [9], $Ba_3YB_3O_9$ (BYB) [10], and $Sr_3YB_3O_9$ (SYB) [3], have been reported previously (Table 1). These structures have been analyzed using two models. The first model is the "disordered" model: a divalent alkali earth element and a trivalent REE occupy the 3*a* and 3*b* sites, respectively, and the 18*f* site is a chemically disordered site occupied by either the alkali earth element or the REE, as in the case of BDB and BYB (i.e., 3*a*:3*b*:18*f* = A:R:A/R). The other model is the "ordered" model: the 3*a* and 3*b* sites are both occupied by trivalent REEs, and the 18*f* site is occupied solely by a divalent alkali earth element, as in the case of SSB and SYB [3] (i.e., 3*a*:3*b*:18*f* = R:R:A). The most important differences between these models are the valence states of the 3*a* site and whether the 18*f* site is chemically ordered or disordered. An investigation of the lengths of the bonds between the cation and the oxygen anions suggests that the 3*a* site may have a lower valence than that of the 3*b* site because the bond lengths at the 3*b* sites are smaller than those at the 3*a* sites. When the cation elements have significantly different atomic numbers, the site occupancy can be easily determined by ordinary X-ray structural analysis. However, in the case of Sr and Y, which have similar atomic numbers, the difference between the elements cannot be detected by ordinary X-ray structural analysis. Thus, we attempted to confirm the structural model of SYB using the AXS method as follows.

Table 1: Characteristics of $R\bar{3}$ type $A_3RB_3O_9$ crystals

	3*a** (VI)	3*b** (VI)	18*f** (VIII)	Average bond length of 3*a* site (nm)	Average bond length of 3*b* site (nm)	Average bond length of 18*f* site (nm)	Ref.
BDB	Ba^{2+}	Dy^{3+}	Ba^{2+}/Dy^{3+}	0.2713	0.2446	0.2761	[8]
SSB	Sc^{3+}	Sc^{3+}	Sr^{2+}	0.2142	0.2077	0.2690	[9]
BYB	Ba^{2+}	Y^{3+}	Ba^{2+}/Y^{3+}	0.2749	0.2343	0.2767	[10]
SYB	Y^{3+}	Y^{3+}	Sr^{2+}	0.2518	0.2238	0.2660	[3]
SYB	Sr^{2+}	Y^{3+}	Sr^{2+}/Y^{3+}	-	-	-	This study

* The average bond lengths, *d*, were calculated from the distance between the central cation and the coordinated oxygen anions at the 3*a*, 3*b*, and 18*f* sites. Note that the bond lengths calculated from the effective ionic radii [4] for Ba^{2+}(VI)-O, Sr^{2+}(VI)-O, Dy^{3+}(VI)-O, Y^{3+}(VI)-O, and Sc^{3+}(VI)-O are 0.273, 0.256, 0.229, 0.228, and 0.213 nm, respectively.

3.2. AXS-based X-ray diffraction of the SYB crystal

The AXS method has been described in detail in previous research [6-7]. When the X-ray diffraction intensities of the sample are measured using two different X-ray incident energies (16.083 and 15.958 keV) in the close vicinity of the absorption *K*-edge of Sr, the detected variation in the structural factors is attributable to the anomalous scattering terms of Sr ($\Delta f' \approx 1.86, \Delta f'' \approx 0$[11]). This is because the anomalous dispersion effects arising from the other elements such as Y, B, and O are insignificant at these energy levels. The distribution map of Sr, $\rho_{Sr}(x,y,z)$, can be obtained by following Fourier transformation:

$$\rho_{Sr}(x,y,z) = \frac{1}{V}\sum_h\sum_k\sum_l\{|F_{hkl,15.958keV}|e^{i\alpha_{hkl,15.958keV}} - |F_{hkl,16.083keV}|e^{i\alpha_{hkl,16.083keV}}\}e^{-2\pi i(hx+ky+lz)}$$

where V is the volume of a unit cell, $|F_{hkl,EkeV}|$ and $e^{i\alpha_{hkl,EkeV}}$ are the structural factor of an absolute unit and the phase term for an *hkl* reflection measured at the energy *E* keV, respectively.

4. Results and discussion

The isosurfaces of electron density obtained by ordinary single-crystal X-ray diffraction at 15.958 keV and the isosurfaces of electron density obtained by AXS at the Sr *K*-edge are shown in Fig. 2. In the isosurfaces shown in Fig. 2(b), elements other than Sr are omitted by the AXS analysis, and the electron density from Sr is observed only at the 3*a* and 18*f* sites. This implies that the disordered model is suitable for modeling the structure of SYB.

The maximum intensity in the distribution map of Sr at the 18*f* site was about five-sixth of that at the 3*a* site, suggesting that the 18*f* site was a mixed site of Sr and Y, the two elements present in a ratio of 5:1. Consequently, the crystal structure of SYB could be constructed using three types of coordination polyhedrons: three Sr-centered octahedrons at the 3*a* site, three Y-centered octahedrons at the 3*b* site, and eighteen Sr/Y(=5:1)-centered irregular-shaped dodecahedrons at the 18*f* site, together with eighteen BO_3 triangles in the unit cell.

For laser and scintillator applications, a small amount of a photogenic REE is doped in SYB. Judging from the structure of SYB determined in this study, photogenic REEs that have ionic radii greater than that of Y^{3+} (0.09 nm), such as Ce^{3+}(0.101 nm), Pr^{3+}(0.099 nm), and Nd^{3+}(0.0983 nm), among others, can be assumed to occupy the 18*f* site (d = 0.266 nm), which is larger, rather than the 3*b* site (d = 0.224 nm).

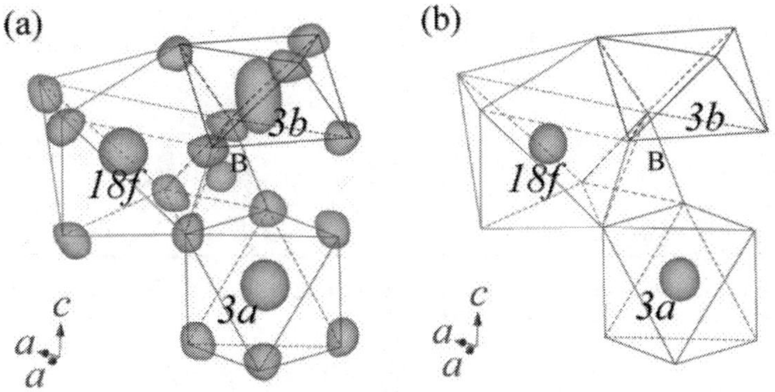

Figure 2. (a) Isosurface three-dimensional (3D) image obtained by ordinary single-crystal X-ray diffraction analysis. (b) Isosurface 3D image obtained by AXS-based analysis at the Sr *K*-edge. The intensity thresholds used for the isosurfaces in (a) and (b) were 5.0×10^{-3} electron/nm^3 and 1.0×10^{-3} electron/nm^3, respectively.

5. Conclusion

AXS-based analysis at the Sr absorption *K*-edge revealed clearly the location of Sr in the $Sr_3YB_3O_9$ crystal. It was found that Sr was distributed at the 3*a* site as well as it was at the chemically disordered 18*f* site. Hence, the combination of single-crystal X-ray diffraction analysis and AXS-based analysis can be considered an indispensable tool for determining the site occupancies of elements having similar atomic numbers.

Acknowledgments

We would like to express gratitude to Mr. S. Suzuki, Mr. S. Kawai, and Mr. T. Sugawara for their invaluable support during FZ crystal growth and AXS-based measurements. This research was partially supported by a Grant-in-Aid for Young Scientists (B) (2376003) from the Japan Society for the Promotion of Science (JSPS) and partially by the research collaboration program (Proposal #2011G680) with the Institute of Materials Structure Science, High Energy Accelerator Research Organization, Tsukuba, Japan.

References

[1] Chenais S, Duron F, Balembois F, Georges P, Gaume R, Haumesser P H, Viana B, Aka G P and Vivien D 2002 *J. Opt. Soc. Am.* B **19** 1083
[2] Simura R, Kawai S, Sugiyama K, Yanagida T, Sugawara T, Shishido T and Yoshikawa A 2013 *J. Cryst. Growth* **362** 2013
[3] Maczka M, Waskowska A, Majchrowski A, Kisielewski J, Szyrski W and Hanuza J 2008 *J. Solid State Chem.* **181** 3211.
[4] Shannon R D 1976 *Acta Cryst.* A **32** 751
[5] Tanaka M and Sasaki S 1993 *KEK Internal* 93-4
[6] Genba M, Sugiyama K, Hiraga K and Yokoyama Y 2002 *J. Alloy. Compd.* **342** 143
[7] Sugiyama K, Genba M, Hiraga K and Waseda Y 2010 *J. Alloy. Compd.* **494** 98
[8] Ilyukhin A B and Dzhurinskii B F 1993 *Zhur. Neorg. Khimii.* **38** 1625
[9] Thompson P D and Keszler D A 1994 *Chem. Mater.* **6** 2005
[10] Li X Z, Chen X L, Jian J K, Wu L, Xu Y P and Guo Y G 2004 *J. Solid State Chem.* **177** 216
[11] Waseda Y 1984 *Novel Application of Anomalous X-ray Scattering for Structural Characterization of Disordered Materials* (Springer-Verlag, Heidlberg)

A structural study of liquid Te-rich Si-Te alloys

Fujio Kakinuma[1], Shinya Hosokawa[2], Hiroyuki Ikemoto[3], Shinji Kohara[4] and Yoshimi Tsuchiya[5]

[1] Department of Information and Electrical Engineering, Niigata Institute of Technology, 1719 Fujihashi, Kashiwazaki, 945-1195, Japan
[2] Graduate School of Science and Technology, Kumamoto University, 2-39-1 Kurokami, Chuo-ku, Kumamoto, 860-8555, Japan
[3] Department of Physics, Faculty of Science, University of Toyama, 3190 Gofuku, Toyama, 930-8555, Japan
[4] JASRI/SPring-8, 1-1-1 Kouto, Sayo-cho, Sayo-gun, Hyogo 679-5198, Japan
[5] Department of Physics, Faculty of Science, Niigata University, 2-8050 Ikarashi, Nishi-ku· Niigata 950-2181, Japan

E-mail: kakinuma@iee.niit.ac.jp

Abstract. X-ray diffraction measurements using synchrotron radiation on liquid Si_xTe_{1-x} ($x \leq 0.2$) were carried out at several temperatures. At constant temperature, the first peak, which is composed of Si-Te and Te-Te correlations, of the pair distribution function gradually shifts to the shorter distances with increasing Si content. These shifts indicate the substitution of Si-Te bond for the Te-Te bonds in the first peak. The total coordination number shows little temperature variation and remains about two in the temperature interval of a few hundreds of degrees. The composition and temperature behaviour of the Te-rich side of Si-Te liquid alloys has already been investigated, among which thermodynamic behaviour is a quite contrast to that of the Ge-Te alloys in the overlapping composition-temperature range of the present investigation. We clarify the evolution of the local structure of the liquid Si-Te alloys in comparison with the liquid Ge-Te alloys.

1. Introduction

Liquid Te-rich alloys exhibit unique temperature dependence of the physical properties [1, 2]. The thermodynamic susceptibilities (specific heat, isothermal compressibility and thermal expansion coefficient) for liquid $Ge_{15}Te_{85}$ which is the eutectic alloy show the sharpest extremum so far found for liquid semiconductors [3, 4]. It has been clarified that the anomalous thermodynamic behaviour is a sign of the steep structural changes in the liquid with temperature [5].

Si and Ge belong to the 14th family of the periodic table and it is assumed that the liquid Si-Te alloy at the Te-rich side has the similar covalent bonding mechanism to the liquid Te-rich Ge-Te alloy and both physico-chemical properties would be much the same. The presumptions are in evidence in some experiments. The electrical conductivity of liquid Te-rich Si-Te alloy has a temperature dependence of a type of a liquid semiconductor [6]. The specific heat, compressibility and thermal expansion coefficient show an extremum in the temperature domain where the electrical transport properties change from semi-conductive to weak metallic state. The composition and temperature behaviour of the Te-rich side of Si-Te liquid alloys is a quite contrast to that of the Ge-Te alloys in the overlapping composition-temperature range, however. For example, the peak value of the specific heat and the amount of volume contraction with temperature of $Si_{15}Te_{85}$ are, respectively, about half and 1/3 of those of $Ge_{15}Te_{85}$, while their temperature variation takes place over much wider range. In other words the rate of structural changes with temperature in the Si-Te alloy is about halftimes as large as the one in the Ge-Te alloy.

In the present study we clarify the evolution of the local structure of the liquid Si-Te alloys in comparison with the liquid Ge-Te alloys.

2. Experimental

The Si-Te alloy samples were made from Si and Te shots of purity (99.999%) by direct mixing. The raw materials were sealed in a molten quartz tube in 3 mm diameter with 0.5 mm wall thickness under a vacuum of 2×10^{-6} Torr, and kept at 1243 K for 24 hours. The liquid alloy was shaken a few times during heating to make thorough mixing. After heating, the sample was quenched in iced water, and then the alloy was ground to grains. Finally the powder sample was encapsulated in a silica glass capillary in 2 mm diameter with 0.02 mm wall thickness for measurements under a vacuum of 2×10^{-6} Torr.

X-ray diffraction experiments were carried out with the angle dispersive method at the BL04B2 beam line [7] of the SPring-8, Japan. Incident photon energy was 113 keV obtained by the third harmonics reflection of Si(111) monochromator. The structure factor, $F(q)$, where $q = 4\pi\sin\theta/\lambda$, 2θ is the diffraction angle, and λ the X-ray wavelength, was obtained from the net scattering intensities from the sample after standard corrections for the background and empty cell intensities, absorption and polarization [7, 8]. The measured temperatures were 763 K, 873 K, 923 K and 973 K for Si_xTe_{1-x} with x = 0.05, 0.10, and 0.15, 873 K, 923K and 973 K for $Si_{0.2}Te_{0.8}$.

3. Results and Discussion

Figure 1 shows the structure factors, $F(q)$, for Si_xTe_{1-x} alloys at 873 K. It can be seen that the profile of $F(q)$ changes systematically with Si concentration. With increasing Si concentration, the first peak shifts to smaller value of q, while the second peak to the larger value. The second peak height increases and the first minimum becomes deeper. A small hump, which is the indication of the development of the medium range order, appears around $q = 0.8$ Å^{-1} in $F(q)$ for x = 0.15 and 0.2.

The total pair distribution function, $g(r)$, is obtained from the Fourier transform of $F(q)$ [8],

$$g(r) = 1 + \frac{1}{2\pi^2 \rho r}\int (F(q)-1)q\sin qr dq \ , \tag{1}$$

where ρ is the average number density of atoms in the alloy. The values of ρ used in this analysis are calculated using the molar volume data in [9]. Figure 2 shows the total pair distribution functions of Si_xTe_{1-x} alloys at 873 K. With increasing Si concentration very large changes in $g(r)$ appear up the second peak. The first peak position shifts to the shorter distance from 2.86 Å for x = 0.05 to 2.75 Å for x = 0.2, accompanying the decrease in the height. The second peak grows steeply and becomes prominent beyond x = 0.15 while its position remains almost unchanged. Corresponding to the appearance of hump in $F(q)$ around $q = 0.8$ Å^{-1} a new peak becomes apparent in around 7.8 Å in $g(r)$.

For a Si-Te alloy, the total structure factor, $F(q)$, is given by equation (2) in terms of Si-Si, Si-Te, Te-Si and Te-Te partial structure factors, $F_{ij}(q)$.

$$F(q) = w_{SiSi}(q)F_{SiSi}(q) + w_{SiTe}(q)F_{SiTe}(q) + w_{TeSi}(q)F_{TeSi}(q) + w_{TeTe}(q)F_{TeTe}(q) \ , \tag{2}$$

where $w_{ij}(q) = x_i x_j f_i(q) f_j(q)/(x_i f_i(q) + x_j f_j(q))^2$ in which $f_i(q)$ and x_i are, respectively, the atomic form factor and the concentration of Si or Te. Since the q-dependence of $w_{ij}(q)$ is weak, the total pair distribution function, $g(r)$, obtained by putting equation (2) into equation (1), can be approximated to equation (3) to a good approximation.

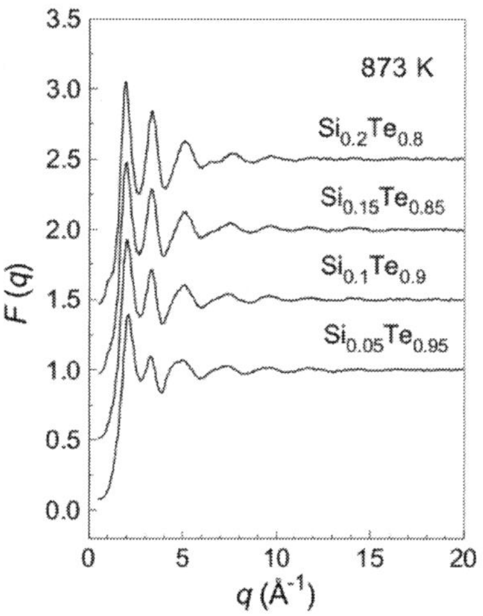

Figure 1. The structure factors for liquid Si_xTe_{1-x} alloys. The curves for $Si_{0.1}Te_{0.9}$, $Si_{0.15}Te_{0.85}$ and $Si_{0.2}Te_{0.8}$ are shifted 0.5, 1.0 and 1.5, respectively.

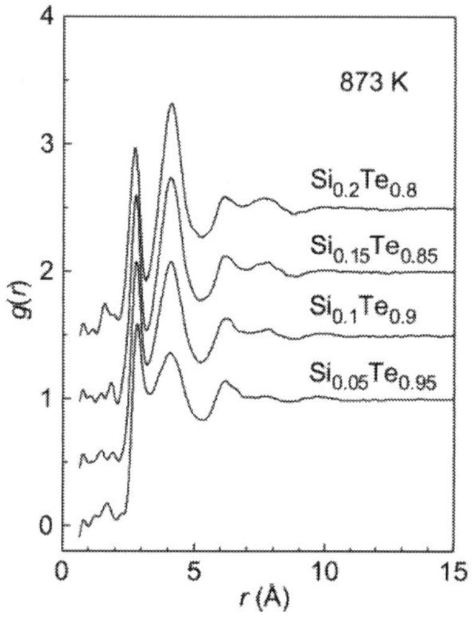

Figure 2. The total pair distribution functions for liquid Si-Te alloys. The curves for $Si_{0.1}Te_{0.9}$, $Si_{0.15}Te_{0.85}$ and $Si_{0.2}Te_{0.8}$ are shifted 0.5, 1.0 and 1.5 , respectively.

$$g(r) \approx c_{SiSi}g_{SiSi}(r) + c_{SiTe}g_{SiTe}(r) + c_{TeSi}g_{TeSi}(r) + c_{TeTe}g_{TeTe}(r)\,, \tag{3}$$

where the total pair distribution function is a weighted sum of the four partial distribution functions, $g_{ij}(r)$, of Si-Si, Si-Te, Te-Si and Te-Te pairs, defined by equation (4). The weighting factors c_{ij} are $x_i x_j Z_i Z_j/(x_i Z_i + x_j Z_j)^2$ with the atomic number of an atom i, Z_i. The partial pair distribution function for an i-j atom pair is the probability to find j atoms at the distance r when an i atom is at the origin.

$$g_{ij}(r) = 1 + \frac{1}{2\pi^2 \rho r}\int (F_{ij}(q) - 1) q \sin qr dq \tag{4}$$

Intuitive insight to the structure of liquid Si-Te alloy may be obtained from the total coordination numbers n for the definite peaks in the radial distribution function $G(r) = 4\pi r^2 \rho g(r)$.

$$n = \int_{r_1}^{r_2} G(r)dr \approx (c_{SiSi} + c_{SiTe})n_{Si} + (c_{TeSi} + c_{TeTe})n_{Te} \tag{5}$$

For the total coordination number, r_1 and r_2 were chosen to be 2.2 Å and the first minimum position of $G(r)$, respectively. The results are listed in table 1. The total coordination number remains about two though a slightly decreasing tendency with increasing Si fraction can be seen. Further the depth of the first minimum in $g(r)$ becomes deeper as can be seen in figure 2, which indicates that overlap between the first and second coordination shell decreases. Since the atomic number of Si and Te is 14 and 52, respectively, the weighting factors, c_{SiSi} and c_{SiTe}, for the partial radial distribution function of a Si atom at the origin are 0.00398 and 0.0591, respectively, for the Si-richest $Si_{0.2}Te_{0.8}$ alloy. Therefore, the composition-temperature dependence of the coordination number around a Te atom, n_{Te}, dominates the present results.

The coordination number in pure Te at the melting temperature, 450 °C, is about 2.7, and decreases to about two in the deeply under cooled liquid state, which has been interpreted as a liquid-liquid crossover transition in liquid Te. At the melting point the transition is almost finished and liquid Te is in a high temperature form in which inter-chain bonding caused by bond defects in a chain structure brings about a local three-dimensional structure. By adding Si, the present results indicate that the local structure around Te transforms back to a chain structure. A Si-Te alloy in the composition range from x = 0.15 to 0.25 easily forms bulk glass of which the structure is shown to be Te chains interconnected with Si atoms tetrahedrally coordinated to four Te atoms. The results are in good agreement of Refs. [10, 11]. The latter structure is the structural unit of Si_2Te_3 that is the only

Table 1. The total coordination number of Si_xTe_{1-x} alloys.

	763 K	873 K	923 K	973 K
$Si_{0.05}Te_{0.95}$	2.00±0.02	2.21±0.02	2.17±0.02	2.13±0.02
$Si_{0.1}Te_{0.9}$	2.07±0.02	2.08±0.02	2.08±0.02	2.06±0.02
$Si_{0.15}Te_{0.85}$	1.95±0.02	1.96±0.02	1.98±0.02	2.02±0.02
$Si_{0.2}Te_{0.8}$	—	1.92±0.02	1.84±0.02	1.85±0.02

stable compound in the Si-Te alloy. It is supposed, therefore, that added Si atoms compensate bond defects in Te chains to form tetrahedral unit with surrounding four Te atoms, which would decrease bond defects in a liquid and change the electrical transport properties from the semi-metallic to semi-conductive state. The measurements by Gasser *et al* support this conjecture [6].

The structural evolution in the Si-Te system summarized above comparatively contrasts with the results for the Te rich Ge-Te system in harmony with the difference in the behaviour of thermodynamic susceptibilities. For example, the total coordination number of $GeTe_6$ is as large as the one for pure Te and increases apparently with increasing temperature from 623 to 796 K [12]. The origin of liquid-liquid transitions in both systems might be somewhat different, therefore.

4. Conclusion

The X-ray scattering measurements were performed for liquid Si_xTe_{1-x} alloys with x = 0.05, 0.1, 0.15 and 0.2 at several temperatures. The total pair distribution functions for the liquid Si-Te alloys exhibit characteristic behaviour. The variation of the first minimum in the total pair distribution function with Si concentration suggests that the bond defects in liquid Te decreases with addition of Si, and Si atom connects four Te chains and forms the tetrahedral unit. The total coordination number obtained is about two and decreases with increasing Si concentration. The structure of liquid Te by addition of Si atoms transforms to a chain structure similar to that of liquid Se. The $Si_{0.15}Te_{0.85}$ and $Si_{0.2}Te_{0.8}$ alloys have a hump at around 0.8 $Å^{-1}$ in the structure factor, which suggests that the medium range order still remain around near the melting temperature.

Acknowledgement

This work has been performed under the approval of the Japan Synchrotron Radiation Research Institute (JASRI) (proposal no. 2012A1097). The authors thank Dr. K Ohara for his useful support on experiments.

References

[1] Tschirner H -U, Pazdrij I P, Weikart R 1983 *Wiss. Z. d. Techn. Hochsch. Karl-Marx-Stadt* **25/2** 219.
[2] Tsuchiya Y 1993 *J. Non-Cryst. Solids* **156-158** 704
[3] Tsuchiya Y, Saitoh K, Kakinuma F 2005 *Monatshefte für Chemie* **136** 1963
[4] Narita M, Konishi I , Tsuchiya Y 2011 *EPJ Web of Conferences* **15** 01021
[5] Bergman C, Bichera C, Gaspard J -P, Tsuchiya Y 2003 *Phys. Rev. B* **67** 104202
[6] Gasser J -G, Halim H, Wax J -F, Vinckel J 1996 *J. Non-Cryst. Solids* **205&206** 120
[7] Kohara S, Itou M, Suzuya K, Inamura Y, Sakurai Y, Ohashi Y, Takata M 2007 *J. Phys. Condens. Matter* **19** 506101
[8] Waseda Y 1980 *The Structure of Non-Crystalline Materials* (McGraw-Hill, New York)
[9] Tsuchiya Y 1991 *J. Non-Cryst. Solids* **136** 37
[10] Schoening F R L 1979 *J. Mater. Sci.* **14** 2397
[11] Bartsch G E A, Bromme H, Just T 1975 *J. Non-Cryst. Solids* **18** 65
[12] Otjacques C, Raty J -Y, Gaspard J -P, Tsuchiya Y, Bichara C 2011 *Collection SFN* **12** 233

Photocreation of a quantum domain and its detection by inelastic X-ray scattering and X-ray CARS

Kaoru Iwano

Photon Factory, Institute of Materials Structure Science, High Energy Accelerator Research Organization (KEK), 1-1 Oho, Tsukuba, Ibaraki 305-0801, Japan

E-mail: `kaoru.iwano[at-mark]kek.jp`

Abstract. We briefly argue the concept of a quantum domain and propose its detection using X-ray inelastic scattering and X-ray coherent anti-Stokes Raman scattering (CARS). The quantum domain is defined as a spatial region of which the phase state is converted to a different one from that in the background. In a case where photoinduced phase transitions are allowed to exhibit, such a domain has a relatively low excitation energy and is expected to be detected experimentally. Especially the X-ray inelastic scattering and the CARS are attractive methods, since they can give information of both the momentum and the energy.

1. Introduction

The realization of material-phase switching is now a hot issue as often mentioned as "photoinduced phase transition." In particular, attention is paid to its ultrafast aspect, namely, electronic phase transition in a time scale of femto or subfemto seconds. The recent advance in synchrotron technology is now making it partially possible to detect such ultrafast phenomena. Needless to say, the advantage of using synchrotron radiation is in the full search of $(q,\ \omega)$ space. In particular, the q-scan gives a great merit, because a created domain of a photoinduced phase has a spatial internal structure and must be analyzed either in the real space or in the reciprocal space. Note that the real-space method is impossible here because of the translational invariance.

In a work performed a few years ago [1], we proposed the detection of a domain of an ionic phase in the neutral-phase background through stationary inelastic X-ray scattering. Following this work, we extended the theory to include the dissipation originating from scattering by phonons and found the effect of "quantum friction," that is, the deceleration of domain growth [2]. The current interest is in the detection of such a dissipating process by time-resolved experiment and we propose a detection based on coherent anti-Stokes Raman spectroscopy (CARS) in the X-ray region.

Below, we start our explanation from a model that describes the quantum domain, and briefly review the possible detection using the inelastic X-ray scattering. We also describe a proposed scheme of the X-ray CARS and give some calculated results for demonstration.

2. Model of quantum domain

First, we introduce our effective model hamiltonian, which is defined as

$$H = H_D + H_{D-ph} + H_{ph}\ , \tag{1}$$

where the part for the bare domain in one dimension is expressed as

$$H_D = -t_{eff} \sum_{jp} \{(|j,p+1\rangle + |j-1,p+1\rangle)\langle j,p| + H.c.\} + \sum_{jp} E(p)|j,p\rangle\langle j,p| \, , \tag{2}$$

the part of domain-phonon interaction as

$$H_{D-ph} = g \sum_{i=j}^{j+p-1} (b_i + b_i^\dagger)|j,p\rangle\langle j,p| \, , \tag{3}$$

and the bare phonon part as

$$H_{ph} = \omega \sum_j (b_j^\dagger b_j + \frac{1}{2}) \, . \tag{4}$$

Regarding the details of the definitions, the readers are encouraged to see Ref. [2]. Instead, we describe our idea very briefly in the following. Namely, a one-dimensional domain as depicted in Fig. 1 is assumed to be photoexcited. Here, the state of the background phase is a type of electronic state, that is, a charge-density wave state, while that inside the domain is regarded as a Mott-insulator state. Next, we consider the eigenstates of the bare domain part, H_D, for

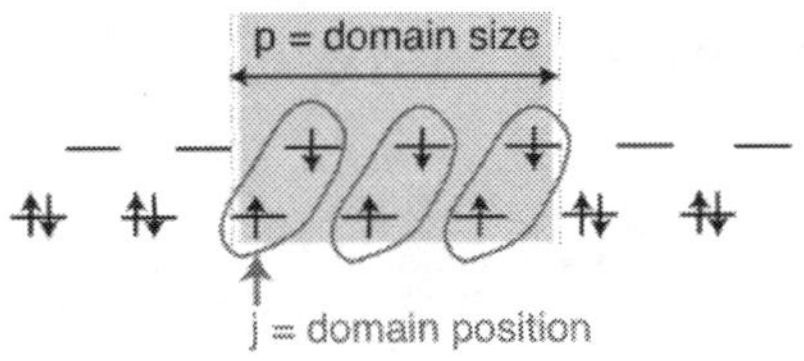

Figure 1. Schematic picture of the quantum domain. The position of the domain is counted at its left end.

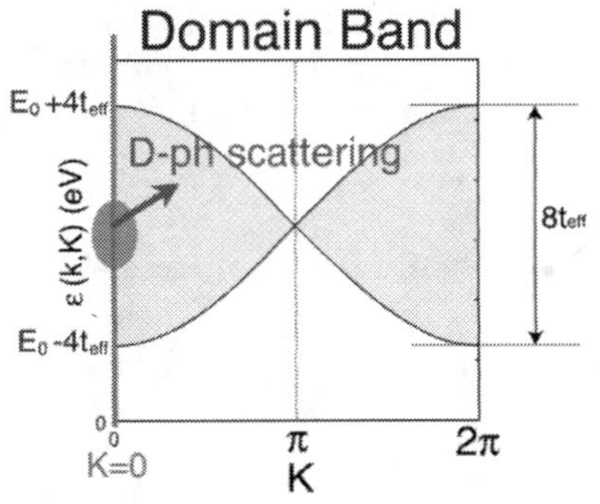

Figure 2. Band structure for the quantum domain.

the special case of $E(p) \equiv E_0$. This case corresponds to the photoexcitation just at the phase boundary, which means that the formation energy is constant, E_0, irrespective of its size p. In this case, the domain energy does not change even when it expands. Considering also the apparent fact that the domain has a translational symmetry, these basic properties lead to a picture that the two domain wall move freely to each other. As a result, we find a continuum band for the domain eigenstate as a function of its total momentum, K, as shown in Fig. 2.

3. Inelastic X-ray scattering

In this section, we briefly discuss the inelastic X-ray scattering expected only for the H_D part. Since the part is already diagonalized for the case of $E(p) = E_0$, as $\epsilon(k,K) = -4t_{eff}\cos(K/2)\cos(k) + E_0$ with k being the internal momentum, we easily obtain expressions for the spectra. Namely, for K=π, which corresponds to zero-momentum transfer, the spectrum becomes $I(\Omega,\pi) = \delta(\omega - E_0)$. Meanwhile, it becomes a very broad spectrum expressed as

$I(\Omega,0) = (1/\pi)\sqrt{((1-(\Omega-E_0)^2/(4t_{eff})^2}$, as a normalized spectrum. Moreover, when $E(p)$ is not constant, we have the results that were reported in Ref. [1], although an analytical treatment is impossible for this case. By the way, the validity of the present effective model was checked for the latter case, by the direct comparison between the result by this model and that by the original microscopic electronic hamiltonian. As the last remarks for this measurement itself, we point out the following two points. One is the large spectral width particularly for K=0. Considering that the t_{eff} is not so different from the bare electron transfer energy, the width as large as $8t_{eff}$ is rather outstanding, because even the bare electron has a band width of only $4t_{eff}$. Another point is the drastic band shrinking from K=π to 0, which suggests the experimental feasibility to detect such a feature.

4. X-ray CARS

In the previous section, we only considered the effect from H_D. While such effect is very basic and predicts a coherent growth for the quantum domain, we also expect other effects that hinder such a growth. The domain-phonon interaction that was introduced in Eq. (3) plays such a role. Here, we assume an intramolecular type of vibrations, of which the boson operators are defined as b_j and $b_j^\dagger$ at the jth site. After the photoexcitation by visible light, the initial total momentum is almost zero. However, in the course of the dynamics, the domain-phonon scattering occurs and hence the domain acquires a finite center-of-gravity momentum, K, as depicted in Fig. 2. At the same time, such incoherence manifests itself as a deceleration in the domain growth [2]. What is important here is that such domain growth can never been observed by ordinary real-space imagines, since every process is translationally invariant except for special cases where some crystal imperfections or defects play some role. In this sense, it is required to to "visualize" such an evolution by using X-rays, not in the real space but in the (q,ω) space. From here on, we

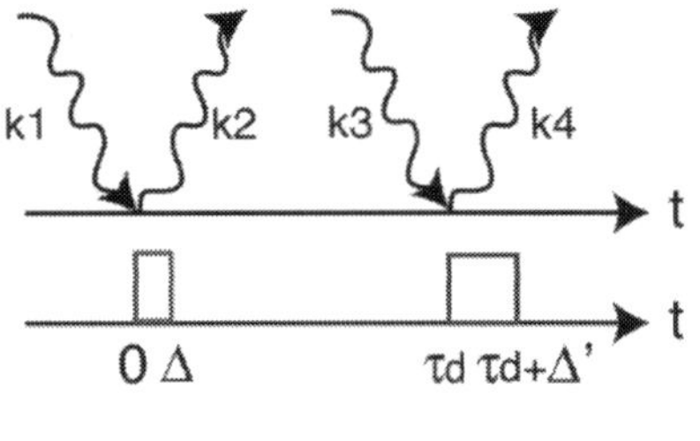

Figure 3. Feynman diagram for the CARS. Rectangle shapes are assumed for the X-ray pulses.

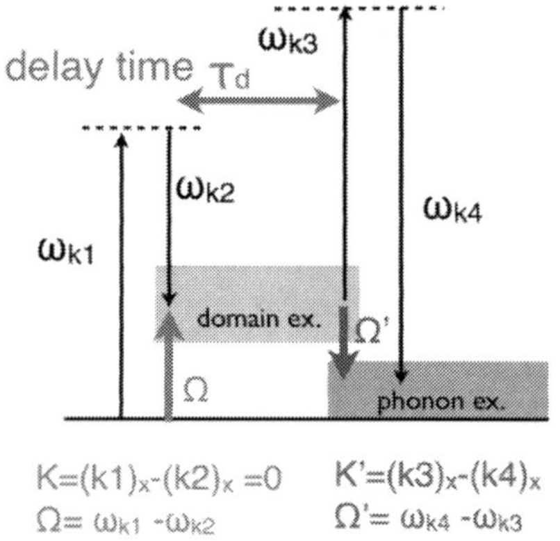

Figure 4. Level scheme for the CARS. "x" is the direction of the one-dimensional axis.

think about the case that two photons come in and other two photons come out, as illustrated in Figs. 3 and 4. To drive such a process, we utilize induced emission, which corresponds to the injection of k_2. Assuming the light pulse widths of Δ and Δ' and defining the delay time as τ_d, we find the scattering intensity as a function of $K' \equiv (k3)_x - (k4)_x$ and $\Omega' \equiv \omega_{k4} - \omega_{k3}$,

$$S(\Omega',K') = \sum_\nu |_D\langle K', p=1|_{ph}\langle\nu|\frac{e^{i(\Omega'-H+E_\nu)\Delta'}-1}{\Omega'-H+E_\nu}|\psi(\tau_d)\rangle|^2\,, \quad (5)$$

where $\psi(\tau_d)$ is the state that evolved up to $t = \tau_d$, $|K', p = 1\rangle_D$ is the domain state with $K = 0$ and the size of p, and $|\nu\rangle_\nu$ and E_ν are the νth phonon state and its energy, respectively.

As the initial condition, we equal the x-component of $k1$ and $k2$, with x being the one-dimensional direction, which corresponds to a preparation of the excited domain state with K=0 for the initial state. The probability of the evolved domain state as a function of K and the energy then changes depending on the time, as a result of the domain-phonon scattering. Here, we calculate the time evolution exactly within a restricted Hilbert space of one domain and several phonons. In Fig. 5, we show a density map traced out for the phonon states, which is taken at t=140 in the unit of $\hbar$/eV $\sim$ 0.66 fs. Note that Δ' is set at 2Δ and Δ is 30 in the same unit. The other parameter values are the same as those in Ref. [2]. Although it is natural that the densities are distributed at the available domain eigenstates, the high densities around the band edges and, especially, at the crossing point should be paid attention, because they are related to the deceleration of the domain growth [2]. Next, we calculate the CARS intensities using the same final state, i. e., $\psi(\tau_d)$. In this study, we make one approximation, that is, the use of $H - H_{D-ph}$, instead of H in the denominator of Eq. (5), neglecting the domain-phonon interaction in the short time of Δ'. The obtained result is shown in Fig. 6. Although it is somewhat smeared out due to the finite observation time Δ', we recognize the same tendency as that in Fig. 5, which indicates the feasibility of this type measurement.

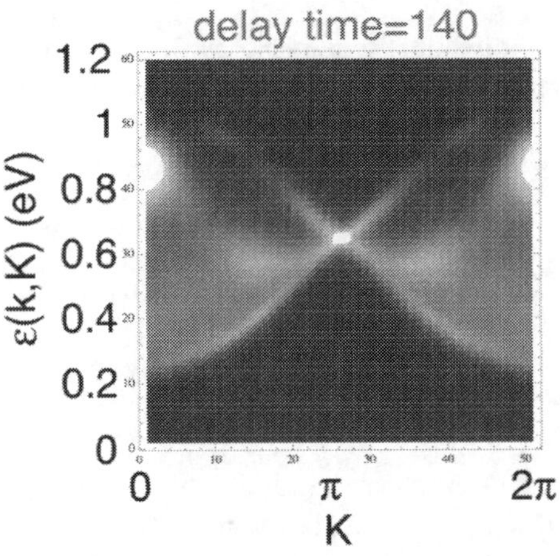

Figure 5. Probability density for the domain state with respect to ϵ and K at τ_d=140.

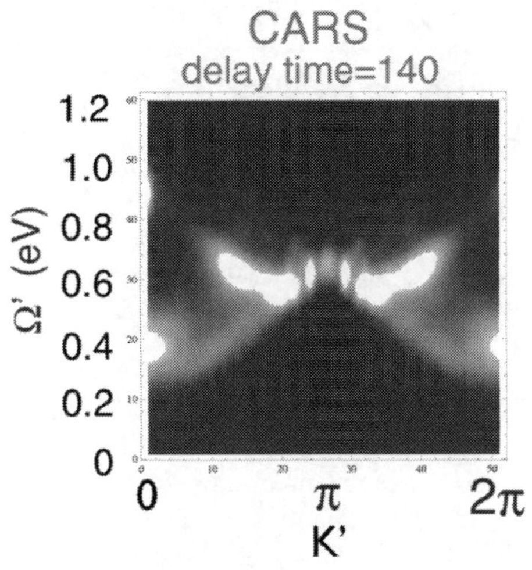

Figure 6. Two-dimensional map for the CARS at τ_d=140.

5. Future Prospect

Through the detections discussed above, a new possibility will be opened in both the future X-ray spectroscopy and the research of the photoinduced phase transition. For those purposes, a new brilliant coherent light source is essential and its construction is highly waited to promote the science in these fields substantially.

References

[1] K. Iwano, Phys. Rev. Lett. 102, 106405 (2009).
[2] K. Iwano, Phys. Rev. B 84, 235139 (2011).

Journal of Physics: Conference Series **502** (2014) 012031 doi:10.1088/1742-6596/502/1/012031

Polarized XAFS study of Al *K*-edge for *m*-plane AlGaN films

T Miyanaga[1], T Azuhata[1], K Nakajima[1], H Nagoya[1], K Hazu[2], S F Chichibu[2]

[1]Department of Advanced Physics, Hirosaki University, Hirosaki, Aomori 036-8561, Japan

[2]Institute of Multidisciplinary Research for Advanced Materials and Department of Applied Physics, Tohoku University, Aoba, Sendai 980-8577, Japan

Takafumi Miyanaga, email: takaf@cc.hirosaki-u.ac.jp

Abstract. Local structures around Al atoms in high-quality *m*-plane $Al_xGa_{1-x}N$ films (x=0.32 and 0.58) deposited on *m*-plane GaN substrates by the NH_3 source molecular beam epitaxy method were investigated by Al *K*-edge X-ray absorption fine structure (XAFS) for the first time. XAFS spectra were measured using a linearly-polarized X-ray source from synchrotron radiation for three different directions; along the *c*-, *a*-, and *m*-axes. The interatomic distances along the *a*-axis are close to Ga-Ga distance in GaN, indicating that the local structures are strongly affected by GaN substrates. The localization of Al atoms was observed for the $Al_{0.32}Ga_{0.68}N$ film.

1. Introduction

Group III-nitride semiconductors have attracted much attention as materials for blue and ultraviolet light emitters [1]. The most studied member of this group, wurtzite GaN, has a direct band gap of 3.42 eV, while pseudo-binary alloys, $Al_xGa_{1-x}N$, provide a direct gap up to 6.0 eV. It is attractive that it covers the wide range of ultraviolet rays. There are several XAFS studies of Ga *K*-edge for $Al_xGa_{1-x}N$ (x=0~0.4) [2] and (x=0~0.6) [3] films, where the interatomic distances and the coordination numbers of Ga-Al and Ga-Ga pairs are discussed. On the other hand, no Al *K*-edge XAFS study has been reported for AlGaN films. Recently, it was pointed out that the problem is the reduction in oscillator strength of electron-hole pairs in quantum wells due to the quantum-confined Stark effect caused by the spontaneous and piezoelectric polarization discontinuity at *c*-plane heterointerfaces. To avoid this effect in AlGaN heterostructures, epitaxial growth in nonpolar orientation such as *m*- or *a*-plane has been attracting attention. However, the local structure and the basic physical properties of nonpolar AlGaN films are not fully understood. In this paper, polarized XAFS studies have been carried out for Al *K*-edge of *m*-plane $Al_xGa_{1-x}N$ thin films. Especially, we discuss the interatomic distance of the second nearest neighbour (2NN) Al-Al and Al-Ga atomic pairs and the localization of Al atoms in *m*-plane $Al_xGa_{1-x}N$.

2. Experimental and Data analysis

140-nm-thick *m*-plane $Al_xGa_{1-x}N$ films (x=0.32 and 0.58) have been grown by the NH_3 source molecular beam epitaxy method on *m*-plane GaN substrates. It is reported that the $Al_{0.32}Ga_{0.68}N$ film is pseudomorphic but the $Al_{0.58}Ga_{0.42}N$ film is partially or nearly fully relaxed [4]. Thus, it is important to study the local structure around Al atoms in the films by EXAFS.

Al *K*-edge (1560 eV) XAFS was measured on BL11A at Photon Factory, KEK, in fluorescence mode using a Silicon Drift Detector (SDD). To measure the polarization dependent XAFS, we set the sample in three configurations with the electric field vector of X ray parallel to the (1) *a*-axis [11-20], (2) *c*-axis [0001], and (3) *m*-axis [1-100].

XAFS analyses were performed using XANADU code [5] and FEFF8.01 code [6]. Figure 1 shows the schematic structure for $Al_xGa_{1-x}N$. The black center circle is an Al atom. Gray and blue circles are N atoms and Al or Ga atoms, respectively.

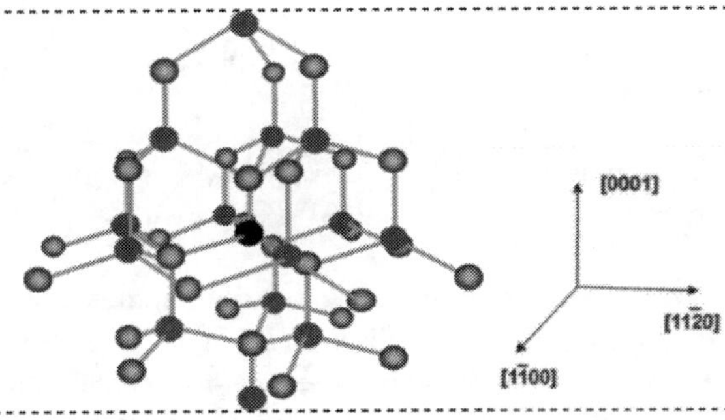

Figure 1. Schematic structure for $Al_xGa_{1-x}N$ with crystal axes. Black, gray, and blue circles are Al, N, and Al or Ga atoms, respectively.

3. Results and discussion

Figure 2 shows EXAFS $k^2\chi(k)$ spectra of Al *K*-edge for the *m*-plane $Al_{0.32}Ga_{0.68}N$ and $Al_{0.58}Ga_{0.42}N$ films, where '*a*-', '*c*-', and '*m*-' denote that the polarization direction of the incident X ray is parallel to the *a*-, *c*-, and *m*-axes, respectively. The data quality is acceptable. Although it is not so easy in general to analyze Al *K*-edge EXAFS because of low energy of *K*-edge and narrow energy range of EXAFS region, we tried the EXAFS curve fitting analysis.

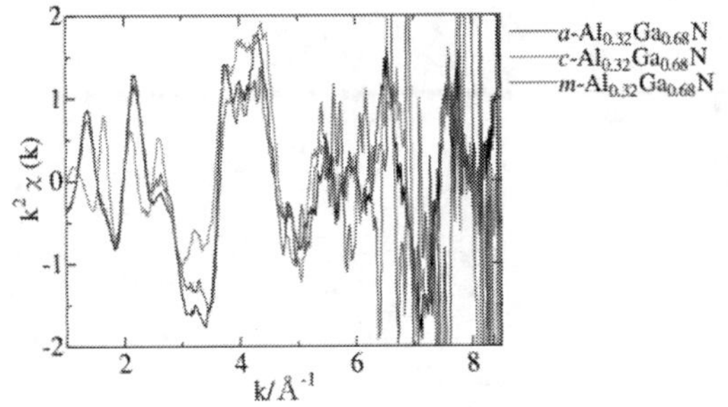

Figure 2(a). EXAFS $k^2\chi(k)$ spectra of Al *K*-edge for the $Al_{0.32}Ga_{0.68}N$ film.

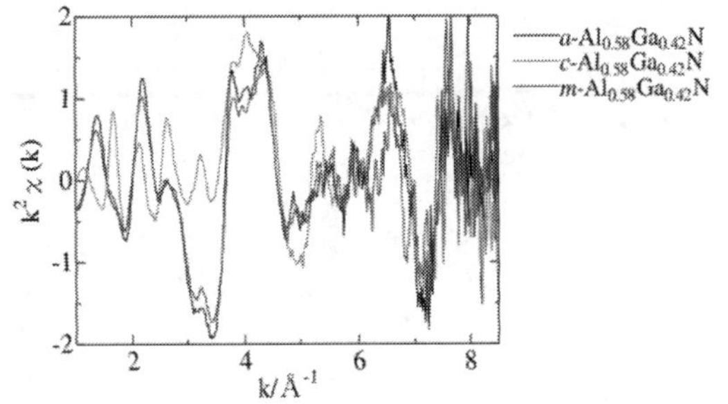

Figure 2(b). EXAFS $k^2\chi(k)$ spectra of Al *K*-edge for the $Al_{0.58}Ga_{0.42}N$ film.

Figures 3 (a) and (b) show Fourier transforms for Al *K*-edge of the *m*-plane $Al_{0.32}Ga_{0.68}N$ and $Al_{0.58}Ga_{0.42}N$ films, respectively. The *k*-range for the Fourier transforms is 2.1-7.75 Å^{-1}. We analyzed polarization depending XAFS data individually. The step-by-step procedure has been proposed for more precise polarization analysis [7], but in this paper we applied simple two-shell fitting for Al-Al and Al-Ga pairs in each direction.

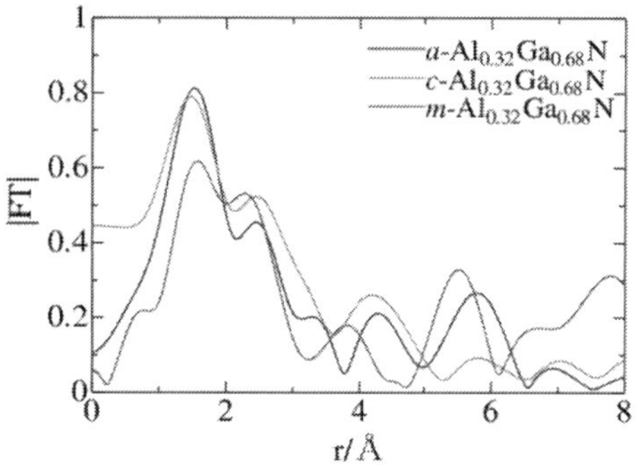

Figure 3(a). EXAFS Fourier transforms for the $Al_{0.32}Ga_{0.68}N$ film.

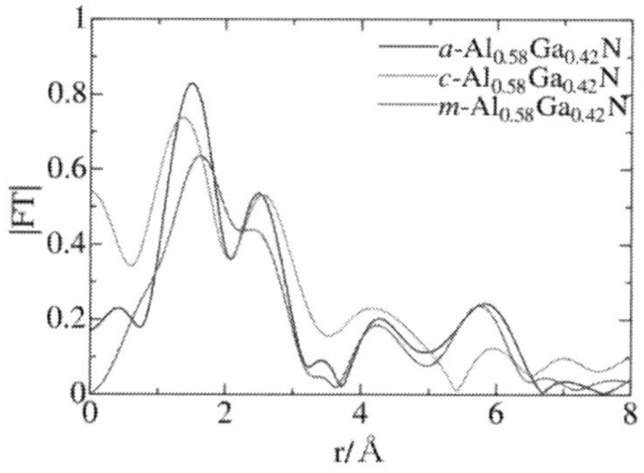

Figure 3(b). EXAFS Fourier transforms for the $Al_{0.58}Ga_{0.42}N$ film.

The structural parameters obtained from the curve fitting analysis are listed in Table 1. We performed *k*-space fitting after extracting the contribution of the second nearest atoms (Al and Ga) by Fourier filtering. At first, we tried to fit the first and the second nearest contribution at once using 3-shell atoms of N, Al, and Ga, but the result was quite unstable because of many parameters. A typical fitted spectrum is shown in Fig.4.

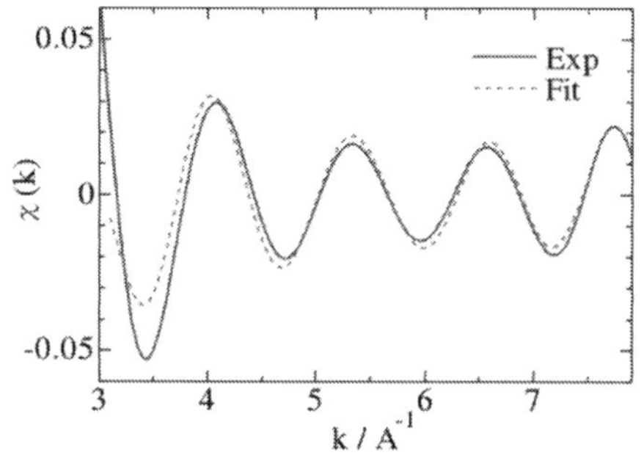

Figure 4. Typical fitted spectrum. The solid line is the experimental a-$Al_{0.58}Ga_{0.42}N$ spectrum and the dashed line is fitted one.

For both the films, the interatomic distances of Al-Al and Al-Ga for the *a*-axis-polarized spectra are about 3.17±0.02 Å, which is close to Ga-Ga distance in GaN. On the other hand, those for the *m*-axis spectra are about 3.10±0.02 Å, which is close to Al-Al distance in AlN. It is suggested that local structures along the *a*-axis of the films are strongly affected by GaN substrates. This result is similar to that for *m*-plane $In_{0.06}Ga_{0.94}N$ XAFS [7], but in that case the 2NN interatomic distance (In-Ga) is shortened by the effect of GaN substrates. The interatomic distances of Al-Ga for the *c*-axis spectra are 3.18±0.02 Å for x=0.32 and 3.11±0.02 Å for x=0.58, respectively. This difference may be caused by the fact that the $Al_{0.32}Ga_{0.68}N$ film is pseudomorphic but that the $Al_{0.58}Ga_{0.42}N$ film is partially or nearly fully relaxed. Yu *et al.* [3] studied Ga *K*-edge XAFS for $Al_xGa_{1-x}N$ (x=0~0.6) films and reported that the interatomic distance of Ga-Al is almost constant to be 3.10 Å and that of Ga-Ga changes linearly from 3.18 Å for GaN to 3.14 Å for $Al_{0.6}Ga_{0.4}N$. The present polarization dependent analysis shows that the interatomic distances of Al-Al and Al-Ga for the *m*-axis spectra are close to their Ga-Al distance (3.10 Å) and those for the *a*-axis spectra are close to their Ga-Ga distance for GaN (3.18 Å). The reason of these discrepancies may be that our samples were grown on *m*-plane GaN substrates, while their ones on *c*-plane sapphire substrates.

Next, we evaluated the proportion of the 2NN Al atoms around Al atoms in the films (N_{Al}/N_{Al+Ga}). Those for the $Al_{0.32}Ga_{0.68}N$ and $Al_{0.58}Ga_{0.42}N$ films are around 0.63±0.05 for each direction, suggesting that Al atoms are almost randomly distributed in the $Al_{0.58}Ga_{0.42}N$ film but are localized in the $Al_{0.32}Ga_{0.68}N$ film within the simple model. Polarization direction dependence of the localization of Al atoms was not observed. Yu *et al.* [3] reported that Ga and Al atoms are randomly distributed in the concentration range of x=0~0.6. Our result shows that Al atoms are localized in the $Al_{0.32}Ga_{0.68}N$ film. Hazu *et al.* have studied spatio-time-resolved cathode luminescence for an *m*-plane $Al_{0.25}Ga_{0.75}N$ film and suggested the localization of Al atoms [8]. The present EXAFS results support their result.

Table 1. Structural parameters for $Al_xGa_{1-x}N$ obtained from EXAFS curve fitting.

	$r_{Al\text{-}Al}$/Å	$r_{Al\text{-}Ga}$/Å	N_{Al}/N_{Al+Ga}
a-$Al_{0.32}Ga_{0.68}N$	3.17	3.17	0.63
c-$Al_{0.32}Ga_{0.68}N$	3.11	3.18	0.67
m-$Al_{0.32}Ga_{0.68}N$	3.09	3.10	0.67
a-$Al_{0.58}Ga_{0.42}N$	3.17	3.18	0.58
c-$Al_{0.58}Ga_{0.42}N$	3.10	3.11	0.67
m-$Al_{0.58}Ga_{0.42}N$	3.11	3.12	0.59

4. Conclusion

The polarization dependent XAFS study revealed the local atomic structures in the *m*-plane $Al_xGa_{1-x}N$ films (x=0.32 and 0.58). The in-plane atomic distances are strongly affected by GaN substrates for both the films. The Al atom localization was observed in the $Al_{0.32}Ga_{0.68}N$ film but Al atoms are randomly distributed in the $Al_{0.58}Ga_{0.42}N$ film.

Acknowledgements

The synchrotron radiation experiments were performed at the Photon Factory in KEK under Proposal 2009G554 and 2011G587.

References

[1] Morkoc F H and Mohammad S N, 1995, Science **267**, 51.
[2] Miyano K E, Woicik J C, Robins L H, Bouldin C E, and Wickenden D K, 1997 Appl. Phys. Lett. **70**, 2108.
[3] Yu K M, Glover C J, Ridgway M C, Wong W S, and Yang W, 1999 Appl. Phys. Lett. **75**, 4097.
[4] Hazu K, Hoshi T, Kagaya M, Onuma T, and Chichibu S F, 2010 J. Appl. Phys. **107**, 033701.
[5] Sakane H, Miyanaga T, and Watanabe I, Matsubayashi N, Ikeda S, and Yokoyama Y, 1993 Jpn. J. Appl. Phys. **32**, 4641.
[6] Ankudinov A L and Rehr J J, 2000 Phys. Rev. B. **62**, 2437.
[7] Miyanaga T, Azuhata T, Mikami S, Nitta K, Onuma T, and Chichibu S F, unpublished.
[8] Hazu K, Kagaya M, Hoshi T, Onuma T, and Chichibu S F, 2011 J. Vac. Sci. Technol. **B 29**, 21208.

Magnetic and local structure analysis for Fe/Cr multilayer films

Takafumi Miyanaga, Koki Takasugi, Tetsuro Ohba, Takahiro Aida, Teiko Okazaki

Department of Advanced Physics, Hirosaki University, Hirosaki, Aomori, 036-8561, Japan

Takafumi Miyanaga, email: takaf@cc.hirosaki-u.ac.jp

Abstract. XMCD of Fe $L_{2,3}$-edge and EXAFS of Fe and Cr K-edge measurements were carried out for Fe/Cr multilayer films which were deposited with several conditions. Magnetic moment of Fe was weakly depending on the film thickness. Relation between the disorder of the local structure and the coercivity was investigated. From EXAFS Debye-Waller analysis, it was found that the thinner films show the smaller Debye-Waller factor, or the more stable local structure. The stable local structure presents large coercivity.

1. Introduction

In 1989, a GMR (giant magnetic resistivity) effect was first observed in metal multilayer films as Fe/Cr/Fe [1]. The thin films of Fe and Cr deposited alternatively on the substrate changes the reluctance as a function of the external magnetic field. It becomes indispensable now to bring about a new field in nanotechnology called as spin electronics (spintronics). To reveal the mechanism of the change of the reluctance, the local structure study is important. Coercivity is the important magnetic parameters for the industrial application of the multilayer magnetic films. The coercivity increases as the crystalline size or film thickness decreases for the intermediate range (100nm-0.1mm) [2]. EXAFS is unique tool to investigate the local structure and magnetic structure for such a multilayer structure, because of its element selectivity. In this paper, we report the XMCD and EXAFS analyses for Fe/Cr multilayer films and discuss the relation between the coercivity and the atomic disorder.

2. Experimental and Data analysis

The Fe/Cr multilayer films were prepared by RF high frequency magnetron sputtering method. The sample thicknesses were controlled by the deposition time. Fe and Cr of 10nm, 25nm, 50nm, 100nm, were deposited on the glass and MgO crystal, as the total film thickness was set to 300nm. The magnetic characterization (saturated magnetization and coercivity) of the Fe/Cr multilayer films were performed by VSM. XMCD spectra were measured at BL7A at KEK. Applied magnetic field was 0.2 T. XAFS was measured on BL9C at KEK by use of fluorescence mode with Lytle detector. XAFS analyses were performed by XANADU code [3] and FEFF8.01 code [4].

3. Results and discussion

3.1. XMCD results for Fe/Cr multilayer films

First we show the XMCD and magnetic measurement results. Figure 1 (a) shows the XMCD spectra for Fe $L_{2,3}$-edge for several multilayer films. They show almost the same spectral structure as the bulk Fe. Figure 1 (b) shows the plot of the magnetic moment of Fe atom obtained from $L_{2,3}$ XMCD by sum rule as a function of saturation magnetization (M_s) for these samples.

The value of M_s increases as the film thickness decreases but the change of the magnetic moment of Fe atom as the film thickness is rather small. This result means that the magnetic moment of Fe in the top layer shows almost constant or weak dependence on the total magnetic moment of film which is strongly dependent on film thickness.

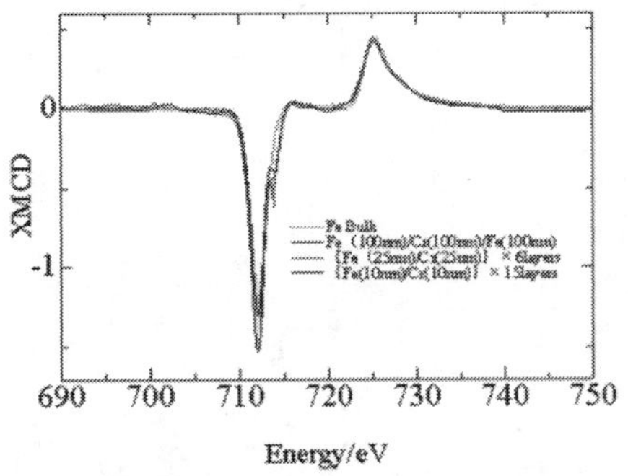

Figure 1(a). XMCD spectra for Fe $L_{2,3}$-edge for Fe, 100nmFe/100nmCr/100nmFe, $[25nmFe/25nmCr]_6$, $10nmFe/10nmCr]_{15}$.

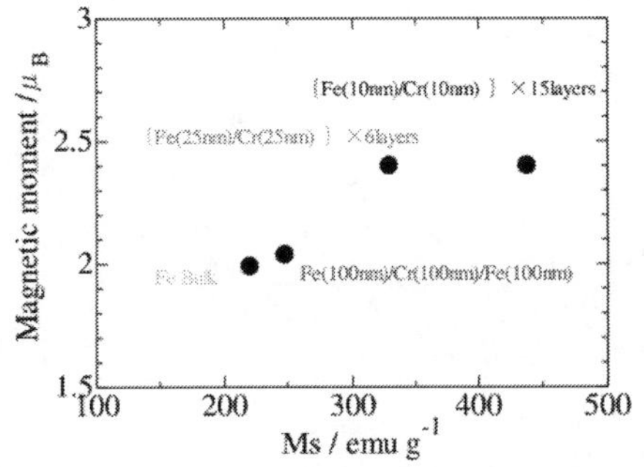

Figure 1(b). Plot of the magnetic moment of Fe as a function of saturation magnetization (M_s) for these samples.

3.2. *EXAFS results for Fe/Cr multilayer films*

Figure 2 shows EXAFS $k^3\chi(k)$ spectra for (a) Fe *K*-edge and (b) Cr *K*-edge for several Fe/Cr multilayer films.

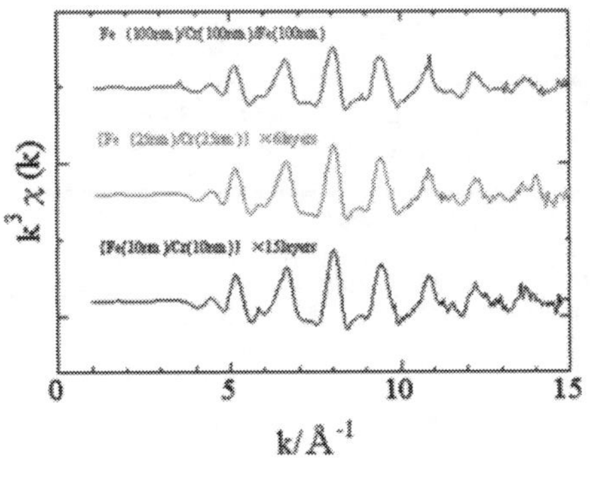

Figure 2(a). EXAFS $k^3\chi(k)$ spectra of Fe *K*-edge for 100nmFe/100nmCr/100nmFe, $[25nmFe/25nmCr]_6$, $10nmFe/10nmCr]_{15}$.

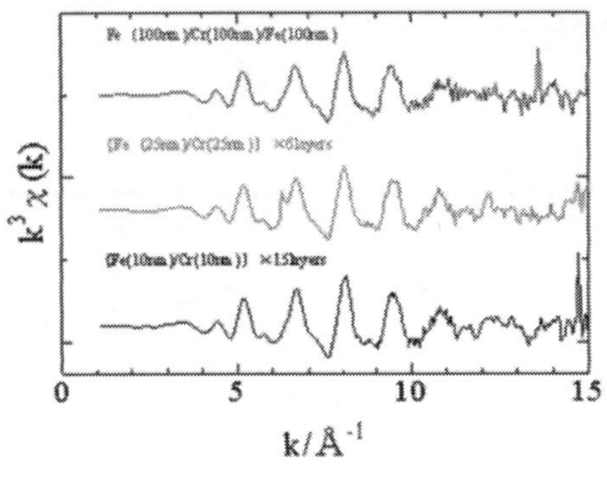

Figure 2(b). EXAFS $k^3\chi(k)$ spectra of Cr *K*-edge for 100nmFe/100nmCr/100nmFe, $[25nmFe/25nmCr]_6$, $10nmFe/10nmCr]_{15}$.

Figure 3 shows the Fourier transforms (FT) for (a)Fe *K*-edge and (b) Cr *K*-edge of several Fe/Cr multilayer films. Both edges showed the bcc structures in Fe and Cr.

Non-linear least square fitting was applied to Fourier filtered data in *k*-space to obtain the structural parameters, interatomic distances and Debye-Waller factors. We applied two-shell fitting for the first nearest peak. The fitting spectra is shown in Fig.4 for 10nmFe/10nmCr]$_{15}$. Generally the coercivity decreases depending on the film thickness or particle size (*T*) as T^{1} in the case that they are longer than coupling length. This study shows the qualitatively that the coercivity decreases as the film thickness.

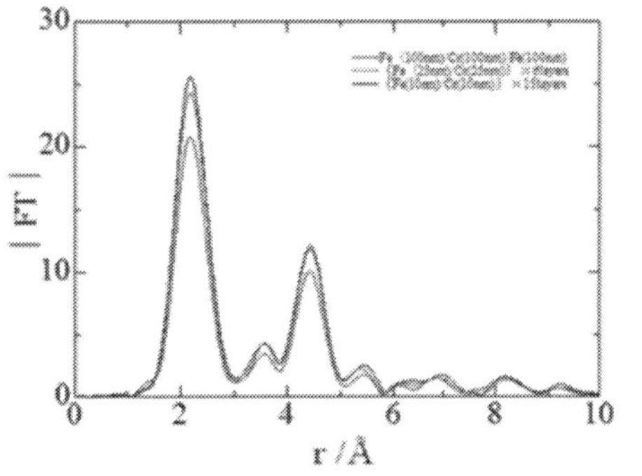

Figure 3(a). EXAFS FT spectra of Fe *K*-edge for 100nmFe/100nmCr/100nmFe, [25nmFe/25nmCr]$_{6}$, 10nmFe/10nmCr]$_{15}$.

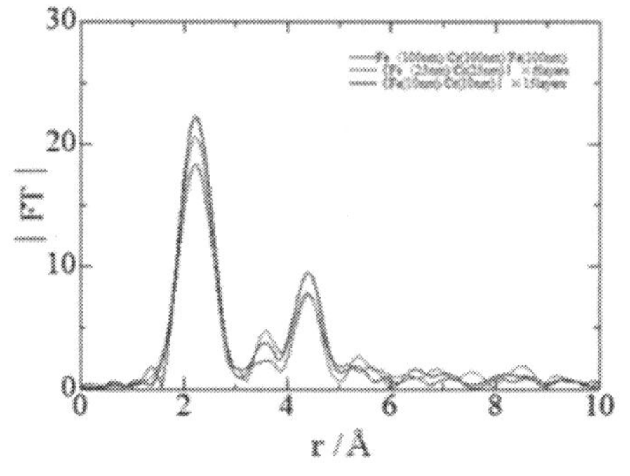

Figure 3(b). EXAFS FT spectra of Cr *K*-edge for 100nmFe/100nmCr/100nmFe, [25nmFe/25nmCr]$_{6}$, 10nmFe/10nmCr]$_{15}$.

Figure 5 shows the plots of the EXAFS Debye-Waller factor for the first nearest neighbor atom as a function of the coercivity of the sample. The numbers in the figure show the film thickness. Except for 50nm and 100nm films on glass, thinner multilayer film shows larger coercivity. It can be seen a strong correlation between the Debye-Waller factor and the coercivity. Thus, the sample which has a small Debye-Waller factor shows the large coercivity. Because the Debye-Waller factor is an index of the disorder of local structure, this result indicates that the multilayer film sample having a stable local structure shows the large coercivity. Furthermore, it is suggested that the thinner multilayer films shows the smaller Debye-Waller factor, or the more stable local structure.

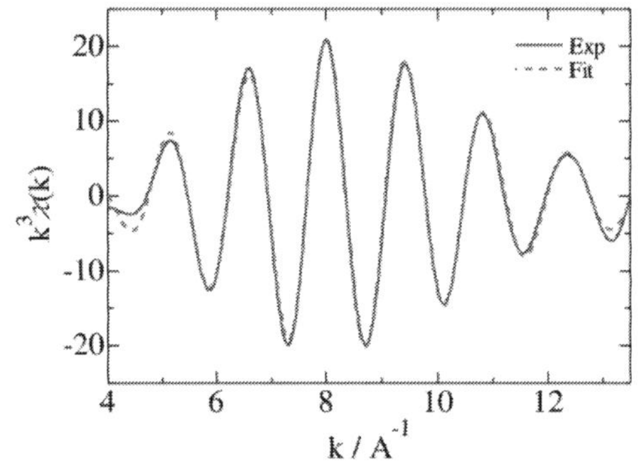

Figure 4. The example of fitting spectra for [10nmFe/10nmCr]$_{15}$ film. Solid line is experimental and dashed line is fitting data.

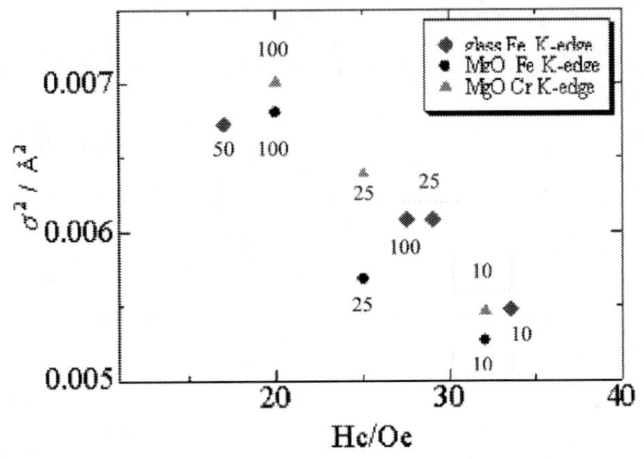

Figure 5. Plots of the EXAFS Debye-Waller factor for the first nearest neighbour atom as a function of the coercivity of the sample. The numbers in the figure are film thickness in nm.

4. Conclusion

XMCD of Fe $L_{2,3}$-edge and EXAFS of Fe and Cr *K*-edge measurements were carried out for Fe/Cr multilayer films. The relation between disorder of the local structure and the coercivity was investigated. It was found that the thinner films show the smaller Debye-Waller factor, or the more stable local structure. The stable local structure shows the large coercivity. The reason is that the thinner film composes the smaller number of domains which relates the magnetic domains and that shows smaller disorder.

Acknowledgements

The synchrotron radiation experiments were performed at the Photon Factory in KEK under Proposal 2007G621, 2009G565.

References

[1] Baibich M N, Broto J M, Fert A, Nguyen Van Dau F, Petroff F, Eitenne P, Creuzet G, Friederich A, and Chazelas J, 1988 *Phys. Rev. Lett.* **61** 2472.
[2] Herzer G, 1989 *I.E.E.E.Trans. Mag., MAG.* **25,** 3327.
[3] Sakane H, Miyanaga T, Watanabe I, Matsubayashi N, Ikeda S, and Yokoyama Y, 1993 J*pn. J. Appl. Phys.* **32**, 4641.
[4] Ankudinov A L and Rehr J J, 2000 *Phys. Rev.* B. **62**, 2437.

In-situ XAFS study of Ag clusters in Ag-type zeolite-A

Atsushi Nakamura, Masataka Narita, Sho Narita, Yushi Suzuki, Takafumi Miyanaga

Department of Advanced Physics, Hirosaki University, Hirosaki, Aomori, 036-8561, Japan

Takafumi Miyanaga: takaf@cc.hirosaki-u.ac.jp

Abstract. We studied the relation between structure of the Ag clusters produced in Ag-type zeolite-A and properties of photo luminescence (PL) by use of in-situ X-ray absorption fine structure (XAFS). Ag clusters are generated in the cavity of Ag-type zeolite-A which is cooled to room temperature after heated at 500°C for 24 hours under vacuum or atmosphere. We investigated the change of the Ag clusters in the zeolite cavity when various gases are introduced: oxygen, nitrogen, water vapor, and their mixture. The Ag clusters were breakdown when H_2O gas was introduced and the fact of the breakdown of the cluster is necessary for appearing the strong PL band. It was suggested that the structural change in the zeolite induced by the formation and breakdown of the Ag clusters is key point of the PL mechanism.

1. Introduction

Zeolites are crystalline aluminosilicates showing unique properties due to the presence of cavities and cages in the structure. Some cations are easily exchange to Na^+ ions in the original Na-type zeolites and they show various unique physical and chemical properties [1]. The Ag-type zeolite-A cooled to room temperature after heated at 500°C for 24 hours under the vacuum or in the atmosphere shows a strong PL band around 2.1 eV [2]. It was known that the intensity of PL band was influenced by conditions of the heating steps and the kinds of atmospheric gases introduced after cooling. However, the detail mechanism of PL still has not been clear. On the other hand, when the Ag-type zeolite is heated, the lattice water molecules are removed and then the Ag^+ ion is reduced to Ag^0, the "Ag clusters" are produced inside the cavity of zeolite [2]. In the cavity of the zeolite, a part of Ag ions are still connected to oxygen atoms at the zeolite lattice, but we call such "Ag aggregation" as "Ag cluster" according to our previous paper [2].

In this paper we investigated the relation between the existence of Ag clusters and the intensity of PL band to reveal the PL mechanisms using X-ray absorption fine structure (XAFS), which has advantages for the local structure analyses in the field of nano scaled sciences. In the result of previous study, it was concluded that the PL of the Ag-type zeolite-A is attributed to the Ag clusters [2]. However the conclusion may be somewhat skeptical because the measurement conditions of PL and XAFS are not completely the same. The strong PL band was appeared even at 24 hours passed after the air was introduced. In order to obtain a further exact result, we carried out in-situ XAFS measurement at the same condition of PL measurements. And further we tried to introduce various gases; oxygen, nitrogen, water vapor, and their mixtures into zeolite after heating to study the origin of the structural change of the Ag clusters.

Journal of Physics: Conference Series **502** (2014) 012033 doi:10.1088/1742-6596/502/1/012033

2. Experimental

2.1. Sample preparations

The fully Ag^+-exchanged zeolite A (hydrated 12Ag-A) powder samples were prepared by immersing the hydrated 12Na-A in 0.1 M $AgNO_3$ solution for 24 hours at 25°C. The solution was stirred every an hour. After careful filtration the 12Ag-A powder was dried under air at RT in the darkroom.

Ag-type Zeolite-A was heated at 500°C in vacuum to produce the Ag clusters [2]. After keeping at 500°C for 24 hours, they were cooled to room temperature, and then, various gases: oxygen, nitrogen, water vapor, their mixtures and air are introduced. In-situ XAFS measurements were performed on each process.

2.2. PL measurements

The PL measurements were performed on a UV-VIS spectrometer (SPM-002; KLV) and a 405 nm violet laser (SU-61-405; audio-technica) light source was used. The chamber for high vacuum (RVX-3) was used in order to PL spectra measurement at various atmospheric conditions. PL measurements were performed 24 hours after the gases were introduced.

2.3. XAFS measurements

X-ray absorption spectra of *K*-edge of Ag (25.5 keV) were measured at NW10A at Photon Factory in KEK with transmission mode. A Si (311) monochrometer was used. The EXAFS oscillation function extracted from the X-ray absorption spectra and Fourier transformed by XANADU code [3]. In order to obtain the structural parameters, the EXAFS function was fitted in *k*-space by a non-linear least-squares method using the theoretical parameters calculated by FEFF 8.10 [4].

3. Results and discussion

First, we discuss the result of the PL measurements. Figure 1 shows the PL curves for various introduction gases. The strongest intensity of PL band around 2.1 eV was observed when the air or the mixture of the water vapor and nitrogen are introduced. It is suggested that the combination of the water vapor and the nitrogen is important for the strong PL band. On the other hand, PL is not observed when the oxygen was introduced. In the present stage the reasons of these phenomena are unclear.

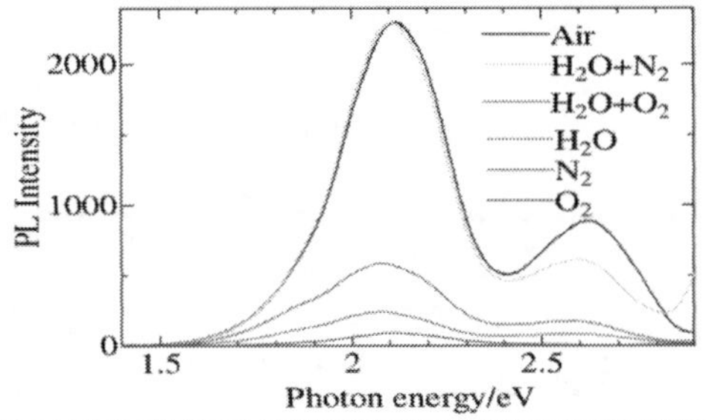

Figure 1. PL curves for various introduced gases; air, O_2, N_2, H_2O, H_2O+O_2, H_2O+N_2.

Figure 2 shows the Ag-*K* EXAFS $k^2\chi(k)$ spectra for the Ag-type zeolite-A at room temperature in atmosphere (black solid line) and that at room temperature in vacuum after heated (red dashed line). It is found that the structure around 3.5 $Å^{-1}$ is especially changed after heating process, which is the characteristic feature by producing the Ag clusters [2]. Figure 3 shows the EXAFS $k^2\chi(k)$ spectra for Ag-type Zeolite-A after following gases were introduced; air (black solid line), H_2O (blue slid line) and O_2 (red dashed line). The structure around 3.5 $Å^{-1}$ is returned to initial state when air or H_2O+N_2

were introduced. This reversible change in $k^2\chi(k)$ spectra indicates the breakdown of the Ag clusters. On the other hand, the $k^2\chi(k)$ spectra was not changed when O_2 gas was introduced.

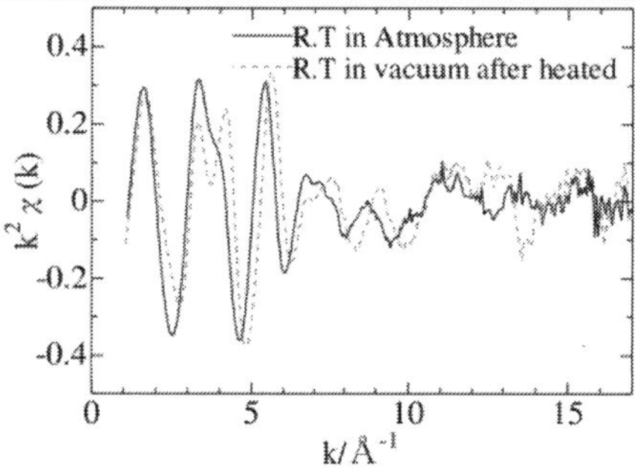

Figure 2. The EXAFS $k^2\chi(k)$ spectra for Ag type Zeolite-A at room temperature in atmosphere and that at room temperature in vacuum after heated.

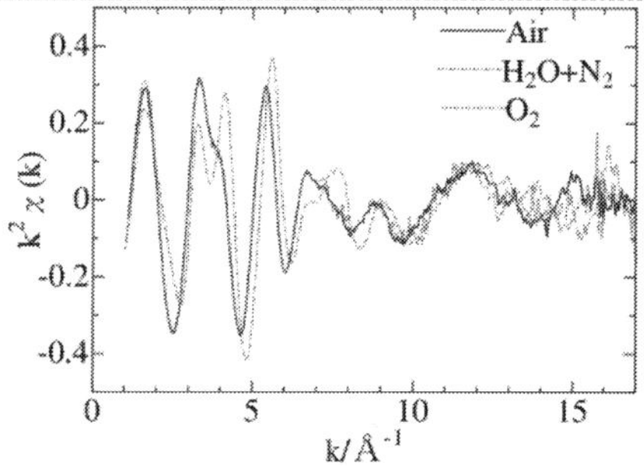

Figure 3. The EXAFS $k^2\chi(k)$ spectra for Ag type Zeolite-A after following gases were introduced; Air, H_2O+N_2, O_2.

Figure 4 shows the Ag *K*-edge Fourier transform spectrum for Ag zeolite-4A measured at room temperature in atmosphere and that after following gases were introduced: Air, O_2, H_2O, N_2, and H_2+O_2. The peak around 1.7 Å corresponds to the first nearest neighbor O1 atom, the second peak around 2.7 Å corresponds to the second nearest neighbor O2 and Ag atom. We carried out the 3-shell fitting and obtained the structural parameters; the atomic distance *r* and the coordination number *N* for each shell.

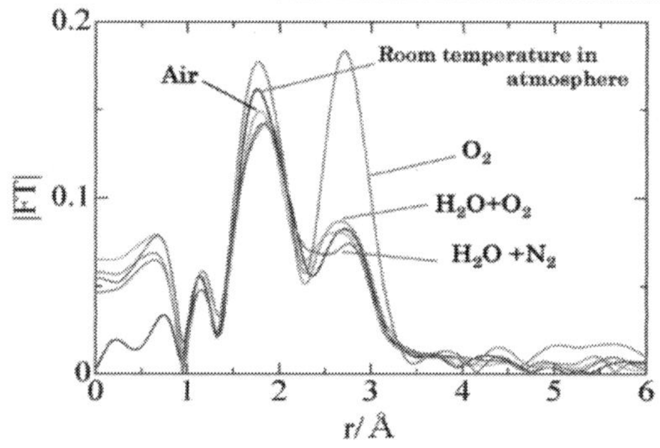

Figure 4. Fourier transform spectrum of Ag *K*-edge; unheated in atmosphere, introduction Air, O_2, H_2O+O_2, H_2O+N_2.

Table 1. The structural parameters; *r* and *N* for the process before the gases are introduced.

	r_{O1}(Å)	N_{O1}	r_{O2}(Å)	N_{O2}	r_{Ag}(Å)	N_{Ag}
Atmospheric pressure at RT	2.38	5.51	2.85	3.92	2.88	1.12
Vacuum at RT	2.30	3.70	2.82	3.89	2.83	1.98
Vacuum at 500°C	2.29	3.40	2.74	3.49	2.80	2.58
Vacuum at 500°C after 24h heated	2.29	3.10	2.74	3.20	2.80	2.55
Vacuum at RT after 24h heated	2.29	3.14	2.81	3.61	2.81	1.93

In Table 1, the results of the curve fittings at the processes before introducing the gases are summarized. When the sample was evacuated N_{Ag} increases from 1.12 to 1.98, which indicates producing Ag clusters. Then it was observed that the Ag clusters were grown by heating in vacuum (N_{Ag} ~2.58), and they remain after cooling to room temperature (N_{Ag} ~1.93).

Now we discuss the structure of the Ag clusters after the various gases are introduced. The structural parameters after the gases are introduced are summarized in Table 2. As shown in Fig. 1, the strong PL band is observed when the air is introduced. After introducing the air, the Ag clusters were disappeared. The values of r_{Ag} and N_{Ag} are nearly same as those at the room temperature in atmospheric pressure. The air leads to breakdown of the Ag clusters, but it takes long time enough. Judging from the results of the PL and the XAFS, the existence of the Ag clusters is not necessary to make the strong PL band. The PL band is not observed when the oxygen is introduced. The r_{O1} and r_{Ag} have not been changed after the oxygen gas is introduced, and the N_{Ag} is still large (N_{Ag} ~2.35). This indicates that the Ag clusters exist after oxygen gas is introduced. Finally we discuss the results after the mixed gases of (1) water and oxygen, and (2) water and nitrogen were introduced. Feeble PL was observed when two gases of (1) were introduced. On the other hand, strong PL band was observed when two gases of (2) were introduced. The results of curve fitting indicate the breakdown of the Ag clusters as same as the case of the introduction of the air for both cases (1) and (2).

Table 2. The structural parameters; r and N for the process after the various gases are introduced.

	r_{O1}(Å)	N_{O1}	r_{O2}(Å)	N_{O2}	r_{Ag}(Å)	N_{Ag}
Air	2.38	4.57	2.86	2.96	2.86	1.40
Oxygen	2.30	4.14	2.85	4.85	2.84	2.35
H_2O+O_2	2.39	6.73	2.86	5.19	2.88	0.94
H_2O+N_2	2.39	5.75	2.86	4.23	2.87	1.23

4. Conclusion

We studied the in-situ XAFS for the structural changes of Ag cluster after introduction of various gases. It was found that the Ag clusters were broken when the water vapor is introduced. It is suggested that the deformation of the Ag clusters after the introduction of the water vapor as primary and the nitrogen gas as secondary plays an important role of generation of the strong PL band. Ag clusters might be not direct species for the PL. We predict as one possibility that the structural change in the zeolite cavity induced by the formation and breakdown of Ag clusters is key point of the PL mechanism.

Acknowledgements

The synchrotron radiation experiments were performed at the Photon Factory in KEK under Proposals 2009G564 and 2011G586.

References

[1] Kim Y and Seff K, 1978 *J. Chem. Soc.* **100:22**, 6986.
[2] Hoshino H, Sannohe Y, Suzuki Y, Azuhata T, Miyanaga T, Yaginuma K, Itoh M, Shigeno T, Osawa Y and Kimura Y, 2008 *J. Phys. Soc. Jpn* **77**, 064712 .
[3] Sakane H, Miyanaga T, Watanabe I, Matsubayashi N, Ikeda S and Yokoyama Y, 1993 *Jpn. J. Appl. Phys.* **32**, 4641 .
[4] Ankudinov A L, Rehr J J, 1998 *Phys. Rev.* **B 58**, 7565.

Bond cleavages of adenosine 5'-triphosphate induced by monochromatic soft X-rays

K. Fujii, A. Narita, and A. Yokoya

Advanced Science Research Center, Japan Atomic Energy Agency, 2-4 Shirakata-Shirane, Tokai, Naka, Ibaraki, JAPAN

fujii.kentaro@jaea.go.jp

Abstract. To investigate which type of bond is likely to be cleaved by soft X-ray exposure to an adenosine 5'-triphosphate (ATP), we observed spectral changes in X-ray absorption near edge structure (XANES) around nitrogen and oxygen *K*-edge of an ATP film by soft X-ray irradiation. Experiments were performed at a synchrotron soft X-ray beamline at SPring-8, Japan. The XANES spectra around the nitrogen and oxygen *K*-edge slightly varied by exposure to 560 eV soft X-rays. These changes are originated from the cleavage of C-N bonds between a sugar and a nucleobase site and of C-O, P-O or O-H bond of sugar and phosphate site. From the comparison between the change in XANES intensity of σ^* peak at nitrogen and that at oxygen *K*-edges, it is inferred that the C-O, P-O or O-H bond of sugar and phosphate is much efficiently cleaved than the C-N of N-glycoside bond by the exposure of 560 eV soft X-ray to ATP film.

1. Introduction

Nucleotide molecules composed of nucleobases, pentose (sugar) rings, and phosphate groups, have been known to play a variety of roles in a biological system. Among those nucleotides, adenosine 5'-triphosphate (ATP) is one of the most universal molecules distributed in a living system. When a living cell is exposed to ionizing radiation from soil or space, ATP molecules presumably sustain damage by ionization or excitation, or reaction with OH radicals resulting from water radiolysis. Although it is inferred that the damage of ATP causes an abnormal condition in the irradiated cell, there have been very few studies of the radiation effect on ATP. Using an HPLC product analysis method, Watanabe *et al.* [1] analyzed ATP decomposition induced by soft X-ray (2.15 keV) and ^{60}Co γ-ray irradiation. They found that adenosine 5'-diphosphate (ADP), and adenosine 5'-monophosphate (AMP) residue are produced by the soft X-ray irradiation, in addition to production of adenine residue which was also observed by γ-irradiation. Although this study has clarified the final products as a result of complex chemical reaction of ATP and water molecules, little has been known yet about the molecular changes in ATP just after photoabsorption before the chemical reaction with water included in an atmosphere. Measurement of X-ray Absorption Near Edge Structures (XANES) is one of the powerful tools to give us information on electric states of a target molecule, particularly unoccupied states [2], or modification of the electric states by irradiation before the chemical reaction of the product species with solvent molecules [3, 4]. In the present paper, we report spectrum changes in XANES of ATP induced by soft X-ray irradiation.

2. Materials and methods

A thin film of adenosine 5'-triphosphate (ATP) (Sigma Aldrich Co. Ltd., Tokyo, Japan) was used as sample in the experiments. An aqueous solution (pH ~3.0) of ATP (225 µL, 4.0 mg/mL) prepared in distilled water was spread onto a Si(100) plate (30 mm × 30 mm; KN Platz Co. Ltd., Aira-gun, Kagoshima, Japan) and dried at room temperature under normal atmospheric conditions to obtain the thin film samples. A part of the sample substrate (10 mm × 10 mm) was clipped from the prepared Si substrate and attached to the sample holder. All experiments were performed at a soft X-ray beamline BL23SU at SPring-8, Japan [5]. The energy resolution of the beamline E/ΔE was ~10^4 (~40 meV at 400 eV and ~50 meV at 530 eV, respectively). The photon energy was calibrated by measuring 1s to π* resonant peak of nitrogen and oxygen gas absorption spectra. Reproducibility of scanning of photon energy is less than 5 meV. The detail of the experimental procedures were described in the previous report [3]. We chose 560 eV soft X-rays, which correspond to energy levels above phosphorus *L*-edge, carbon *K*-edge, nitrogen *K*-edge, and oxygen *K*-edge, for irradiation. The XANES spectra were obtained before and immediately after irradiation. The sample position was never changed during an experiment to ensure that XANES measurements were performed at exactly the same position for the irradiation. A distinct ATP sample was prepared for 10 min exposure from that for the 5 min exposure. To eliminate the variability of the XANES intensity originated from a sample variation, the estimation of the spectral changes induced by the exposure was performed using the spectra obtained before and after the irradiations at the exact same sample position.

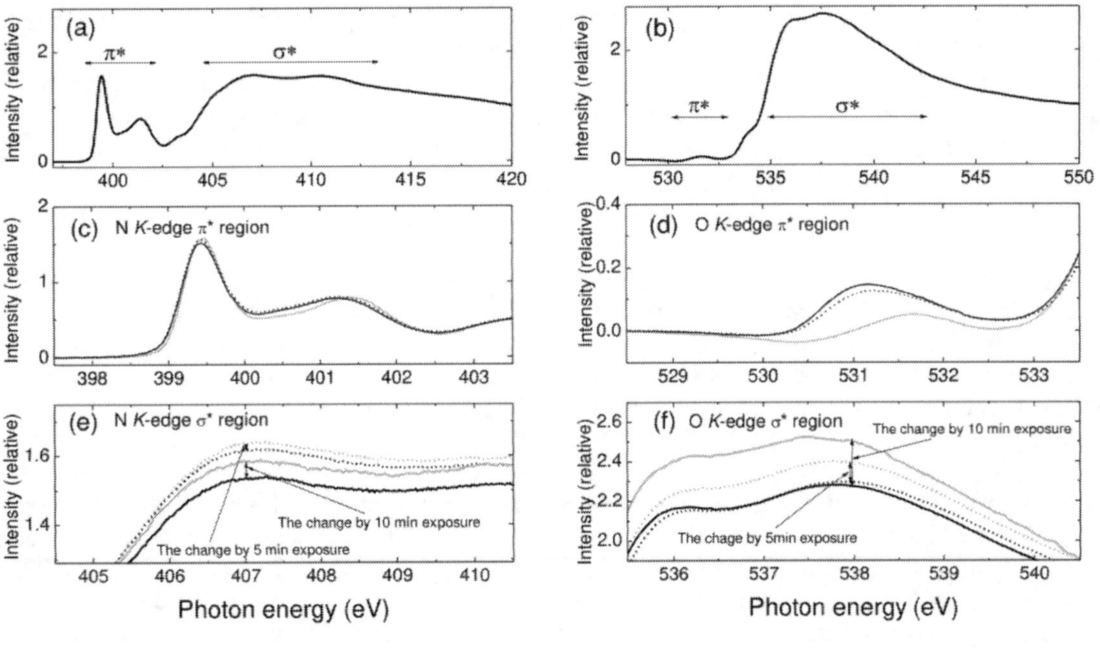

Figure 1. XANES spectra around nitrogen (a), and oxygen *K*-edge (b) of an ATP film. The extended spectra of π* ((c) and (d)) and σ* ((e) and (f)) regions of nitrogen and oxygen *K*-edge obtained after exposure to 5 min (dashed line) and 10 min (solid line) of 560 eV soft X-rays, respectively. The spectra obtained before the irradiation showed gray line. The averaged spectrum of six different samples was used as a control spectrum of π* regions ((c) and (d)) to reduce the statistical variability.

3. Results

Figure 1(a) shows XANES spectrum of the ATP film around nitrogen K-edge region. Figure 1(c) and 1(e) show the XANES spectra obtained by following exposure to 560 eV soft X-rays for 5 min (3.3 × 10^{15} photons) and 10 min (6.6 × 10^{15} photons) around π* and σ* region, respectively. Baselines were subtracted from the obtained XANES spectra. The spectral intensities were normalized above nitrogen

Journal of Physics: Conference Series **502** (2014) 012034 doi:10.1088/1742-6596/502/1/012034

K-ionization potential, 420 eV, around which no resonant peak structure was observed. The XANES spectra can be divided into two regions, which are called π^* transitions located around 400 eV and σ^* in 405-413 eV, respectively. The fine structures of XANES spectrum around 400 eV were similar with that of gas phase adenine [6]. Recently Kelly *et al.* [7] observed XANES spectra of ATP aqueous solution and thin films around nitrogen *K*-edge region. They attributed the fine peak structures around 399.4 eV and 401.4 eV to originating from the transitions of nitrogen 1s to π^*(N-C=N) and $\pi^*(NH_2)$, respectively. A broad peak located in 407-412 eV attributes the transition of nitrogen 1s to σ^* resonance due to C-N bonds [8-10]. We observed a slight reduction and broadening in the π^* peak at 399.4 eV and slightly low-energy peak shift of the π^* peak from 401.4 eV to 401.2 eV around the nitrogen *K*-edge caused by the irradiation. The energy of π^* peak of XANES spectrum of nitrogen *K*-edge did not show any remarkable differences between the spectra obtained after exposure of 5 min and that of 10 min. Figure 1(b) shows the XANES spectrum of the ATP film around oxygen *K*-edge region. Figure 1(d) and 1(f) show the XANES spectra obtained by following exposure to 560 eV soft X-rays for 5 min and 10 min around π^* and σ^* region, respectively. Baselines were subtracted from the obtained XANES spectra. The spectral intensities were normalized above oxygen *K*-ionization potential, 550 eV, around which no resonant peak structure was observed. A weak π^* peak which at 531.5 eV of the XANES spectrum of non-irradiated and irradiated samples is thought to be attributed to the transition from O 1s to π^*(P=O) because this peak was also reported in that of sodium phosphate [10]. The intensity of this peak increased with the exposure of soft X-rays. A new π^* peak around 531 eV appeared by the irradiation. A broad peak located in 536-543 eV is attributed to the transition of oxygen 1s to σ^* resonance due to oxygen containing bonds [9-11]. The values at the peaks (407 eV and 538 eV) were used for the estimation of the changes of the σ^* intensity induced by the exposure for nitrogen and oxygen K-edge XANES, respectively. The changes caused by 5 min and 10 min exposure were 1 % and 3 % for nitrogen σ^*, and 4 % and 10 % for oxygen σ^*, respectively.

4. Discussion

A new π^* peak around 531 eV of the spectrum was not observed in the spectrum of non-irradiated sample. For several organic molecules, Yang *et al.* [12] suggested to attribute the peak appeared in this energy to originating from ketone (RCOR') or aldehyde (RCHO) which contains unsaturated bonds. There are three kinds of C-O bonds in ATP molecule as follows: C-O-C in pentose ring, C-O-H at hydroxyl group, and P-O-C located between sugar and phosphate. One of the most probable candidates to produce a C=O bond is the cleavage C-O-C bond in the pentose ring as known to be a reaction scheme of DNA single strand break (SSB) [13]. These results suggest that soft X-ray irradiation causes a C-O bond cleavage in the pentose ring to produce a SSB. The second notable point is the changes of π^* peaks in nitrogen *K*-edge. From the similarity of the peak energies of π^* excitations with those of adenine or AMP [10], the observed changes of the spectrum might show that an adenine or an AMP molecule was released from an ATP by the irradiation. The AMP release is induced by a bond cleavage of P-O bond between a sugar and a phosphate; an adenine release, a bond cleavage of C-N of N-glycoside bond. The result of the increment of intensity of the π^* peak located at 531.5 eV suggests that some new P=O bonds are presumably produced by the exposure of soft X-rays, but the yield is not much high. Decrements of σ^* intensity of oxygen and that of nitrogen *K*-edge seem to be appeared as the result of bond cleavages of C-O or P-O, and C-N, respectively. The cleavage of O-H bond was also expected to be produced by the soft X-ray exposure to deoxyribose molecule in the recent study [14]. The irradiation induces about three times higher decrease of the intensity at oxygen σ^* energy than that at nitrogen σ^* energy. These evidences also suggests that C-O, P-O, or O-H bond is much efficiently cleaved than C-N of N-glycoside bond by the exposure of 560 eV soft X-rays.

Conclusions

Spectral changes of XANES were observed for an ATP film after an exposure of 560 eV monochromatic soft X-rays. The C-O bond cleavage of the pentose ring in the sugar site is produced by the irradiation, and then the C=O containing substitutes such as aldehyde or ketone are newly formed in the sugar site. Adenine or AMP release is expected from the spectral changes of the nitrogen *K*-edge XANES spectrum by the irradiation of 560 eV soft X-rays. It is expected that the bonds which contain oxygen atom more efficiently cleaved than that of C-N bond by the exposure of 560 eV soft X-rays to ATP molecules.

References

[1] Watanabe R, Ishikawa M, Kobayashi K and Takakura K 1992 Damage to Adenosine-triphosphate Induced by Monochromatic X Rays around The K Shell Absorption Edge of Phosphorus *AAPM Symposium Series* No. 8

[2] Stöhr J ed 1992 *NEXAFS Spectroscopy* (Berlin: Springer)

[3] Fujii K, Fukuda Y and Yokoya A 2012 Observation of cleavage in DNA and nucleotides following oxygen K-shell ionization by measuring X-ray absorption near edge structure *International Journal Radiation Biology* 88 888-94

[4] Fujii K and Yokoya A 2009 Spectral change in X-ray absorption near edge structure of DNA thin films irradiated with monochromatic soft X-rays *Radiation Physics and Chemistry* 78 1188-91

[5] Saitoh Y, *et al.* 2012 Performance upgrade in the JAEA actinide science beamline BL23SU at SPring-8 with a new twin-helical undulator *Journal of Synchrotron Radiation* 19 388-93

[6] MacNaughton J, Moewes A and Kurmaev E Z 2005 Electronic structure of the nucleobases *The Journal of Physical Chemistry B* 109 7749-57

[7] Kelly D N, Schwartz C P, Uejio J S, Duffin A M, England A H and Saykally R J 2010 Communication: Near edge x-ray absorption fine structure spectroscopy of aqueous adenosine triphosphate at the carbon and nitrogen K-edges *Journal of Chemical Physics* 133 101103

[8] Kirtley S M, Mullins O C, Chen J, Vanelp J, George S J, Chen C T, Ohalloran T and Cramer S P 1992 Nitrogen Chemical-Structure in DNA and Related Molecules by X-Ray Absorption-Spectroscopy *Biochimica Et Biophysica Acta* 1132 249-54

[9] Kummer K, Vyalikh D V, Gavrila G, Preobrajenski A B, Kick A, Bönsch M, Mertig M and Molodtsov S L 2010 Electronic Structure of Genomic DNA: A Photoemission and X-ray Absorption Study *The Journal of Physical Chemistry B* 114 9645-52

[10] Samuel N T, Lee C Y, Gamble L J, Fischer D A and Castner D G 2006 NEXAFS characterization of DNA components and molecular-orientation of surface-bound DNA oligomers *Journal of Electron Spectroscopy and Related Phenomena* 152 134-42

[11] Fujii K, Akamatsu K, Muramatsu Y and Yokoya A 2003 X-ray absorption near edge structure of DNA bases around oxygen and nitrogen K-edge *Nuclear Instruments & Methods in Physics Research Section B-Beam Interactions with Materials and Atoms* 199 249-54

[12] Yang L, Ågren H, Carravetta V and Pettersson L G M 1996 Static exchange and quantum defect analysis of X-ray absorption spectra of carbonyl compounds *Physica Scripta* 54 614-24

[13] von Sonntag C 1987, *The Chemical Basis of Radiation Biology* (London: Taylor and Francis)

[14] Ha D T, Huels M A, Huttula M, Urpelainen S and Kukk E 2011 Experimental and ab initio study of the photofragmentation of DNA and RNA sugars *Physical Review A* 84 033419

Acknowledgement

The authors thank Mr. Y. Fukuda, Dr. Y. Takeda and Dr. Y. Saitoh of SPring-8, JAEA for tuning the soft X-ray beamline and for their technical support during the XANES measurement. The authors gratefully acknowledge Prof. K. Tanaka in JASRI, and Dr. I. Shimoyama in JAEA, for helpful discussions regarding X-ray absorption spectrum. This work was supported by JSPS KAKENHI Grant Numbers 24651057 and 24310074, and was also supported by REIMEI project of JAEA. The synchrotron radiation experiments were performed at the BL23SU beamline of SPring-8 with the approval of the JASRI (Proposal No. 2009B3813, 2009B3814).

Development of surface sensitive DXAFS measurement method by applying Kramers-Kronig relations to total reflection spectra

Hitoshi Abe, Yasuhiro Niwa, Hiroaki Nitani and Masaharu Nomura
Institute of Materials Structure Science, High Energy Accelerator Research Organization, 1-1 Oho, Tsukuba, Ibaraki 305-0801, Japan

E-mail: hitoshi.abe@kek.jp

Abstract. Total re ection XAFS spectra are obtained combining with dispersive XAFS (DXAFS) con guration. A total re ection DXAFS spectrum of 30 nm Co layer on Si substrate was measured in 4 ms. Kramers-Kronig analysis was applied to extract XAFS signal from total re ection spectra. Spectra obtained by this method are comparable to usual XAFS spectra in terms of signal-to-noise ratios and peak positions in Fourier transformed EXAFS functions. The development of this Kramers-Kronig relations based DXAFS method is useful to study the variation of the surface state with the time resolution of millisecond.

1. Introduction

X-ray absorption ne structure (XAFS) is a useful technique to study local structures and electronic states of elements. XAFS is one of the important methods to examine catalysts especially under the conditions of reactions, and time-resolved XAFS techniques are widely used to observe them in a real time manner. Quick XAFS (QXAFS)[1, 2] and dispersive XAFS (DXAFS)[3, 4] are the two methods usually used to perform time-resolved XAFS measurements.

QXAFS techniques are widely used at XAFS experimental stations in the world. It is relatively easy to install the system to perform QXAFS because the same hardwares as usual step scanning XAFS can be used. Dispersive XAFS (DXAFS) is a powerful technique to obtain time-resolved XAFS spectra, though the setup and detection would be di cult relative to QXAFS. Since a DXAFS spectrum is recorded at once without scanning the x-ray energies by moving monochromators, it has a better capability of time resolution and is suitable to study time evolutions of chemical reactions and transient species. Chemical reactions and other fast processes have been studied by DXAFS methods[5, 6, 7].

We would like to expand this DXAFS technique to the eld of surface science, for example, in order to study fundamental mechanisms of surface chemical reactions. We have tried to obtain surface sensitive spectra by detecting signals of total reflection. A total reflection spectrum is so surface sensitive that bulk signals should be eliminated. The total reflection measurements are key to obtain information on surface. Absorption spectra are required to be analyzed by XAFS method, and reflection spectra are transformed to "absorption" spectra by Kramers-Kronig relations.

This kind of method was formerly named ReflEXAFS by G. Martens and P. Rabe[8, 9], and some groups had tried to develop the method[10, 11, 12, 13]. R. Frahm's group reported lm deposit processes observed by the method[14, 15].

We have combined this method with the DXAFS technique. The DXAFS system provides a better time resolution to the total reflection XAFS method, which has a surface sensitivity.

2. Experiment

Usual step scanning XAFS measurements were performed at BL-9C of the Photon Factory at the Institute of Materials Structure Science, High Energy Accelerator Research Organization. A 30 nm Co lm deposited on a Si wafer was used as a test sample, and a Co foil was used as a standard. Reflection spectra were measured by using stages for grazing incident measurements. The sample was aligned to realize a total reflection condition. The total reflection critical angle of Co is $\sim$7 mrad in this energy range. The incident angle was set to be $\sim$3 mrad, which is well below the critical angle. Total reflection XAFS experiments combined with a DXAFS system were carried out at NW2A of PF-AR. All the measurements were performed under the air atmosphere.

The total reflection spectra were converted to "absorption" spectra via Kramers-Kronig relations. Kramers-Kronig relations are de ned with integration of unlimited regions. It is, however, impractical to integral unlimited regions, and the integrations in our analyses of Kramers-Kronig relations were performed in the limited regions as wide as possible for the data until the oscillations can be almost null. It is helpful and usual to adopt this approximation in practice. All the spectra were processed through a standard XAFS analysis procedure.

3. Results and Discussion

The validity of the Kramers-Kronig relations based XAFS method was con rmed by measuring usual step scanning XAFS spectra at BL-9C. A standard Co foil was measured by a standard transmission mode. "XAFS" spectra via Kramers-Kronig relations were obtained from total reflection spectra of the Co(30 nm)/Si lm. The reflection spectrum of Co(30 nm)/Si is shown in Fig. 1(a) with that of Co foil.

EXAFS oscillations $k^2\chi(k)$ are shown in Fig. 1(b). The Kramers-Kronig analysis was applied to the total reflection spectrum. The smooth EXAFS oscillations were observed at least until 15 A ^{1}. The structures of the oscillations are roughly similar to each other, and they contain structural information on metallic cobalt. There's a characteristic di erence between the two around 2–4 A ^{1}. The di erence would reflect some speci c features at anomalous dispersion at the absorption edge. The features at the edge are generally complicated, and discussions on the point are beyond this paper. We would like to focus on the similar oscillations above $\sim$5 A ^{1}. The same frequency corresponds to the same local structure, such as atomic distances.

The EXAFS oscillations were processed through Fourier transformation as shown in Fig. 1(c). The characteristic peak structures of the two[16] resemble well, which are the rst peak at 2.2 A ^{1} and the double peak structure around 4–5 A ^{1}. Though further analyses and careful treatments are required, the whole structures are in good agreement each other.

We had combined the Kramers-Kronig relations based XAFS method with a DXAFS system. The rst trial spectra of reflection DXAFS are shown in Fig. 2. They were measured in 4 ms every 10 ms. The speci c features are seen in the spectra, and the good reproducibility was observed. We have developed a Kramers-Kronig relations based DXAFS method. It is con rmed that the reflection DXAFS measurements are useful to obtain surface sensitive XAFS spectra via Kramers-Kronig relations. This surface sensitive technique can be applied to study the eld of surface science with a good time resolution.

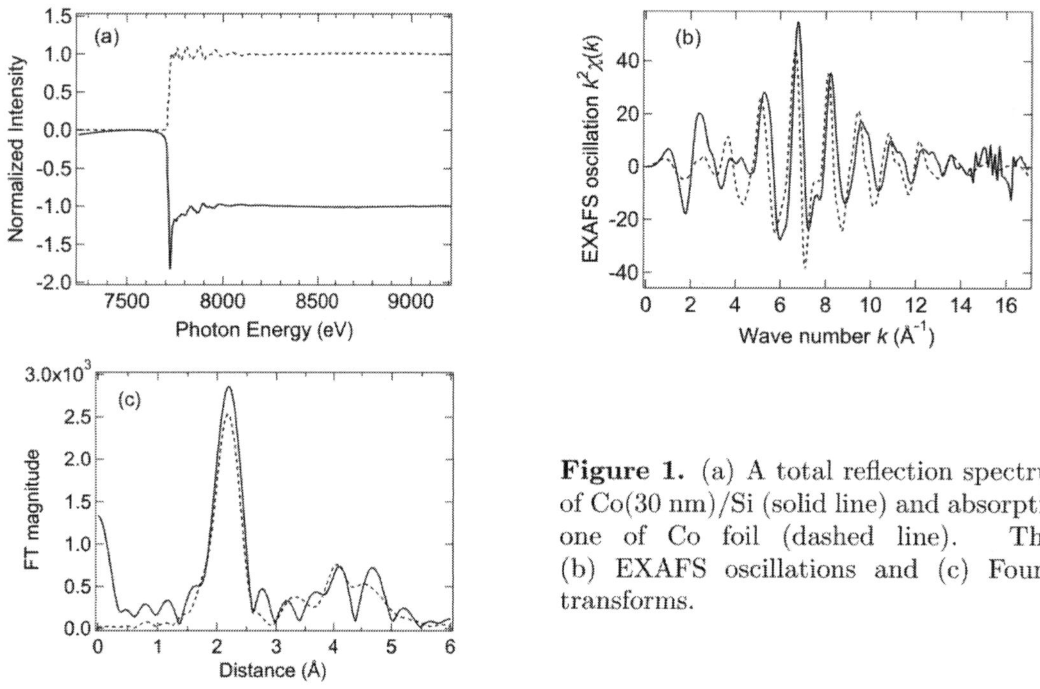

Figure 1. (a) A total reflection spectrum of Co(30 nm)/Si (solid line) and absorption one of Co foil (dashed line). Their (b) EXAFS oscillations and (c) Fourier transforms.

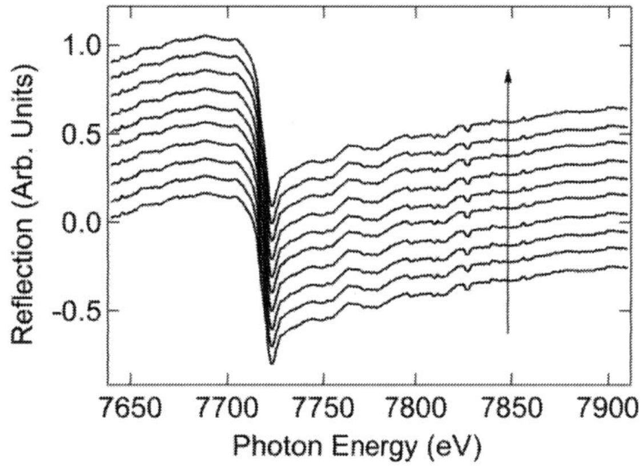

Figure 2. Reflection spectra of Co(30 nm)/Si measured by DXAFS method. Each spectrum was recorded for 4 ms.

4. Summary

Total reflection XAFS spectra are obtained combining with DXAFS con guration in order to study the variation of the surface state. Total reflection DXAFS spectra of Co(30 nm)/Si were measured in 4 ms every 10 ms. Kramers-Kronig analysis was applied to extract XAFS signal from total reflection spectra. XAFS spectra obtained from the analysis were comparable to usual XAFS spectra in terms of signal-to-noise ratios and peak positions in Fourier transformed EXAFS functions. The obtained spectra showed a possibility to apply this Kramers-Kronig relations based DXAFS method to the eld of surface science with the time resolution of millisecond.

Acknowledgments
Prof. H. Kondoh is appreciated for fruitful discussions. We are grateful to Mr. T. Nakayama for his support. This work was performed under the approval of PF-PAC No. 2011G116. The authors are grateful for the nancial support Grant-in-Aid for Young Scientists (B) (No. 24710102).

References

[1] Frahm R 1989 *Physica B* **158** 342–343
[2] Frahm R, Weigelt J, Meyer G and Materlik G 1995 *Rev. Sci. Instrum.* **66** 1677–1680
[3] Matsushita T and Phizackerley R P 1981 *Jpn. J. Appl. Phys.* **20** 2223–2228
[4] Flank A M, Fontaine A, Jucha A, Lemonnier M and Williams C 1982 *J. Phys. Lett.* **43** 315–319
[5] Iwasawa Y 2003 *J. Catal.* **216** 165–177
[6] Tada M, Murata S, Asakoka T, Hiroshima K, Okumura K, Tanida H, Uruga T, Nakanishi H, Matsumoto S, Inada Y, Nomura M and Iwasawa Y 2007 *Angew. Chem. int. Ed.* **46** 4310–4315
[7] Nagai Y, Dohmae K, Teramura K, Tanaka T, Guilera G, Kato K, Nomura M, Shinjoh H and Matsumoto S 2009 *Catalysis Today* **145** 279–287
[8] Martens G and Rabe P 1980 *J.Phys. C: Solid St. Phys.* **13** L913–L918
[9] Martens G and Rabe P 1981 *J.Phys. C: Solid St. Phys.* **14** 1523–1534
[10] Poumellec B, RCortes, FLagnel and CTourillon' 1989 *Physica B* **158** 282–283
[11] Pizzini S, Roberts K J, Greaves G N, Harris N, Moore P, Pantos E and Oldman R J 1989 *Rev. Sci. Instrum.* **60** 2525–2528
[12] Borthen P and Strehblow H H 1995 *J. Phys.: Condens. Matter* **7** 3779–3787
[13] L-Hecht D and Frahm R 2000 *Physica B* **283** 108–113
[14] L-Hecht D and Frahm R 2001 *J. Synchrotron Rad.* **8** 478–480
[15] L-Hecht D and Frahm R 2005 *Thin Solid Films* **493** 67–76
[16] Miyawaki J, Matsumura D, Nojima A, Yokoyama T and Ohta T 2007 *Surf. Sci.* **601** 95–113

Journal of Physics: Conference Series **502** (2014) 012036 doi:10.1088/1742-6596/502/1/012036

Molecular electronic states in charge transfer complex studied by x-ray absorption spectroscopy

Y Takahashi[1,2], H Nakao[1,2], R Kumai[1,2,3], S Ishibashi[2,4], S Horiuchi[2,3], M Kohyama[5], K Kobayashi[1], Y Yamasaki[1], J Okamoto[1], T Sudayama[1], Y Murakami[1] and Y Tokura[6,7]

[1]Condensed Matter Research Center and Photon Factory, Institute of Materials Structure Science, High Energy Accelerator Research Organization (KEK), Tsukuba, 305-0801, Japan
[2]CREST, Japan Science and Technology Agency (JST), Tokyo, 102-0076, Japan
[3]Flexible Electronics Research Center (FLEC), National Institute of Advanced Industrial Science and Technology (AIST), Tsukuba, 305-8562, Japan
[4]Research Institute for Computational Sciences (RICS), AIST, Tsukuba, 305-8568, Japan
[5]Research Institute for Ubiquitous Energy Devices (UBIQEN), AIST, Osaka 563-8577, Japan
[6]Department of Applied Physics, University of Tokyo, Tokyo, 113-8656, Japan
[7]RIKEN Center for Emergent Matter Science (CEMS), Wako, 351-0198, Japan

E-mail: yukari.takahashi@kek.jp

Abstract. The electronic states of tetrathiafulvalene (TTF: TTF = $C_6H_4S_4$) molecule in organic ferroelectric TTF-*p*-bromanil (TTF-BA: BA = $C_6Br_4O_2$) and TTF crystals have been investigated by x-ray absorption spectroscopy (XAS) measurement at S *K*-edge. We elucidated that the peak structure at 2470.5 eV directly reflects the existence of hole in the highest occupied molecular orbital (HOMO) state of the TTF molecule in TTF-BA; that is consistent with the ionic TTF molecule (TTF^+). The XAS of TTF-BA was evaluated on the basis of first-principles calculations, and the calculated spectra reproduce well the shape of experimental spectrum and the peak energy of the HOMO state.

1. Introduction

In the study of charge transfer (CT) complexes, many multi-functional materials have been discovered to date. Organic CT complex, $D^{+\rho}A^{-\rho}$ (ionicity ρ is the degree of CT), composed of electron donor (D) and acceptor (A) molecules forms a π-π stack with alternating $D^{+\rho}$ and $A^{-\rho}$ molecules. This is characterized by quasi-one-dimensional electronic structure where electrons are easy to move in the stacking axis direction with an overlap of π orbitals. Among them, tetrathiafulvalene-*p*-chloranil (TTF-CA: TTF = $C_6H_4S_4$, CA = $C_6Cl_4O_2$) [1] and TTF-*p*-bromanil (TTF-BA: BA = $C_6Br_4O_2$) [2] have been attracting attention for a long time as CT complexes with ferroelectricity. The crystal structure of TTF-BA is shown in Fig. 1(b), in which two one-dimensional column structures composed of TTF and BA molecules align parallel to the *a* and *b* axes. TTF-BA undergoes a magnetic phase transition at around 50 K and it has been revealed that this is a spin-Peierls transition forming spin-singlet state of dimerized D^+A^- pair along the stacking direction. The ferroelectricity emerges due to the dimerization below the transition temperature [2, 3], as TTF and BA molecules remain almost ionic ($\rho \sim 1$). In contrast with TTF-BA, TTF-CA shows ferroelectric transition at T_{NI} = 81 K associated with neutral-ionic transition in which the ρ jumps from ~0.3 for the quasi-neutral phase to ~0.6 for the quasi-ionic phase [4, 5]. It is very interesting that the different type of ferroelectricity emerges in TTF-CA and TTF-BA, although these are composed of common TTF molecules and CA/BA molecules having similar molecular formula and chemical properties (Fig. 1(a)). We have been focusing on the electronic states of molecules in order to discuss the mechanism of the ferroelectricity in CT complexes.

X-ray absorption spectroscopy (XAS) technique is possible to detect the unoccupied orbitals electronic state of the target element, and the spectrum reflect the valence and coordination structures. In the case of organic system, the electronic state of the unoccupied molecular orbital can be observable. Only the electronic state of the TTF molecule can be selectively proved in TTF-BA by XAS at S *K*-edge, because only the TTF molecule contains sulfur ions and the BA molecule does not. The highest occupied molecular orbital (HOMO) state of TTF molecule is π orbital as shown in Fig. 1(a). On the basis of the selection rule for the dipole transition, a strong absorption reflecting the HOMO state is expected to be observed, when the electric vector, ε, is parallel to the extended direction of the π orbital. Namely the absorption reflecting the HOMO state can be observed, when the ε aligns in the *ab* plane. Utilizing this XAS technique, we have attempted to examine the electronic state of TTF molecule in TTF-BA.

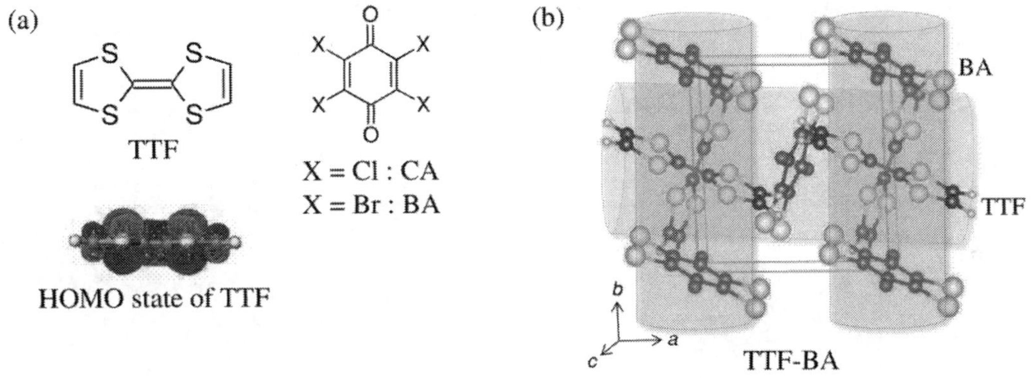

Figure 1. (a) Molecular structures of TTF, CA and BA. Molecular orbital of HOMO of TTF. (b) Crystal structure of TTF-BA.

2. Experimental and theoretical details

High-quality single crystals of TTF-BA and TTF were used in this experiment. The paraelectric phase of TTF-BA has a triclinic structure with the lattice parameters, a = 8.3904(5) Å, b = 8.6996(6) Å, c = 11.6305(7) Å, α = 93.628(5) °, β = 97.169(5) °, γ = 91.929(5) ° at T = 293 K [6]. XAS experiments were performed at S *K*-edge using a high vacuum two-circle diffractometer at BL-11B in Photon Factory, KEK [7]. The incident beam was monochromatized by Si(111) double crystals and the energy resolution $E / \Delta E$ was about 5000 [7]. The incident x-ray was linearly polarized in horizontal plane. He flow cryostat was used for cooling of the sample. The XAS was measured by fluorescence yield mode. Silicon drift detector was used for x-ray detector with high energy resolution to reduce the background of fluorescence from other elements. The azimuthal angle dependence of XAS was measured by rotation of theta (θ) angle as shown in Fig. 3(c). The electric vector of x-ray (polarization), ε, is parallel to the b axis at θ = 90 °. X-ray absorption spectra were simulated theoretically by using the computational code QMAS (Quantum MAterials Simulator) [8], which utilizes the projector augmented-wave (PAW) method [9] and the planewave basis set. The PAW potential for the absorbing atom was generated introducing half a core-hole and the standard supercell approach was adopted [10, 11]. As for the exchange-correlation energy for electrons, the generalized gradient approximation (GGA) [12] was used.

3. Results and Discussion

The XAS of TTF-BA were measured near S *K*-edge at $T = 130$ K (paraelectric phase) for $\boldsymbol{\varepsilon} // b$, since the absorption reflecting the HOMO state is observable at the $\boldsymbol{\varepsilon} // b$ as discussed in the introduction. As shown in Fig. 2(a), the clear peak structure was observed at 2470.5 eV (E_S: dotted line) near the absorption edge energy. XAS of simple TTF crystal with an ideal neutral phase was also measured at $\boldsymbol{\varepsilon} // b$. The *b* axis is the molecular stacking direction in the TTF crystal. The spectrum reflects the valence of the TTF molecule and the local structure around the molecule. Hence the spectrum of simple TTF is different from that of TTF-BA as shown in Fig. 2(c). Especially the peak structure at E_S was not observed, although the $\boldsymbol{\varepsilon} // b$ is a possible setting to detect the absorption reflecting the HOMO state. It corresponds that the TTF molecule of simple TTF is neutral but the TTF molecule of TTF-BA is ionic. In other words, the HOMO in TTF-BA has a hole, whereas the HOMO in simple TTF is fully occupied. Finally, we concluded that the peak at E_S reflects the HOMO state of TTF molecule, and the peak intensity is expected to be proportional to the valence of the TTF molecule.

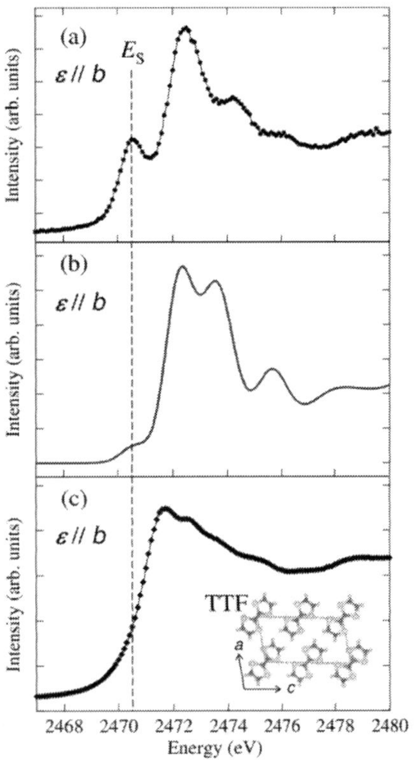

Figure 2. (a) Experimental result and (b) theoretical calculation of XAS near S *K*-edge of TTF-BA. The energy of theoretical calculation is calibrated by the experimental result. (c) XAS near S *K*-edge of simple TTF.

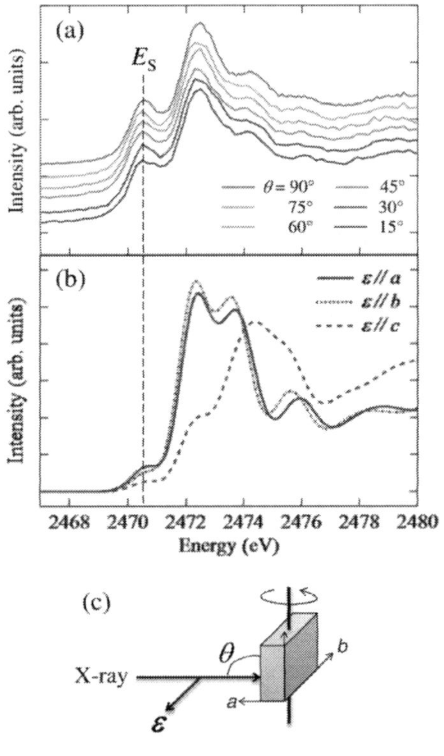

Figure 3. (a) Azimuthal angle dependence of XAS near S *K*-edge of TTF-BA, and (b) the theoretical calculations. (c) Schematic view of the experimental configuration. θ is defined as the angle between incident x-ray and the *b* axis of the TTF-BA.

Figure 2(b) shows the theoretical calculation of XAS near S *K*-edge of TTF-BA ($\boldsymbol{\varepsilon}$ // *b*). The calculated XAS reproduces the peak structures at E_S and above 2472.5 eV. The shape of the spectrum near the edge energy shows a good agreement with the experimental spectrum, although the intensity ratio of these peaks has a difference between the experiment and calculation. Theoretical calculation also makes clear that the peak at E_S reflects the HOMO state of TTF molecule.

Next, the azimuthal angle dependence of XAS near S *K*-edge of TTF-BA was measured as shown in Fig. 3(a). The spectra show very small azimuthal angle dependence. This reflects that the molecular stacking structures along the *a* and *b* axes are similar to each other, as shown in Fig. 1(b). The observation of the E_S peak for any azimuthal angle is consistent with that the absorption reflecting the HOMO state can be observable when the electric vector $\boldsymbol{\varepsilon}$ aligns in the *ab* plane.

The azimuthal angle dependence of XAS near S *K*-edge was also calculated theoretically as shown in Fig. 3(b). The energy spectrum at $\boldsymbol{\varepsilon}$ // *a* is almost the same as that of $\boldsymbol{\varepsilon}$ // *b*, it is good agreement with the experimental spectra. Moreover, the peak intensities at E_S when $\boldsymbol{\varepsilon}$ // *a* and $\boldsymbol{\varepsilon}$ // *b* are much stronger than that at $\boldsymbol{\varepsilon}$ // *c*. This is consistent with that the π orbital of the HOMO state of TTF molecule aligns in the *ab* plane. Utilizing the XAS technique, we succeeded in detecting the electronic state of HOMO of TTF molecule in TTF-BA.

4. Conclusion

We have examined the electronic state of TTF molecule in TTF-BA in comparison with the neutral TTF crystal utilizing XAS measurements near S *K*-edge. It was found that the peak at E_S reflects the HOMO state of TTF in TTF-BA. The experimental results were supported by theoretical calculation with good agreement. This XAS measurement near the S *K*-edge energy became a first step to investigate the electronic state of molecules in organic compounds.

Acknowledgments

This study has been carried out under approval of the Photon Factory Program Advisory Committee (Proposal Nos. 2009S2-008 and 2012S2-005). This work was partly supported by JSPS Grant-in-Aid for Scientific Research (Nos. 21224008 and 24224009) and (B) (No. 25286090) from the Ministry of Education, Culture, Sports, Science, and Technology, Japan, by the Japan Society for the Promotion of Science (JSPS) through "Funding Program for World-Leading Innovative R&D on Science and Technology (FIRST Program)" initiated by the Council for Science and Technology Policy (CSTP), and by CREST, Japan Science and Technology Agency (JST).

References

[1] Cointe Le M, Lemee-Cailleau H M, Cailleau H, Toudic B, Toupet L, Heger G, Moussa F, Schweiss P, Kraft H K and Karl N 1995 *Phys. Rev. B* **51** 3374
[2] Kagawa F, Horiuchi S, Tokunaga M, Fujioka J and Tokura Y 2010 *Nature Phys.* **6** 169
[3] Bewick A S and Soos G Z 2006 *J. Phys. Chem. B* **110** 18748-18757
[4] Girlando A, Marzola F, Pecile C and Torrance B J 1983 *J. Chem. Phys.* **79** 1075
[5] Horiuchi S, Okimoto Y, Kumai R and Tokura Y 2000 *J. Phys. Soc. Jpn.* **69** 1302
[6] Girlando A and Pecile C 1985 *Solid State Commun.* **54** 753
[7] Kitajima Y 1996 *J. Elec. Spec. Relat. Phenom.* **80** 405
[8] http://www.qmas.jp.
[9] Blöchl P 1994 *Phys. Rev. B* **50** 17953
[10] Tailleufumier M, Cabaret D, Flank A and Mauri F 2002 *Phys. Rev. B* **66** 195107
[11] Iannuzzi M and Hutter J 2007 *Phys. Chem. Chem. Phys.* **9** 1599
[12] Perdew P J, Burke K and Ernzerhof M 1996 *Phys. Rev. Lett.* **77** 3865

Journal of Physics: Conference Series **502** (2014) 012037 doi:10.1088/1742-6596/502/1/012037

Buried Interface between N-alkane Thin Film and Monolayer Graphene Studed by Depth-Dependent C K-NEXAFS

Osamu Endo

Department of Organic and Polymer Materials Chemistry, Faculty of Engineering, Tokyo University of Agriculture and Technology, Koganei, Tokyo 184-8588, Japan

E-mail: oendo@cc.tuat.ac.jp

Masashi Nakamura

Department of Applied Chemistry and Biotechnology, Faculty of Engineering, Chiba University, Inage-ku, Chiba 263-8522, Japan

Kenta Amemiya

Photon Factory, IMSS, KEK, Tsukuba, Ibaraki 305-0801, Japan

Abstract. Polarization-dependent X-ray absorption fine structure spectroscopy at the carbon near K-edge (C K-NEXAFS) is developed for a layer-by-layer analysis, by measuring the distribution of electrons emitted by X-ray absorption as a function of emission angles. The C K-NEXAFS spectra for multilayer n-$C_{12}H_{26}$ grown on a Pt(111) surface modified with monolayer graphene are separated on the basis of the depth of the layers. The spectral signal of n-$C_{12}H_{26}$ molecules in the first layer, which is in contact with the graphene substrate, is detected in the separated spectra corresponding to the lower layer in addition to the spectra of the monolayer graphene. The energy and polarization dependence of σ^*_{CH}/R (R: Rydberg state) resonances indicate that the orientation of the carbon plane of flat-lying chains with respect to the graphene sheet in the first layer is perpendicular (edge-on) below the multilayer, and is parallel (flat-on) in the monolayer.

1. Introduction

In the field of materials science, organic and carbon-based thin films are increasingly attracting attention for the development of flexible nanodevices. The electronic properties of these nanodevices are controlled via the interface between the film and substrate; therefore, non-destructive observation of the interface, which is buried after film growth, is critically important. Polarization-dependent X-ray absorption fine structure spectroscopy at the carbon near K-edge (C K-NEXAFS) is a powerful technique for studying the molecular orientation at surfaces and interfaces [1]. In this study, the first layer of multilayer n-$C_{12}H_{26}$, which was buried and in contact with a graphene substrate grown on a Pt(111) surface, was observed by depth-analysis C K-NEXAFS method that we developed recently [2, 3]. The orientation of the carbon plane with respect to the substrate was then determined.

Journal of Physics: Conference Series **502** (2014) 012037 doi:10.1088/1742-6596/502/1/012037

2. Experimental

C K-NEXAFS measurements were conducted at the soft X-ray beamline BL-7A at the Photon Factory, Institute of Materials Structure Science, High Energy Accelerator Research Organization (KEK-PF), Japan. A mechanically polished single crystal of Pt(111) (MaTecK) was cleaned under an ultrahigh vacuum by annealing at an O_2 pressure of 2×10^{-7} Torr at 700 K, followed by vacuum annealing at 1100 K. After dosing with 18 L of n-$C_{12}H_{26}$ at a pressure of 1×10^{-7} Torr for 180 s at room temperature, the Pt(111) substrate was heated at 1000 K for 1 h under the same pressure to obtain monolayer graphene (MG). n-$C_{12}H_{26}$ was then dosed to the graphene-covered Pt(111) surface in increments of 1 L at temperatures below 270 K, pressure of 1×10^{-7} Torr, and period of 10 s (for each increment). C K-NEXAFS spectra were measured at temperatures between 120 and 300 K by the partial electron yield method, using a microchannel plate (MCP) at a retarding voltage of 200 V. The fluorescence screen image obtained at the back was monitored using a CCD camera. Figure 1 shows a schematic illustration of the detector setup. The incidence angle with respect to the surface normal was set to 0° (normal incidence, NI) and 75° (grazing incidence, GI), to predominantly observe the orbitals directed in and out of plane, respectively. Details of the depth-analysis method are described elsewhere [3].

3. Results and Discussion

Figure 2(a) shows the C K-NEXAFS spectra for the MG grown on the Pt(111) surface. The π^*_{CC} resonance at 285.5 eV and σ^*_{CC} resonance at 292 eV predominantly observed in the GI and NI spectrum, respectively, are characteristic of a flat-lying graphene sheet [4–6]. Figure 2(b) shows the C K-NEXAFS spectra of multilayer n-$C_{12}H_{26}$ grown on the MG with a 5 L dosage at 120 K. The σ^*_{CH}/R resonances of the n-$C_{12}H_{26}$ layers (R: Rydberg state) are enhanced in both the NI and GI spectra at around 288 eV. The spectra were separated for two layers on the basis of their depth by fitting the experimental decay curves of photoelectron intensity, which were obtained by utilizing channel-dependent intensity as shown in the bottom panel of Fig. 1.

Figure 3 shows the separated C K-NEXAFS spectra for the NI (left side panel) and GI (right side panel) conditions. The spectra corresponding to the lower layer are mostly for the MG with the π^*_{CC} and σ^*_{CC} resonances in the GI and NI spectra, respectively, as shown in Fig. 2(a). In addition, the σ^*_{CH}/R resonance of the buried first layer of n-$C_{12}H_{26}$ is observed in the lower-layer spectra. The spectra corresponding to the rest of the n-$C_{12}H_{26}$ layers are observed as upper-layer spectra. The σ^*_{CC} resonance observed at around 292 eV for the upper-layer NI spectrum is consistent with the flat-lying chains in the upper layer, because the transition dipole moment of this resonance is parallel to the chain axis [7, 8]. The chains are considered to be flat-lying as well as present in the first layer because there is no signature of the σ^*_{CC} resonance in the lower-layer GI spectrum, which would indicate standing chains.

Figure 4(a) shows the energy region for the σ^*_{CH}/R resonance of the separated C K-NEXAFS spectra for multilayer n-$C_{12}H_{26}$. In the upper-layer spectra, peaks at 287.6 and 288.2 eV are observed with a shoulder at 287.2 eV for the GI and NI conditions, respectively. A pair of bands at around 287.6 and 288.2 eV is typical for alkyl chains and is assigned to the $\sigma^*_{CH}/R_{\parallel}$ and $\sigma^*_{CH}/R_{\perp}$ resonance, for which the transition dipole moment is directed parallel and perpendicular to the carbon plane, respectively (see inset for the corresponding orbital distribution) [8–14]. Because the $\sigma^*_{CH}/R_{\parallel}$ resonance is comparatively enhanced in the GI spectrum and the $\sigma^*_{CH}/R_{\perp}$ resonance is predominantly observed in the NI spectrum, the carbon plane of n-$C_{12}H_{26}$ molecules in the upper layer is oriented relatively perpendicular to the graphene sheet. In a condensed phase, the σ^*_{CH}/R resonance exhibits a high energy (blue) shift owing to the orthogonality condition of the final-state orbitals with the valence orbitals of the neighboring molecules [8, 10, 11, 13, 14]. The shoulder at 287.2 eV is attributed to the *less*-blue-shifted $\sigma^*_{CH}/R_{\parallel}$ resonance, which appears as a result of decreased interactions with the neighboring molecules [15]. This resonance is attributed to the top-layer molecules of the

Journal of Physics: Conference Series **502** (2014) 012037 doi:10.1088/1742-6596/502/1/012037

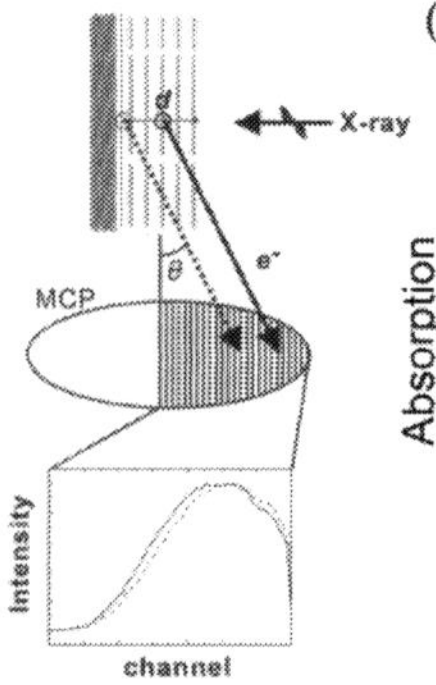

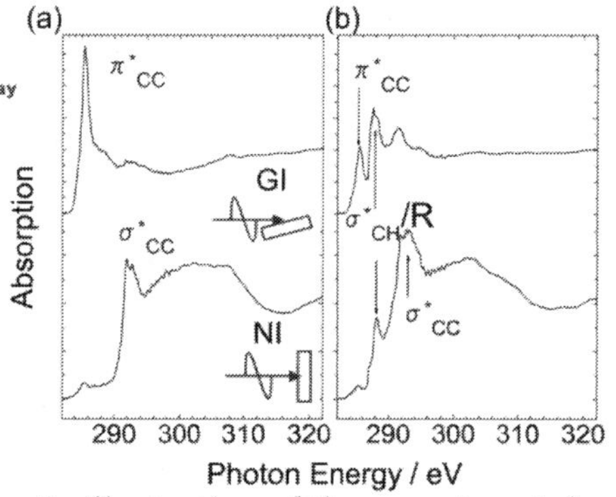

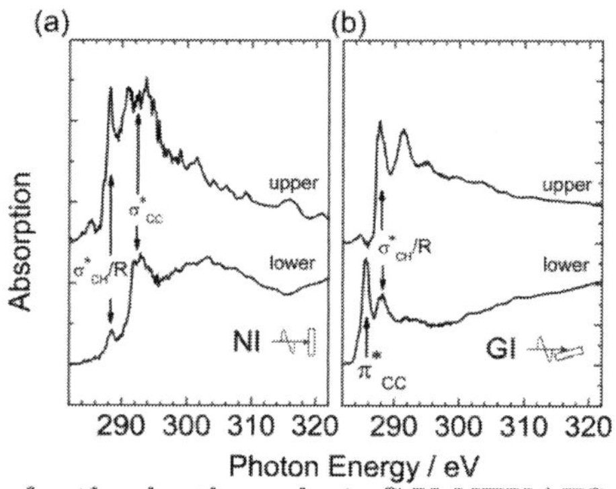

Figure 1. Schematic illustration of the experimental setup for the depth-analysis C K-NEXAFS method.
Figure 2. C K-NEXAFS spectra of (a) MG grown on Pt(111) surface and (b) multilayer n-$C_{12}H_{26}$ on MG covering Pt(111) surface.
Figure 3. Separated C K-NEXAFS spectra of multilayer n-$C_{12}H_{26}$ on MG covering Pt(111) surface.

multilayer n-$C_{12}H_{26}$, which are not surrounded by any other molecules above them. The shoulder at 287.2 eV possibly includes the contribution of the σ^*_{CH}/R_{CH} resonance instead of the $\sigma^*_{CH}/R_{\|}$ resonance. The former resonance corresponds to the top-layer molecules for which the carbon plane is oriented parallel to the substrate; the corresponding orbitals for these molecules are modified by symmetry lowering to be directed along the CH bond and observed in both the NI and GI conditions [15].

In the NI spectrum corresponding to the first layer included in the lower layer, the shoulder at 287.2 eV is reduced and the peak attributed to the $\sigma^*_{CH}/R_{\perp}$ resonance is located at a slightly higher energy of 288.5 eV. Double peaks are observed at 287.6 and 288.2 eV in the GI spectrum, and are assigned to the $\sigma^*_{CH}/R_{\|}$ resonance. Accordingly, the orientation of the carbon plane of the n-$C_{12}H_{26}$ molecules in the first layer is perpendicular to the substrate (edge-on orientation). The reduced shoulder intensity is consistent with the fact that the molecules in the lower layer is surrounded by molecules in the layers above. The $\sigma^*_{CH}/R_{\perp}$ resonance observed at a higher energy in the NI spectrum reflects reduction in the intermolecular distance in the first layer, which results from the periodicity of the graphene lattice [15]. The double peaks of the $\sigma^*_{CH}/R_{\|}$ resonance observed in the GI spectrum are attributed to the difference in the environments of the upper and lower carbon atoms in the molecules with an edge-on orientation. Because the molecule-graphene distance is smaller than the molecule-molecule distance, the carbon atoms in the lower methylene group directed toward the graphene substrate exhibit a higher energy resonance at 288.2 eV, and the carbon atoms in the upper methylene group directed toward the upper layer exhibit a lower energy resonance at 287.6 eV.

Figure 4(b) shows the σ^*_{CH}/R resonance region of the C K-NEXAFS spectra for monolayer n-$C_{12}H_{26}$ obtained with a 1 L dosage at 270 K and 240 K. The chain axis is flat-lying because the σ^*_{CC} resonance is predominantly observed in the NI spectra (not shown). At 270 K, the n-$C_{12}H_{26}$ molecules adopt the parallel (flat-on) orientation, because both the NI and GI spectra exhibit the σ^*_{CH}/R_{CH} resonance at 287.2 eV. On the other hand, the $\sigma^*_{CH}/R_{\perp}$ resonance appears at around 288.2 eV in the NI spectrum at 240 K. Further, the shoulder height at 287.2 eV is increased in the GI spectrum, which is ascribable to the contribution of the $\sigma^*_{CH}/R_{\|}$ resonance. Therefore, the molecules that adopt the edge-on orientation are included in the monolayer at 240 K. The σ^*_{CH}/R_{CH} resonance in the NI spectrum, however, represents that the molecules with the flat-on orientation are also included in the monolayer at 240 K. The orientation change upon cooling is related to the phase transition of the monolayer [15,16]. Figure 4(c) shows a model for molecular orientation of the monolayer n-$C_{12}H_{26}$ at 270 and 240 K, and for multilayer n-$C_{12}H_{26}$ at 120 K. The n-$C_{12}H_{26}$ molecules in the monolayer adopt the flat-on orientation at 270 K, and some molecules change their orientation to edge-on at 240 K. Almost all the molecules at the buried interface between the graphene and multilayer adopt the edge-on orientation. The molecules in the upper layer of the multilayer adopt a slightly tilted orientation. The orientation change below the multilayer is caused by either the temperature or multilayer formation. It should be noted here that multilayer n-$C_{12}H_{26}$ cannot be obtained above the desorption temperature (224 K for the HOPG surface [17], and the desorption temperature of n-alkane for graphene on the Pt(111) surface is nearly equal to this value [18]).

In summary, the orientation of the carbon plane of flat-lying n-$C_{12}H_{26}$ molecules in the buried interface between n-$C_{12}H_{26}$ multilayer and MG covering a Pt(111) surface was determined by depth-analysis C K-NEXAFS spectroscopy. It was found that the molecules in the buried interface adopt an edge-on orientation, after changing their flat-on orientation in the monolayer.

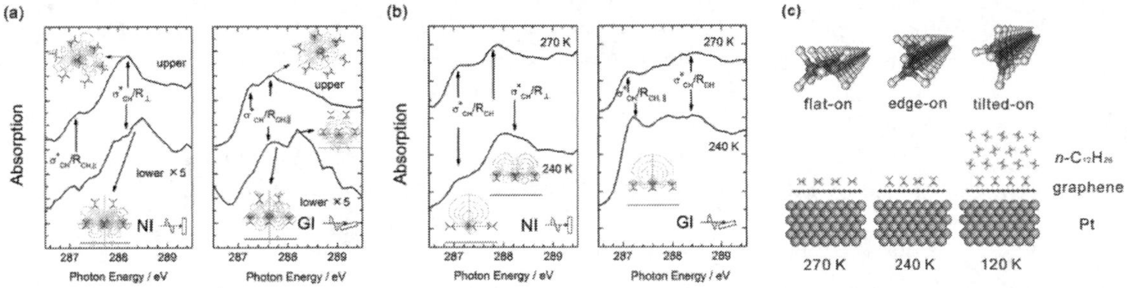

Figure 4. Energy region corresponding to the σ^*_{CH}/R resonance of separated C K-NEXAFS spectra for (a) multilayer n-$C_{12}H_{26}$ at 120 K and (b) monolayer n-$C_{12}H_{26}$ at 270 and 240 K on MG covering Pt(111) surface. (c) Schematic illustration of molecular orientation (upper panel) and structural model (lower panel).

References

[1] Stöhr J 1992 *NEXAFS Spectroscopy* (Berlin Heidelberg New York Tokyo: Springer-Verlag)
[2] Endo O, Nakamura M and Amemiya K 2013 *J. Electron Spectrosc. Relat. Phenom.* **187** 72
[3] Endo O, Nakamura M and Amemiya K 2013 *J. Electron Spectrosc. Relat. Phenom.* **189** 27
[4] Skytt P, Glans P, Mancini D C, Guo J J, Wassdahl N and Nordgren J 1994 *Phys. Rev. B* **50** 10457
[5] Brühwiler P A, Maxwell A J, Puglia C, Nilsson A, Andersson S and Mårtensson N 1995 *Phys. Rev. Lett.* **74** 614
[6] Pacilé D, Papagno M, Rodríguez A F, Grioni M, Papagno L, Girit Ç Ö, Meyer J C, Begtrup G E and Zettl A 2008 *Phys. Rev. Lett.* **101** 066806
[7] Hähner G, Kinzler M, Wöll C, Grunze M, Scheller M K and Cederbaum L S 1992 *Phys. Rev. Lett.* **69** 694
[8] Fu J and Urquhart S G 2005 *J. Phys. Chem. A* **109** 11724
[9] Ohta T, Seki K, Yokoyama T, Morisada I and Edamatsu K 1990 *Phys. Scripta.* **41** 150
[10] Bagus P S, Weiss K, Schertel A, Wöll C, Braun W, Hellwig C and Jung C 1996 *Chem. Phys. Lett.* **248** 129
[11] Weiss K, Bagus P S and Wöll C 1999 *J. Chem. Phys.* **111** 6834
[12] Weiss K, Öström H, Triguero L, Ogasawara H, Garnier M G, Pettersson L G M and Nilsson A 2003 *J. Electron Spectrosc. Relat. Phenom.* **128** 179
[13] Schöll A, Fink R, Umbach E, Mitchell G E, Urquhart S G and Ade H 2003 *Chem. Phys. Lett.* **370** 834
[14] Zou Y, Araki T, Appel G, Kilcoyne A L D and Ade H 2006 *Chem. Phys. Lett.* **430** 287
[15] Endo O, Horikoshi T, Katsumata N, Ohtani K, Fujishima T, Goto H, Minamia K, Akaike K, Ozaki H, Sumii R, Amemiya K, Nakamura M and Kosugi N 2011 *J. Phys. Chem. C* **115** 5720
[16] Hansen F Y, Herwig K W, Matthies B and Taub H 1999 *Phys. Rev. Lett.* **83** 2362
[17] Paserba K R and Gellman A J 2001 *J. Chem. Phys.* **115** 6737
[18] Tait S L, Dohálek Z, Campbell C T and Kay B D 2006 *J. Chem. Phys.* **125** 234308

Nitrogen *K*-edge soft X-ray natural circular dichroism of histidine thin film

Yudai Izumi[1], Masahito Tanaka[2], Yoshiaki Tanigawa[3], Maiko Tanabe[3], Yohei Momoki[3], Miki Iwai[4], Shun Takenaka[4], Kimihiro Ishiyama[4], Misa Kuroki[4], Takayuki Muro[5], and Kazumichi Nakagawa[3,4]

[1] Advanced Science Research Center, Japan Atomic Energy Agency, Ibaraki, Japan
[2] Research Institute of Instrumentation Frontier, National Institute of Advanced Industrial Science and Technology, Ibaraki, Japan
[3] Graduate School of Human Development and Environment, Kobe University, Hyogo, Japan
[4] Faculty of Human Development, Kobe University, Hyogo, Japan
[5] Japan Synchrotron Radiation Research Institute, Hyogo, Japan

E-mail: izumi.yudai@jaea.go.jp

Abstract. We report preliminary natural circular dichroism (NCD) spectra of amino acids, L- and D-histidine, thin films in the nitrogen *K*-edge energy region. NCD peaks were observed at 399.9 eV and were assigned to the $1s \rightarrow \pi^*$ transition of the nitrogen atoms in imidazole ring. Imidazole ring is positioned at the side chain distant from the asymmetric carbon. This result suggests that NCD spectroscopy in the soft X-ray region can selectively pick up the information on local structure around specific atoms, regardless of the position of atoms in molecules. It is interesting in view of application of NCD spectroscopy in the soft X-ray region to investigation of protein structures.

1. Introductions

Natural circular dichroism (NCD) is defined as the difference in the magnitudes of the optical absorption cross-section for left circularly polarized light (LCPL) and that for right circularly polarized light (RCPL). This phenomenon is exhibited in absorption bands of chiral molecules. NCD spectra are typically measured in the ultraviolet or infrared region and used in many different fields. For example, NCD spectroscopy in the ultraviolet region is one of the tools to investigate the molecular structures of chiral materials such as proteins. It is expected that the expansion of NCD measuring region extends applications of NCD.

Our group has been trying to measure NCD spectra in the soft X-ray region [1–4]. Soft X-rays are able to excite *K*-shell electrons of carbon, nitrogen, and oxygen atoms. It is expected that the NCD signals from transitions between core and valence levels can provide information about molecular structure in the vicinity of selected atoms because such transitions are an internal atomic process. Indeed, we measured NCD spectra in the soft X-ray (oxygen *K*-edge energy) region and succeeded to observe the differences reflecting in the local structure around excited oxygen atom which cannot be distinguished using ordinary soft X-ray absorption spectroscopy [4]. However, it has not been cleared yet whether NCD signals originating from core-absorption of atoms located at various sites, as well as next to the asymmetric carbon, in the molecules can be observed selectively.

Journal of Physics: Conference Series **502** (2014) 012038 doi:10.1088/1742-6596/502/1/012038

In this study, we measured NCD spectra of an amino acid, histidine (His; Fig. 1) which has three nitrogen atoms, thin film in the nitrogen *K*-edge energy region, and confirmed that we can pick up the NCD signals originating from each nitrogen atom throughout the molecules.

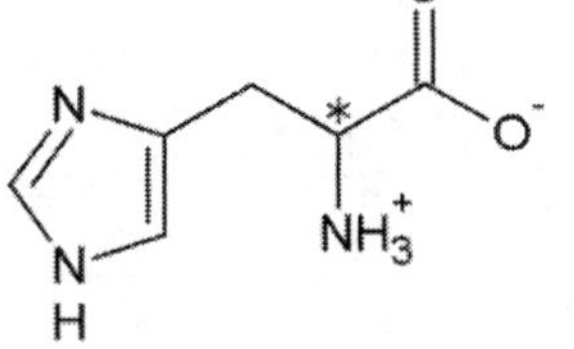

Figure 1. Structural formula of His zwitterion. The asymmetric carbon is marked with an asterisk.

2. Experimental

The detail experimental and analytical procedures have been reported in our previous paper [4]. Thin films of His were prepared by vacuum sublimation. The L- and D-enantiomers of His were purchased from Tokyo Chemical Industries (Japan) and used as amino acid reagents without further purification. The powder of each reagent was sublimated on SiC membranes (membrane size: square with sides of 3.0 mm, thickness 300 nm; MEM-C03003/10M, NTT Advanced Technology Corporation) by heating up to 370 K with a nickel-chrome heater. The base pressure of the vacuum sublimation chamber was about 10^{-3} Pa. The thickness of the films was estimated to be about 400 nm by a quartz crystal deposition controller (CRTM-500, ULVAC, Inc.).

Absorption and NCD measurements in the nitrogen *K*-edge were carried out at beamline BL25SU of SPring-8, Japan [5–7]. Photon energies were calibrated by measuring absorption spectrum of nitrogen gas and comparing it to literature data [8]. Absorption spectra were obtained by using transmission method. Incident light intensity was monitored by measuring the drain current from another SiC membrane positioned in front of the sample. The light through the SiC membrane and the sample reached a gold film sublimated on a stainless steel plate which was positioned at the back of the sample. The drain current from the gold film was recorded as the transmitted light intensity. Absorbance of each His film for LCPL and RCPL were determined. Absorption spectrum for RCPL was subtracted from that for LCPL to give ΔA spectrum. The ΔA spectrum is proportional to the NCD spectrum in ideal case. However, ΔA spectra often include contamination from linear dichroism (LD) and linear birefringence (LB) components in the case of oriented samples. To evaluate the magnitude of contamination, the ΔA spectra were obtained for 8 sets of rotation angles $(\theta, \phi) = (0°, 0°)$, $(0°, 180°)$, $(90°, 0°)$, $(90°, 180°)$, $(45°, 0°)$, $(45°, 180°)$, $(-45°, 0°)$, and $(-45°, 180°)$ for L- and D-samples, where θ and ϕ are rotation angles around axes parallel and vertical to the incident light, respectively. Since contamination of LD and LB signals were found by comparing ΔA spectra, we averaged those spectra for each sample and canceled out LD and LB components. Those averaged spectra are described as ΔA_{av} spectra, hereafter.

3. Results and Discussion

Figure 2(a) shows an absorption spectrum of L-His at the nitrogen *K*-edge energy region. In the case of D-His, similar spectrum was observed. Those spectral shapes is similar to that of reported spectrum [9]. Two sharp peaks observed at 399.9 and 401.5 eV are assigned to the $1s \rightarrow \pi^*$ transition of the nitrogen atoms in C-NH-C and C-N-C of imidazole ring, respectively [9]. Two broad peaks observed at 406.7 and 412.2 eV are mainly assigned to the $1s \rightarrow \sigma^*$ transition of the nitrogen atoms in amino group (NH_3^+) and imidazole ring, respectively [9].

Figure 2(b) shows ΔA_{av} spectra of L- and D-His. The ΔA spectrum of SiN membrane (thickness 100 nm; MEM-N03001/7.5M, NTT Advanced Technology Corporation) measured for $(\theta, \phi) = (0°, 0°)$ is also shown for comparison. In the case of SiN membrane, no significant structure was observed as expected, because SiN is achiral and amorphous, that is, it has neither NCD, LD nor LB. This result shows that no pseudo NCD signal originates from beamline optics and so on.

In the case of L-His (Fig. 2(b)), a positive peak was observed at 399.9 eV, in contrast, a negative peak was observed at the same energy in the case of D-His. Since the NCD spectra of L- and D-type molecules possessed opposite signs each other, these peaks originate from the NCD component. The ratio of ΔA_{av} to absorption intensity was the order to 10^{-3} which is typical value in soft X-ray NCD spectroscopy. At around 401–415 eV, the intensity difference between L- and D-His was observed, while the NCD signs were not opposite. This result indicates that obtained ΔA_{av} spectra in this energy region involve the background (BG) component as well as the true NCD component.

The BG component should originate from His films, because BG component was not observed in the case of SiN membrane. To consider the origin of BG component, the BG component was analytically estimated. Since BG component was observed, the ΔA_{av} spectra of both samples are described as

$$\Delta A_{av}{}^{x}(E) = CD^{x}(E) + BG^{x}(E) \quad (x = \mathrm{L, D}), \qquad (1)$$

where CD is true NCD component and superposition of L and D represent L- and D-His films. Using equation (1), summation of BG^{L} and BG^{D} is described as

$$\Delta A_{av}{}^{\mathrm{L}}(E) + \Delta A_{av}{}^{\mathrm{D}}(E) = BG^{\mathrm{L}}(E) + BG^{\mathrm{D}}(E), \qquad (2)$$

where it is assumed that NCD spectrum of D-His should be opposite of that of L-His in signs; that is, $CD^{\mathrm{L}}(E) = -CD^{\mathrm{D}}(E)$. In Fig. 3, the BG component determined by using equation (2) is shown. The ΔA spectrum of D-His measured for $(\theta, \phi) = (-45°, 0°)$, which shows the largest LD and LB components among obtained ΔA spectra, is also shown for comparison. The spectral shapes below ~410 eV seem similar. It implies that the BG component mainly originated from the LD and LB components which were not perfectly removed in this work. It is supported by the result that BG component was not observed in the case of amorphous SiN membrane. Thus, the ΔA_{av} difference between L- and D-His observed at around 401–415 eV (Fig. 2(b)) might not only originate from NCD but also LD and LB. In order to confirm whether NCD peaks exist at around 401–415 eV or not, remeasurements using randomly oriented His films, which show no LD and LB, are necessary.

Nevertheless, it is quite reliable that the peaks observed at 399.9 eV (Fig. 2(b)) originate from NCD, because they were opposite in signs each other as mentioned above. Those are assigned to the $1s \rightarrow \pi^*$ transition of the nitrogen atoms in C-NH-C of imidazole ring which positioned at the side chain. Thus, we concluded that core absorption NCD originating from nitrogen atoms in the side chain distant from the asymmetric carbon is also detectable. This result suggests that NCD spectroscopy in the soft X-ray region can selectively pick up the information in the vicinity of specific atoms, regardless of the position of atoms in molecules. It is interesting in view of application of NCD spectroscopy in the soft X-ray region to investigation of protein structures.

4. Conclusion

We measured NCD spectra of thin films of L- and D-His in the nitrogen *K*-edge energy region. NCD peaks assigned to $1s \rightarrow \pi^*$ transition of the nitrogen atoms in C-NH-C of imidazole ring were observed. We concluded that NCD signals originating from core-absorption of atoms located at various sites in the molecules can be site-selectively observed. This result is expected that structural information at the vicinity of excited atoms will be obtained by using NCD spectroscopy in the soft X-ray region. It is interesting in view of investigation of protein structures.

Acknowledgment

This work was carried out at BL25SU of SPring-8 with the approval of the Japan Synchrotron Radiation Research Institute (proposal numbers 2010A1254, 2010B1522, 2011A1473, 2012A1394, and 2012B1529). Part of this work was supported by Grant-in-Aid for JSPS Fellows (21-3972).

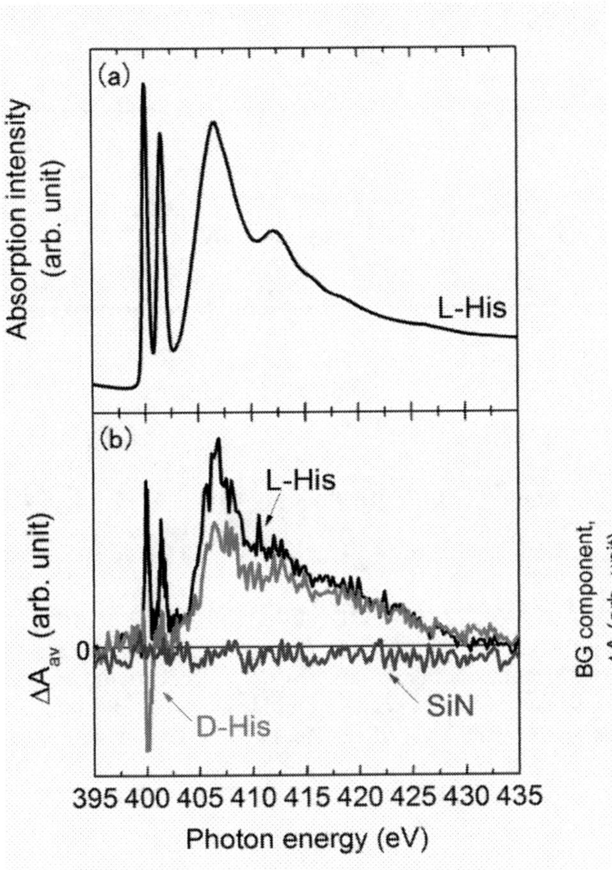

Figure 2. (a) Absorption spectrum of L-His, (b) ΔA_{av} spectra of L- and D-His (black and red lines), and ΔA spectrum of SiN membrane measured for $(\theta, \phi) = (0°, 0°)$ (blue line).

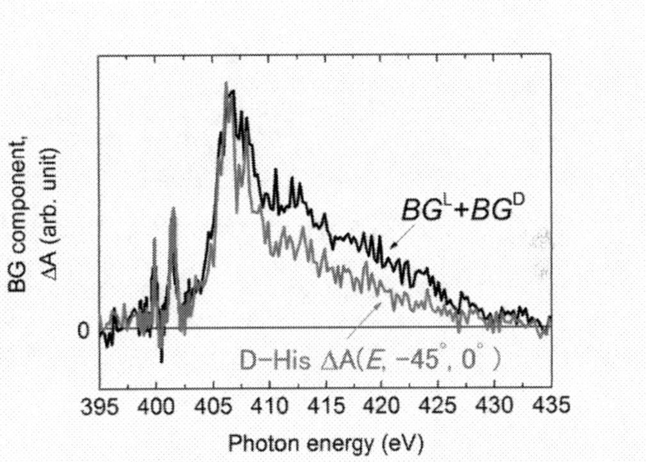

Figure 3. BG component estimated using equation (2) (black line) and ΔA spectrum of D-His measured for $(\theta, \phi) = (-45°, 0°)$ (red line).

References

[1] Tanaka M *et al* 2005 *Phys. Scripta* **T115** 873
[2] Nakagawa K *et al* 2005 *J. Electron Spectrosc. Relat. Phenom.* **144-147** 271
[3] Izumi Y *et al* 2009 *J. Phys.: Conf. Series* **190** 012209
[4] Izumi Y *et al* 2013 *J. Chem. Phys.* **138** 074305
[5] Saitoh Y *et al* 1998 *J. Synchrotron Rad.* **5** 542
[6] Saitoh Y *et al* 2000 *Rev. Sci. Instrum.* **71** 3254
[7] Hara T *et al* 2003 *Nucl. Instr. Meth. A* **498** 496
[8] Sodhi R N S and Brion C E 1984 *J. Electron Spectrosc. Relat. Phenom.* **34** 363
[9] Zubavichus Y *et al* 2005 *J. Phys. Chem. A* **109** 6998

Development of in-situ cell for fluorescence-mode XAFS in BL14B2 of Spring-8

Y-T Cui[1, *], S Hirayama[1, 2], H Oji[1, 2], J-Y Son[1, 2], T Honma[1], I Hirosawa[1]

[1] Industrial Application Division, JASRI / Spring-8, 1-1-1 Kouto, Sayo-cho, Hyogo 679-5198, Japan
[2] SPring-8 Service Co., Ltd., 1-20-5 Kouto, Shingu, Tatsuno, Hyogo 679-5165, Japan

*E-mail: yitaocui@spring8.or.jp

Abstract. An in-situ cell for fluorescence-mode XAFS was developed in BL14B2 of SPring8. It was confirmed that the cell temperature can be up to 850 °C under atmospheric pressures of reaction gases basing on our firm tests in real experiment. The temperature stability of the cell especially at low temperature region is checked with samples of highly dispersed Pt atoms on FeO_x nano-sized supporters by detecting Pt L_3 and Fe K edges. The in-situ cell show very good stabilities and the projected temperature on the test position correspond well with that of TPR results.

1. Introduction

In-situ analysis of the materials under reaction conditions has widely been performed together with synchrotron radiation light sources because it is one of the few techniques to obtain the structural and chemical information of catalysts, fuel cells and secondary batteries under well-controlled reaction conditions such as high temperatures and reactive atmospheres. With this motivation and highly demands of industrial users, we developed of the automotive exhaust gas treatment system together with the auto control system in BL14B2 of SPring-8 [1, 2]. Besides time-resolving analytical techniques such as DXAFS and QXAFS, the design of in-situ cell is also crucial, such as the uniform and reproducible temperature around samples, the efficient gas flow to pass the sample, the corrosive attack to the parts of cell and easy to be used in beam line, etc. Basing on the requirements, the in-situ cells for both transmission and fluorescence modes XAFS were designed with quartz. The in-situ cell for transmission mode XAFS measurements have been tested [2] and can be used for user now. However the in-situ cell for fluorescence mode has not been tested yet due to the difficulties in the temperature uniformity and safety consideration coming from the large fluorescence detecting window. In the present work, the in-situ XAFS cell for fluorescence mode developed in BL14B2 of SPring8 will be described. Its performance was evaluated by firm test in real experimental conditions.

2. Experiment

The in-situ XAFS experiment was performed on BL14B2 of Spring-8 - a hard x-ray bending magnet beam line dedicated for XAFS experiments in the wide energy range from 3.8 to 72 keV. A single pair Si (111) or (311) double-crystal monochromator was used to reduce the harmonic component of the monochrome beam with the photon flux of 10^8~10^{11} photons/sec and energy resolution of 10^{-4} at sample position after focusing with Rh thin films mirrors (standard beam size 5.0mm in horizontal and 0.5mm in vertical directions). For more details please see elsewhere in ref. 1. In the experiment, two kinds of samples were selected, one is 0.2%wt Pt/Fe_2O_3, in which Pt atoms are mostly dispersed atomically, and the other is 2.5%wt Pt/Fe_2O_3, in which Pt atoms aggregate to several-atom-sized clusters [3]. Pt foil and PtO_2 were used as reference samples and measured in the transmission mode. Before the experiment, the samples were pressed into self-supporting disks with a diameter of 10 mm. To obtain a much better signal-to-noise ratio, an Al foil filter was used to reduce the direct reflection and fluorescence of Fe in addition to extending the integral time for fluorescence measurements. IFEFFIT software was used to calibrate the energy scale, to correct the background signal and to normalize the intensity. A similar approach was used to analyse the EXAFS data at the Pt L_3 and Fe K edges.

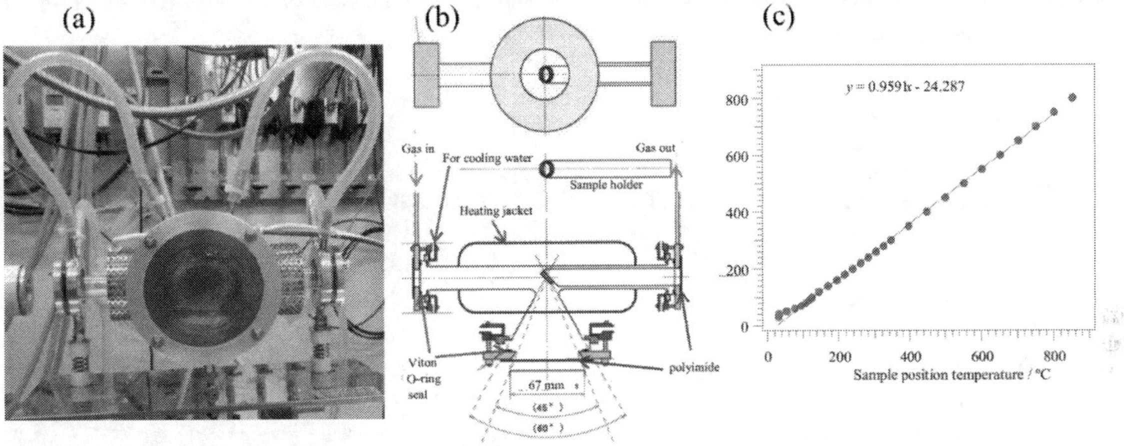

Figure 1 The design of in-situ cell for fluorescence mode: **(a)** the photograph in side view of fully assembled cell for experiment; **(b)** side and top views as well as the sample holder of cell; **(c)** the relationship of setting temperature and the temperature on sample position.

3. Results and discussion

The fully assembled in-situ cell for fluorescence mode XAFS with a catalyst sample inserted in is shown in Figure 1 (a). The main body of the cell is a quartz tube with a conical window opened in the middle of it. The cell is sealed with Viton O-ring and stainless flanges. The windows are designed with polyimide adhered to the exterior of the flanges. In order to avoid over-heating of the Viton O-rings and polyimide windows, water-cooling system is used to the stainless flanges. The heating of the cell is accomplished with heating jacket controlled via a programmable temperature controller. A quartz tube was fabricated as the sample holder with a 45° sample pocket along x ray direction in one side as shown in Figure 1 (b). By the design, the cell can be easily heating to 850 °C with good temperature uniformity in sample position. Indeed, the cell was used to conduct catalyst reductions in hydrogen at 850 °C. It is demonstrated that the correspondence of sample temperature against setting temperature is linearly as shown in Figure 1 (c) with a temperature ramp of 10 °C/min, for example. There are many advantages for this design, which can solve the problems mentioned in part 1. Firstly, quartz tube is inert to most gases and solids typically found in catalysis and it is easy to clean, also, its optically transparent is distinct for detecting the changes of samples in color or shape during experiments. Second, the samples can be easily loaded into sample pocket without any damages. Third, the heating jacket is also convenient cooled quickly from high temperature to room temperature after experiment by using a small fan.

Before experiment, the gas distribution and exhaust gas treatment systems as well as emergency shut-off systems were set up as described in ref. 1. In order to test the whole performance of in-situ cells, firstly, Pt L_3 edge EXAFS with fluorescence mode was measured with and without in-situ cell to evaluate the intensity changes effected by in-situ cell. The fluorescence signal did not show differences with and without in-situ cell. It means no shielding from windows of in-situ cell. Then, we started to test the gas distribution and exhaust system, and no any leakage was detected in 2 hours with the He gas pressure more than one atmosphere. After that, with H_2 flow, we tested the gas distribution and exhaust system to evaluate the gas exchanging time from pure H_2 to pure O_2 detected by a mass spectrometer detector for the gases contribution during the procedure. It was confirmed that 20 minutes is enough for H_2 or O_2 exchanging, which is nearly two times as that of transmission mode owing to the large conical window in fluorescence mode cell [4]. All of the procedures for test were done in nearly 10 hours. The gas flow rate, exhaust gas, QXAFS system and the sample temperature were controlled by the PC with the programs coded by LabVIEW programming language of National Instruments during the procedures [5]. The emergency shut-off system is used during experiments to detect the leak of flammable and/or toxic gasses, fire, earthquake, electric power failure, device failure the abnormal pressure/flow of gases, and so on.

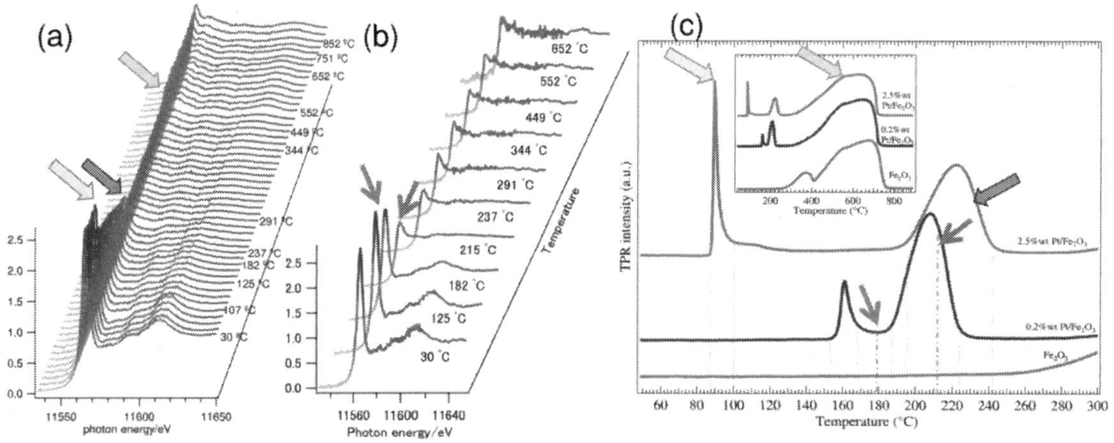

Figure 2 Pt L_3 XANES spectra of 2.5%wt Pt/Fe_2O_3 (a) and 0.2%wt Pt/Fe_2O_3 (b) samples against temperature; H_2-TPR profiles of Fe_2O_3, 0.2%wt and 2.5%wt Pt/Fe_2O_3 (c).

To test the temperature uniformity, the samples were reduced with in-situ cell from room temperature (RT) to 850 °C with a PC program by transition mode at a temperature-increasing rate 10 °C/min for several times to detect the reproducibility of sample temperature. After that, Pt/Fe_2O_3 samples were reduced by H_2/He flow (H_2 10%, He 90%, flow rate 100 ml/min) with a PC programmed temperature from RT to 850 °C as shown in Ref. 4. The Pt L_3 edge XAFS were measured at certain temperatures during reduction by transition mode for 2.5%wt Pt/Fe_2O_3 and fluorescence mode for 0.2%wt Pt/Fe_2O_3 samples respectively. Since the fluorescence signal is weak, the spectra were measured several times by fix temperature at certain values. The exhaust gases contribution is detected with mass spectrometer detector during experiments. Pt L_3 XANES spectra of 0.2%wt and 2.5%wt Pt/Fe_2O_3 were presented in Figures 2 (a) and (b). We can find that Pt L_3 white line suffers gradually changes by variation of temperature and the Pt L_3 white line intensity changes combined well with the H_2 consumption peak and the produce of water [4]. By the knowledge that difference between the area of the L_3 edge resonance (white line) in the 5d pure metals and their compounds can be related to the amount of *d* electron vacancies [6-8], one can detect the amount of charge transferred to the ligands by the changes of white line intensities. So, the abrupt changes of Pt L_3 white line intensity correspond to the reduction or alloying of Pt atoms. To fully understand the changes of Pt L_3 white line intensity, the ex-situ temperature programed reduction (TPR) results (after Ref. 3) of Fe_2O_3, 0.2%wt and 2.5%wt Pt/Fe_2O_3 samples were presented in Figure 2 (c). The H_2 consumption peaks at certain temperature give the information of sample reduction. Such as for Fe_2O_3, the small peak from 300~400 °C indicates the reduction of Fe atoms from Fe_2O_3 to Fe_3O_4, while the broad peak from 420 °C to750 °C is due to the reduction of Fe atoms from Fe_3O_4 to FeO and finally to Fe metal [9]. For Pt/Fe_2O_3 samples, besides the two peaks related with the reduction of Fe mentioned above, an additional sharp peak appears, which should correspond to the reduction of Pt atoms [10]. Besides, the temperatures for reduction of Fe atoms from Fe_2O_3 to Fe_3O_4 are shift to low temperature side due to the Pt interaction with Fe_2O_3 supporters. With the knowledge, now, we can discuss the sample temperatures by combining Pt L_3 white line intensity changes and the H_2-TPR profiles. The abrupt changes of Pt L_3 white line intensity start around 107 °C, 237 °C, 552 °C for 2.5%wt Pt/Fe_2O_3 and 182 °C, 215 °C for 0.2%wt Pt/Fe_2O_3, respectively. While, the H_2 consumption peaks in H_2-TPR profiles, are around 90 °C, 220°C, 580 °C for for 2.5%wt Pt/Fe_2O_3 and 160 °C and 210 °C for 0.2%wt Pt/Fe_2O_3, respectively. It means that the temperature on the samples is nearly the same as that corrected by thermocouples, if we thinking about the temperature-increasing rate of 10 °C/min, the differences of temperature in the different measuring methods are reasonable, since the temperature on the sample should be a little delayed due to the bad thermo conductance of samples during experiments.

4. Conclusions

In this work, the in-situ cell for fluorescence mode EXAFS at BL14B2 was tested successfully. The on-line test of exhaust gas treatment systems and the in-situ cells showed no leakage during experiment even up to 850 °C. The in-situ cell can be used both transmission and fluorescence modes and temperature can be up to 850 °C basing on our firm test in the experiment. The automatic temperature control system can work well with the designed program. But due to the weak signal of 0.2%wt Pt samples, only the temperature-increasing rate of 10 °C/min was tested. Even though, the temperature on real sample and that on the test position correspond well by in-situ test of standard samples comparing with that of ex-situ TPR experiments. This test of in-situ fluorescence mode XAFS will introduce a new method for BL14B2 and compensate the demands of industrial users for in-situ XAFS. After this whole performance test and evaluation, we think that the system can be opened for users of industrial applications now. As described in part 3 of this work, the mass spectrometer detector was used to detect the gases contributions by connected with in-situ cell in the gas outlet. Even though, a true *operando* test is not done for the system, we think that the *operando* mode can be used to monitor the states of the catalyst during the catalytic reactions by determining the conversation and selectivity with the additional mass spectrometer detector to the gas outlet of in-situ cell. And further work will be done for that.

5. Acknowledgements

Thanks to Profs. B Qiao, A Wang and T Zhang of DICP, CAS China for providing the samples. The synchrotron radiation experiments were performed at the BL14B2 of SPring-8 with the approval of JASRI (Proposal No.2011B1974, 2012B1875).

References

[1] Honma T, et al., AIP conf. Proc. **1234**, 13 (2010).
[2] Hirayama S, et al., SPring8 User report, Nos.: 2007B1950, 2008A1832.
[3] Qiao B, et al. Nature Chemistry, **3**, 634, (2011)
[4] Cui Y-T, et al., SPring8 User report, No.: **2011B1974**.
[5] Hirayama S, et al., SPring8 User report, No.: **2012B1875**.
[6] Horsley J A, J. Chem. Phys. **76**, 1451 (1982).
[7] Lytle F W, et al, J. Chem. Phys. **70**, 4849 (1979).
[8] Qi B, et al. Phys. Rev. B, **36**, 2972R (1987).
[9] Hayes P C, et al., Metallurgical and Materials Transactions B, **12**, 319 (1981).
[10] Hurst N W, et al., Catal. Rev. Sci. Eng., **24**, 233 (1982).

DNA damage by soft X-ray exposure at oxygen K-edge

Y Sugaya[1,2] **A Narita**[2], **K Fujii**[2], **and A Yokoya**[2,1]
[1]Ibaraki University, Mito, Ibaraki 310-8512, Japan
[2]ASRC, JAEA, Tokai, Ibaraki 319-1195, Japan

E-mail: yokoya.akinari@jaea.go.jp

Abstract. In order to obtain detailed insights into the physicochemical mechanism of DNA damage induction in terms of photoabsorption modes, we have prepared thin DNA films of closed circular plasmid (pUC18) on a cover slip without any additives. Using this film, we have performed preliminary experiments by exposing to soft X-rays with energies around oxygen K-shell ionization threshold. The DNA damage yields of strand breaks and base lesions or AP sites were quantified by biochemical treatments. We confirmed that the DNA film can work as a specimen irradiation. The DNA damage yields induced by π^* excitation of a K-shell electron of oxygen atoms in DNA were significantly larger those for oxygen K-ionization.

1. Introduction

The radiobiological effects, such as mutation induction, are thought to arise from formation of the DNA damage, e.g. nucleobase lesions, single- or double-strand breaks, or abasic sites (AP sites). In order to obtain more detailed insights into the physicochemical mechanism of the damage induction in terms of photoabsorption-modes, we have tried site selective absorption in DNA using high resolution soft X-rays [1]. Fujii et al. exposed soft X-rays to a thin DNA film in a high vacuum chamber [2]. They suggested that the yields of DNA damage strongly depend on the ionization of nitrogen or oxygen atoms in DNA, particularly much higher yields of base lesions above oxygen K-edge. Their DNA films, however, contained buffer solutes with a few ten % (w/w) to stabilize DNA even in a vacuum. Thus the damage induction by irradiation potentially included some effects from the solutes. The fact makes it difficult to examine specific effects of the resonant photoabsorption-modes, excitation of the *1s* electron to an antibonding π^* or σ^* orbital, because of their overlapping energies with those of the buffer solutes. In order to eliminate the effect from the additives, it is crucial to prepare a thin film that is solely composed of DNA even though the DNA might be destabilized to some extent.

In this study, we have tried to prepare thin DNA films of closed circular plasmid without any additives. We tested the stability of the DNA in the film, and also performed preliminary soft X-ray irradiation experiments. The yield of DNA damage produced by soft X-ray irradiation around oxygen K-shell ionization threshold were determined, and compared with our previous data for DNA films including buffer solutes as additives.

2. Materials and Methods

2.1. Preparation of Plasmid DNA films

[3] Author to whom any correspondence should be addressed.

Plasmid DNA (pUC18, 2686 bp) was purified from *E. coli* strain HB101 using a plasmid purification kit (QIAfilter plasmid Mega kit, QIAGEN, Japan). The plasmid, which is over 95 % in the intact closed circular form, was subsequently stored at -20 °C in TE buffer (10 mmol dm^{-3} Tris, 1 mmol dm^{-3} EDTA, pH 8.0) at a concentration of about 1 mg/ml. Then we removed the solutes from the plasmid DNA solution using a filter kit for PCR purification (QIAquick PCR Purification Kit, QIAGEN), and the plasmid DNA was diluted with sterilized water to give a final DNA concentration of 0.1 μg/μl. Five μl aliquots of this plasmid solution were spotted onto a various substrates and dried for 30 min in air. We tested a cover slip, silicon wafer evaporated with or without gold, cupper plate and graphite plate as substrates for the sample spotting.

2.2. Irradiation of soft X-ray beam to the plasmid DNA films

Irradiation was carried at the soft X-ray beamline (BL23SU) in SPring-8 (Hyogo, Japan). Monochromatic soft X-rays were obtained using a varied line spacing grating. The irradiation of the sample films was carried out at the selected energies at an oxygen K-shell (*1s*) π^* resonance peak (531.6 eV). We also used soft X-rays of 560 eV that is completely larger than the oxygen *1s* binding energy (543.1 eV)[3]. The plasmid DNA film on a substrate was placed in a pre-chamber (~10^{-5} Pa) for 1 h, and was then transferred to a high vacuum irradiation chamber (~10^{-7} Pa). To obtain uniform irradiation, the sample holder was moved up and down, and also right to left in the vacuum chamber using motor-drive manipulators as described our previous report [2]. The photon flux of the beam measured using a photodiode (AXUV-100, International Radiation Detectors Inc., USA) at the sample position was almost constant in the order of 10^{17} photons per sample. The absorbed doses are estimated based on the photo-absorption cross sections and the atomic composition of the sample. Two independent irradiation experiments were performed.

2.3. Treatment of the irradiated DNA samples and gel electrophoresis

After soft X-ray irradiation, the plasmid DNA film was recovered with 15 μl of 40 mM HEPES buffer (pH 7.6) for each sample, and kept at at -20 °C until the biochemical treatment to quantify DNA damage. Sixty μl aliquots of the irradiated samples were mixed with enzymatic reaction buffer (50 μl of HEPES buffer with 0.5 mM dithiothreitol (DTT, Nacalai Tesque, Japan), and 0.21 mg/mL bovine serum albumin (BSA). These samples were then incubated at 37 °C for 0.5 h in the "absence" or "presence" of enzymes. The former treatment was performed to examine the yield of promptly induced SSB, and latter one to examine the yields of base lesions or AP site. The heat treatment did not cause additional SSBs (data not shown). We used three kinds of enzymes, formamidopyrimidine-DNA glycosylase (Fpg), endonuclease three (Nth), or endocuclease four (Nfo) to transform damage to an additional SSB by their glycosylase activities. Nth protein excises mainly ring-saturated pyrimidines (e.g., 5, 6-dihydrothymine [DHT]), thymine glycol, and AP sites. Fpg protein excises mainly 2,6-diamino-4-hydroxy-5-N-methyl formamidopyrimidine, 7,8-dihydro-8-oxo-2'-deoxyguanine (8-oxoGua), and AP sites. Nfo protein excises various AP sites. After the excision of the damage, a SSB is formed at the damage site. Purified enzymes Nth, Fpg, and Nfo were purchased from NEB (New England Biolabs Japan, Japan). The optimal concentrations of the enzymes used for the treatments were determined in our previous study [2].

After incubation with or without enzymes, 2.5 μL of 0.5 M EDTA (pH 8.0) was added to the sample solutions and kept at 5 °C on ice to terminate the enzymatic reactions. For gel electrophoresis, 2.5 μL of the loading buffer (0.1% bromophenol blue, 0.1% xylene cyanol, and 30% glycerol) was added to the samples, which were placed into the wells of a 1% agarose (Type 1-A, Sigma-Aldrich Japan, Tokyo, Japan) gel in TBE buffer (90 mM Tris, 90 mM boric acid, 2 mM EDTA) at pH 7.1 and electrophoresed at 1.4 V/cm for 16.7 h at 5.6 °C.

2.4. Quantification of yields of SSB and DSB in irradiated plasmid DNA

After electrophoresis, the gel was stained with 30 μL of ethidium bromide (Bio-Rad Japan, Tokyo, Japan) (10 mg/mL) in 600 mL of TBE buffer for 1 h. The positions of closed circular, open circular, and linear forms of the plasmid DNA were separately visualized with a charge-coupled device (CCD,

Ettan DIGE, GE Healthcare UK Ltd., England) and analyzed with ImageQuant TL software (GE Healthcare UK Ltd., England). The dose response was determined from the loss of intact closed-circular DNA resulting from the strand breaks promptly induced by irradiation, or enzymatically induced strand breaks at base lesions or AP sites. The D_{37}-values, a dose giving 37% of the remaining DNA in the closed circular form, was obtained from the dose responses. These values, indicators for induction of any kinds of SSBs, reveal the radiation dose required to give an average of one SSB per plasmid DNA. According to the D_{37}-value, a yield of SSB (n[SSB]/Gy/Da) was calculated from equation (1) assuming that the average mass of a base pair is 650 Da and that pUC18 DNA contains 2686 base pairs.

$$n(SSB) = 1/(2686 \times 650 \times D_{37}) \quad (1)$$

The yields of lesions revealed by the enzymatic treatments, described as number of enzyme sensitive sites, n(ESS), observed as additional SSBs were obtained with Equation 2.

$$n(ESS)_{Nth(or\ Fpg,\ Nfo)} = n(SSB)_{Nth\ (or\ Fpg,\ Nfo)} - n(SSB)_{prompt} \quad (2)$$

3. Results and Discussion

Most of the materials tested for the substrate of the sample gave a pronounced degradation effect to plasmid DNA. Over 50 % of the sample plasmid DNA molecules were suffered strand breaks by spotted on the surfaces, or kept in a vacuum for several hours. Only the cover slip gave us sustainable results because over 60 % of the plasmid kept their intact closed circular form in a control condition without exposure as shown in Figure 2. Thus hereafter we adopted the DNA films on the cover slip as a specimen irradiation. A ring-shape film of DNA of 4.0 mm diameter was formed on the cover slip.

X-ray absorption near edge structure (XANES) spectrum determined by transmission of soft X-rays in a calf thymus DNA film [4] is shown in Figure 1 with the energies chosen for exposure in this study. The energies of the oxygen π^* excitation peak (531.6 eV) and above ionization threshold (560 eV) are give the sample DNA a similar photoabsorption cross section. Using an atomic force microscope the thickness of the sample film was determined to be about 200 nm. We also ensured absence of charge up of the sample because the XANES spectrum obtained as a total electron current in the sample film did not get distorted by the soft X-ray exposure.

A typical example of the dependence of the amount of closed circular DNA on absorbed dose is shown in Figure 2 for the samples exposed to soft X-rays at the energy of the π^* excitation. The fractions of the intact plasmid DNA exponentially decreased with increasing dose. These data ensure that soft X-rays considerably penetrated in the sample film and the most of the plasmid molecules

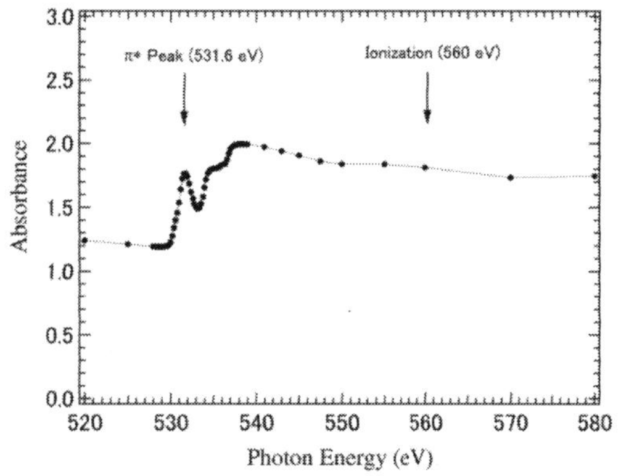

Figure 1. Photoabsorption spectrum of calf thymus DNA film. The arrows show the energy chosen for this study.

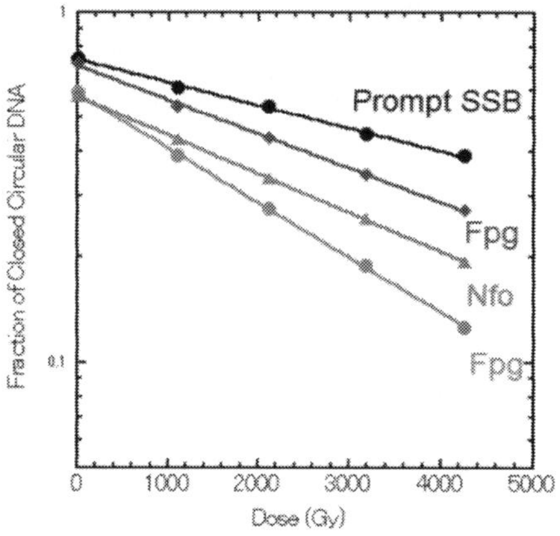

Figure 2. Dependence of the loss of intact (closed circular) plasmid DNA on soft X-ray (531.6 eV) dose after exposure. The straight lines are drawn by the least square method.

were exposed to the soft X-rays uniformly. If the thickness of the sample film much higher than penetration depth of the soft X-rays and the molecules located in the back of the film are not adequately irradiated, resulting that the dose response does show a saturation in the higher dose region. The post-irradiation incubation with the enzymes results in greater loss of closed-circular DNA than seen in the absence of an enzyme treatment for a given dose.

From the obtained dose-dependences, the yields of prompt SSB and base lesions visualized by the biochemical treatments were calculated from the D_{37}-values (see Equation 1 and 2). These yields are tabulated in Table 1 with our previous data [2]. The yields of SSB and Nth sensitive site are larger than those reported by Fujii et al. [2], indicating that the buffer solutes may act to protect DNA chemically active species to produce these DNA lesions. All the yields of DNA damage examined for 531.6 eV X-ray irradiation are significantly larger than those for 560 eV. This indicates that the π^* excitations of an oxygen 1s electron to an antibonding orbital may effectively cause chemical bond breakages resulting induction of certain DNA damage. Impact of ejected photo- or Auger electrons to surrounding DNA is also thought to contribute the damage induction. It is particularly worth noting that the π^* excitation does emit less electrons than oxygen K-ionization, only an Auger electron to be ejected, hence the contribution of the π^* excitation is more apparent in the damage induction than the secondary effect of electron impact. In future the effect of σ^* excitation of oxygen atoms in DNA will be further studied using the experimental procedure established in the present study.

Table 1. Yields of SSB and ESS induced in pUC18 plasmid DNA by soft X-ray irradiation.

	Yield ($\times10^{-11}$ SSB or ESS/Gy/Da)		
DNA Damage	531.6 eV (O-K π^* resonance)	560 eV (O-K ionization)	560 eV[a] (O-K ionization)[a]
n(SSB) Single strand breaks	11.6	7.3	2.8
n(ESS_{Nth}) Pyrimidine lesions and AP sites	12.4	7.4	5.8
n(ESS_{Fpg}) Purine lesions and AP sites	6.3	4.0	4.5
n(ESS_{Nfo}) AP sites	4.5	3.4	-

[a] Fujii et al. 2009.

Reference

[1] Yokoya A, Fujii K, Shikazono N and Ukai M 2011 Spectroscopic study of radiation-induced DNA lesions and their susceptibility to enzymatic repair Charged particle and photon interactions with matter-recent advances, applications and interfaces ed Y Hatano, Y Katsumura and A Mozumder (USA: CRC/Taylor & Francis Group) chapter 20 pp543-574.(2011)
[2] Fujii K, Shikazono N and Yokoya A 2009 J. Phys. Chem. B 113, 16007
[3] X-ray Data Booklet of Lawrence Berkeley National Laboratory: http://xdb.lbl.gov
[4] Oka T, Yokoya A, Fujii K, Fukuda Y and Ukai M 2012 Phys. Rev. Lett. 109, 213001

Acknowledgments

This work was partially supported by JSPS KAKENHI Grant Numbers, 21310041 and 23651049). The soft X-ray irradiation was conducted with the approval of JASRI (2011B3811 and 2012A3812).

Journal of Physics: Conference Series **502** (2014) 012041 doi:10.1088/1742-6596/502/1/012041

Gradual evolution in spin dynamics of $TlCu_{1-x}Mg_xCl_3$ probed by muon-spin-relaxation (μSR) technique

Takao Suzuki[1], Isao Watanabe[2], Fumiko Yamada[3], Yasuyuki Ishii[4], Kazuki Ohishi[5], Takayuki Goto[6], and Hidekazu Tanaka[3]

[1] College of Engineering, Shibaura Institute of Technology, 307 Fukasaku, Minuma-ku, Saitama 337-8570, Japan

[2] Advanced Meson Science Laboratory, Nishina Center, RIKEN, 2-1 Hirosawa, Wako, Saitama 351-0198, Japan

[3] Department of Physics, Tokyo Institute of Technology, O-okayama, Meguro-ku, Tokyo 152-8551, Japan

[4] Department of Physics, Tokyo Medical University, 6-1-1 Shinjuku, Shinjuku-ku, Tokyo 160-8402, Japan

[5] Research Center for Neutron Science and Technology, Comprehensive Research Organization for Science and Society (CROSS), Tokai, Ibaraki 319-1106, Japan

[6] Faculty of Science and Technology, Sophia University, 7-1 Kioi-cho, Chiyoda-ku, Tokyo 102-8554, Japan

E-mail: `t-suzuki@sic.shibaura-it.ac.jp`

Abstract. Longitudinal-field muon-spin-relaxation (LF-μSR) measurements in $TlCu_{1-x}Mg_xCl_3$ were carried out to investigate the spin dynamics in the non-magnetic impurity doped spin gap system. As reported before, in the case of $x \geq 0.006$, LF-μSR time spectra are well fitted by the two components function, which means the phase separation to a spin frozen region and to a spin flucuating region. In this report, we focus on the spin fluctuating region, and discuss the gradual change in its magnetic properties with increasing the Mg-doping.

1. Introduction

$TlCuCl_3$ is well known as a three-dimensionally coupled Cu-3d $S = 1/2$ spin dimer system, and its magnetic ground state is spin-singlet with excitation gap of $\Delta_{gap} = 7.5$ K, which originates from strong intradimer antiferromagnetic interaction [1, 2].

In the nonmagnetic impurity doped $TlCu_{1-x}Mg_xCl_3$ system, it is reported that the magnetic phase transition to an ordered state is observed by magnetization and specific heat measurements, and that neutron elastic scattering measurements identified the impurity-induced ordered state is the antiferromagnetically ordered state of which the magnetic structure is the same with the case of the field-induced phase in $TlCuCl_3$ [3, 4]. The inelastic neutron scattering measurement revealed that a finite spin gap still remains below the transition temperature in $TlCu_{1-x}Mg_xCl_3$ [5].

We have reported gradual changes of spin states probed by zero- and longitudinal-field muon-spin-relaxation (ZF, LF-μSR) measurements as follows [6, 7, 8]. In a slightly doped case of $x = 0.0047$, no evidence for a static internal magnetic field was observed down to 20 mK although the specific heat indicated the magnetic phase transition at $T = 0.70$ K. Above $x =$

Journal of Physics: Conference Series **502** (2014) 012041 doi:10.1088/1742-6596/502/1/012041

0.006, the existence of a static internal magnetic field is confirmed, and with increasing Mg-concentration of x, the internal magnetic field and the volume fraction of a spin frozen region increase simultaneously. These results suggest that the magnetic ground state changes from the spin singlet state to a spin fluctuating state, and to a spin frozen state by the impurity doping. In our previous reports, mainly, the evolution of the static spin state region was discussed. Thus, in this report, we focus on the spin fluctuating region, and discuss the gradual change in its magnetic properties.

2. Experimental

Single crystals used in this study were grown from a melt by the Bridgman method, and the concentration of x was determined by the inductively coupled plasma atomic emission spectrometry (ICP-AES) method. Muon-spin-relaxation (μSR) measurements were carried out at the RIKEN-RAL Muon Facility in the U.K. using a spin-polarized double-pulsed positive surface-muon beam with an incident muon momentum of 27 MeV/c [9], and at the Swiss Muon Source (SμS), Paul Scherrer Institut (PSI), Villigen, Switzerland. In μSR measurements, spin-polarized muons are implanted into samples. The asymmetry was defined as follows:

$$A(t) = \frac{F(t) - \alpha B(t)}{F(t) + \alpha B(t)} \quad (1)$$

Here, $F(t)$ and $B(t)$ were total muon events counted by the forward and backward counters at time t respectively. The α is a calibration factor reflecting relative counting efficiencies between the forward and backward counters. All time spectra in this study were analyzed using the data analysis program named WiMDA [10].

3. Results and Discussions

Figure 1 is summarized data of LF-μSR time spectra of $TlCu_{1-x}Mg_xCl_3$ at lowest temperatures in this study. Data of Fig.1 (a) are quoted from ref.[6], and data of Fig.1 (b) and (c) are from ref.[8]. In Fig. 1 , all the time spectra are plotted as the corrected asymmetry in the same horizontal scale for comparison between spectra quoted from different studies. As for Fig.1 (d), time spectra at T = 0.290 K are not shown in ref.[7], namely, it is the first report. The time spectra for $x = 0.0047$ are well fitted using the function of $A(t) = e^{-\lambda t}G_{\mathrm{KT}}(\Delta, H_{\mathrm{LF}}, t)$. λ is the muon-spin-relaxation rate, and $G_{\mathrm{KT}}(\Delta, H_{\mathrm{LF}}, t)$ is the static Kubo-Toyabe function, where Δ/γ_μ is the distribution width of nuclear-dipole fields at the muon sites [11]. γ_μ is the gyromagnetic ratio of the muon spin ($2\pi \times 13.5534$ kHz/gauss), and H_{LF} is the applied external longitudinal-field. As discussed in ref.[6], this behavior means that no evidence for the existence of a static internal magnetic field is observed down to 20 mK, and that impurity-induced magnetic moments of the Cu-3d spins are slowly fluctuating at 20 mK.

In the case of $x \geq 0.006$, a fast relaxation in the earlier time range below 1 μs of ZF time spectra is observed, and LF dependence is not able to be reproduced using the one component function of $A(t) = e^{-\lambda t}G_{\mathrm{KT}}(\Delta, H_{\mathrm{LF}}, t)$. Thus, in the previous study, LF time spectra are analyzed by a two component function of $A(t) = A_1e^{-\lambda_1 t} + A_2e^{-\lambda_2 t}G_{\mathrm{KT}}(\Delta, H_{\mathrm{LF}}, t)$. Here, $A_1 + A_2 = 1$. The first term and the second term represent a fast and slow relaxation components, respectively. As discussed in ref.[7] and ref.[8], the fast relaxation corresponds to the disappearance of the initial asymmetry A(0), which reveals the existence of the static internal magnetic field at the muon sites. The two components function means that the system is in a phase separated state to a spin frozen state and to a spin fluctuating (paramagnetic) state. In this study, we focus on the spin fluctuating region, and discuss the LF dependence of the muon spin relaxation rate λ_2 of the the spin fluctuating region at low temperatures below 0.3 K.

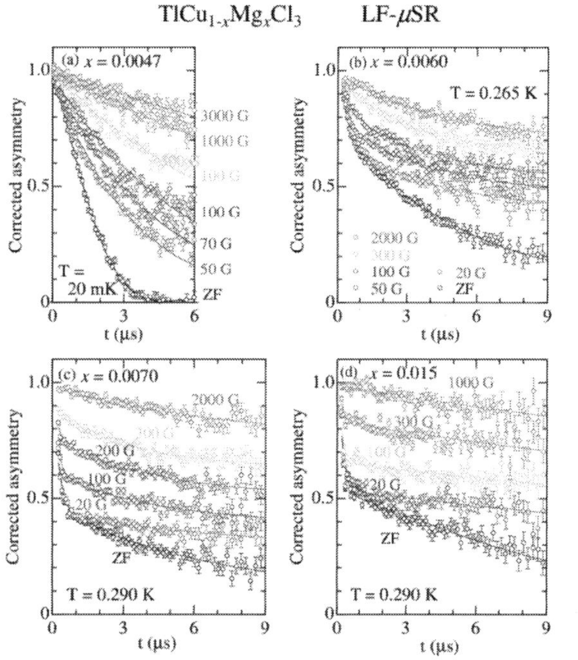

Figure 1. Time spectra of the LF-μSR in $TlCu_{1-x}Mg_xCl_3$. Solid lines are fitted results. The detailed analysis is explained in ref.[6, 7, 8].

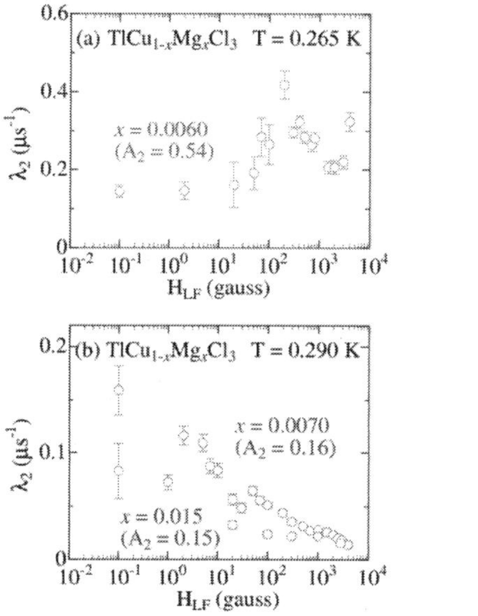

Figure 2. LF dependence of the muon spin relaxation rate λ_2 of the spin fluctuating region in $TlCu_{1-x}Mg_xCl_3$. Zero-field data are plotted at H_{LF} = 0.1 gauss. A_2 corresponds to a volume fraction of the spin fluctuating region.

Figure 2 shows LF dependence of the muon spin relaxation rate λ_2 of the spin fluctuating region in $TlCu_{1-x}Mg_xCl_3$. A value of A_2 corresponds to the volume fraction of the spin fluctuating region. In the case of $x = 0.006$, with increasing H_{LF}, λ_2 begins to increase around 20 gauss, and has a peak around 300 gauss. Such behavior in the LF dependence is quite strange, and can not be explained in usual paramagnetic spin states.

This kind of behavior is observed in the pure (no-impurity) spin singlet state of $KCuCl_3$ which has the same crystal structure with $TlCuCl_3$ [12]. Higemoto *et al.* discussed the possibility of the muon disturbance effect, i.e., implanted muons break spin singlet pairs, and generate free spins near muons. These free spins might cause a strange LF dependence of the muon spin relaxation rate. If so, it is consistent with the measurement results that spin state is in the phase separation to the spin fluctuating region which is in truth the spin singlet state, and to the spin frozen region which causes the disappearance of the initial asymmetry. Here, we emphasize the non-impurity $TlCuCl_3$ is in the spin singlet state at low temperatures. This discussion has not been guaranteed by any theoretical study, but in this stage, this idea is weighty candidate. Similar discussion is reported in the organic triangular-lattice quantum spin magnet κ-(BEDT-TTF)$_2Cu_2(CN)_3$ [13].

In the case of $x = 0.007$ and 0.015, however, the LF dependence of λ_2 shows a linear decrease with increasing the H_{LF}. This means that muon spins feel widely spread distribution of the local magnetic fields H_{loc} and of the frequency of fluctuation f at the muon sites, because when the distribution of the magnitude of H_{loc} and of spin fluctuation frequencies is homogeneous in time and in space, the LF dependence of λ_2 should has a cut-off structure described by the Redfield formula of $\lambda(H_{\rm LF}) = 2\gamma_\mu^2 H_{\rm loc}^2 \tau/(1 + \gamma_\mu^2 H_{\rm LF}^2 \tau^2)$, where $\tau = 1/f$ is the correlation time of muon spins [14, 15, 16]. In other words, the distribution is "white". Thus, the small amount of the spin fluctuating state ($A_2 \sim 0.15$) is in the paramagnetic state for the case of $x = 0.007$ and 0.015.

In the last, we summarize the change of magnetic ground states with Mg-doping to the spin singlet dimer system. In the slightly doped region of $x = 0.0047$, impurity-induced magnetic moments of the Cu-3d spins are slowly fluctuating at 20 mK. In the case of $x = 0.006$, a half of the system is in a spin frozen state, and the other region is in possibly the spin singlet state. Above $x = 0.007$, a static internal magnetic field region extends, and the remained region is in a paramagnetic state.

4. Summary

We carried out zero- and longitudinal-field muon-spin-rotation measurements in the nonmagnetic impuirity doped spin singlet system $TlCu_{1-x}Mg_xCl_3$. Especially in the case of $x = 0.006$, the system goes in the phase separation to a spin frozen region and to possibly a spin singlet region.

Acknowledgments

This work was partially supported by JSPS KAKENHI Grant Numbers 24540350 and 21540344. HT was supported by a Grant-in-Aid for Scientific Research (A) from the Japan Society for the Promotion of Science.

References

[1] W. Shiramura, K. Takatsu, H. Tanaka, K. Kamishima, M. Takahashi, H. Mitamura, and T. Goto, J. Phys. Soc. Jpn. **66**, 1900 (1997).
[2] N.Cavadini, W. Henggeler, A. Furrer, K. Güdel -U H, Krämer, and H. Mutka, Eur. Phys. J. B **7**, 519 (1999).
[3] A. Oosawa, T. Ono, and H. Tanaka, Phys. Rev. B **66**, 020405(R) (2002).
[4] M. Fujisawa, T. Ono, H. Fujiwara, H. Tanaka, V. Sikolenko, M. Meissner, P. Smeibidl, S. Gerischer, and A. H. Graf, J. Phys. Soc. Jpn. **75**, 033702 (2006).
[5] A. Oosawa, M. Fujisawa, K. Kakurai, and H. Tanaka, Phys. Rev. B **67**, 184424 (2003).
[6] T. Suzuki, F. Yamada, I. Watanabe, T. Matsuzaki, T. Goto, A. Oosawa, and H. Tanaka, J. Phys. Soc. Jpn. **78**, 074705 (2009).
[7] T. Suzuki, I. Watanabe, F. Yamada, Y. Ishii, K. Ohishi, Risdiana, T. Goto, and H. Tanaka, Phys. Rev. B **80**, 064407 (2009).
[8] T. Suzuki, M. Yamada, Y. Ishii, I. Watanabe, T. Goto, H. Tanaka, and K. Kubo, Phys. Rev. B **83**, 174436 (2011).
[9] T. Matsuzaki, K. Ishida, K. Nagamine, I. Watanabe, G. H. Eaton, and W. G. Williams, Nuclear Instrumental Methods in Physics Research A **465**, 365 (2001).
[10] F. L. Pratt, Physica B **289-290**, 710 (2000).
[11] R. Kubo and T. Toyabe, in Magnetic Resonance and Relaxation, edited by R. Blinc (North-Holland, Amsterdam, 1967), p. 810.
[12] W. Higemoto, H. Tanaka, I. Watanabe, and K. Nagamine, Phys. Lett. A **243**, 80 (1998).
[13] S. Nakajima, T. Suzuki, Y. Ishii, K. Ohishi, I. Watanabe, T. Goto, A. Oosawa, N. Yoneyama, N. Kobayashi, F. L. Pratt, and T. Sasaki, J. Phys. Soc. Jpn. **81**, 063706 (2012).
[14] C. P. Slichter, *Principles of Magnetic Resonance* (Springer-Verlag, Berlin, Heidelberg, 1978, 1990).
[15] T. Goto, T. Suzuki, K. Kanada, T. Saito, A. Oosawa, I. Watanabe, and H. Manaka, Phys. Rev. B **78**, 054422 (2008).
[16] T. Saito, A. Oosawa, T. Goto, T. Suzuki, and I. Watanabe, Phys. Rev. B **74**, 134423 (2006).

Ultra Slow Muon Microscope at MUSE / J-PARC

J Nakamura[1,2], T Nagatomo[1,2], Y Oishi[3], Y Ikedo[1,2], P Strasser[1,2], N Saito[4], K Miyazaki[4], K Yokoyama[3], K Okamura[4,5], Y Miyake[1,2], S Makimura[1,2], K Nishiyama[1,2], K Shimomura[1,2], N Kawamura[1,2], A Koda[1,2], W Higemoto[1,6], S Wada[4,5], M Iwasaki[3], E Torikai[7]

[1]J-PARC Center, Tokai, Ibaraki 319-1195, Japan
[2]Muon Science Laboratory, IMSS, KEK, Tsukuba, Ibaraki 305-0801, Japan
[3]RIKEN Nishina Center, Wako, Saitama 351-0198, Japan
[4]RIKEN Center for Advanced Photonics, Wako, Saitama 351-0198, Japan
[5]Megaopto Co., LTD., Wako, Saitama 351-0104, Japan
[6]Advanced Science Research Center, Japan Atomic Energy Agency, Tokai, Ibaraki 319-1195, Japan
[7]University of Yamanashi, Kofu, Yamanashi 400-8511, Japan

Email: jumpei.nakamura@kek.jp

Abstract. We report current constructing states of the Ultra Slow Muon Beam at U-line / MUSE / J-PARC, which are supported by thermal muonium (Mu, μ^+e^-) production with the most intense pulsed slow muon beam, laser resonant ionization, and transportation of Ultra Slow Muon Beam. A thermal Mu is produced by a hot tungsten foil in a Mu-production chamber. At the laser resonant ionization process, a thermal Mu is ionized by coherent vacuum ultraviolet radiation and coherent 355-nm radiation. The coherent radiation sources are developed at RIKEN, installed in a laser cabin, and connected via a VUV steering chamber with the Mu-production chamber.

1. Introduction

The Ultra Slow Muon (USM) Beam has been developed at MSL / KEK in Japan [1] and, RIKEN-RAL muon facility / ISIS in UK [2]. But, enough yield of beam, for the study in material science, was not obtained. Now, there are four secondary beam lines (D, U, S, and H-line) at MUSE / J-PARC [3]. Muons are produced via pions (π^+ or π^-) decay by the nuclear reactions between the 3 GeV proton beam and the muon target which made of 20-mm thick graphite [4]. And the muons are extracted from secondary beam line. D-line has been operational for fast muons (decay muons: up to 120 MeV/c, μ^+ or μ^-)) or slow muons (surface muons: 30 MeV/c, 4 MeV, μ^+). And U-line, for Ultra Slow Muon microscope for material science, is planned to be operational with more than 10^4 times higher USM intensity than that at RIKEN-RAL [5]. At U-line, magnetic or electric properties will be investigated for surface or subsurface of condensed or soft matter.

2. Muon U-line

Muon U-line consists of Superomega beam line [6] and Ultra Slow Muon beam line [7]. And USM beam line consists of muonium production chamber and transportation line. (U-line is sometimes

Journal of Physics: Conference Series **502** (2014) 012042 doi:10.1088/1742-6596/502/1/012042

called as Superomega beam line, in a narrow sense) Fig. 1 shows a schematic diagram and a photo of U-line.

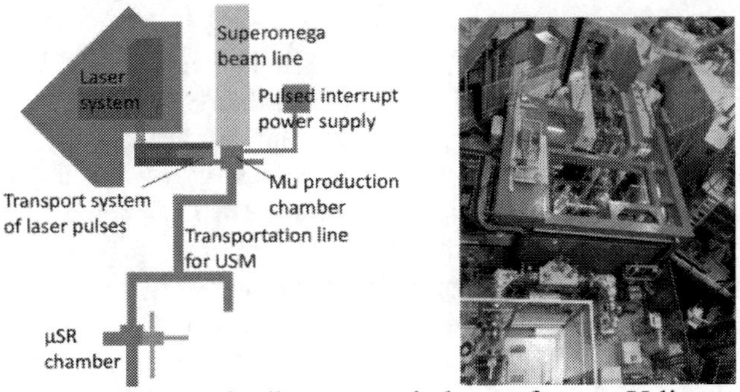

Fig. 1: Schematic diagram and photo of muon U-line

2.1. *Superomega beam line*

Superomega beam line, the slow muon source for the USM beam, was already installed. It consists of the capture solenoid, the curved transport solenoid, and the axial focusing solenoid [6]. And, the positron separators with DC high voltage power supplies were added on the beam line chambers of the axial focusing solenoid [8]. In Nov. 2012, the intensity of slow muons has been achieved 100 times higher than that at RIKEN-RAL [9].

2.2. *Muonium Production chamber and Transportation line for Ultra Slow Muon*

At Mu production chamber, the surface muon beam from Superomega beam line is injected into a hot tungsten (W) foil, which is heated up to 2100-2300 K, and the Mu can escape to the vacuum with thermal velocity (E~0.2-0.3 eV). At this chamber, ultrahigh vacuum (UHV, 10^{-7} Pa~10^{-8} Pa; required for the thermal Mu production) has been achieved with a tandem unit of turbo molecular pumps and a cryopump. The W foil has been heated by a pulsed interrupt power supply [10]. And highly pure W foil (6N, 99.9999%; 50 μm×54 mm× 70 mm) has been developed with mechanical polishing, without hot rolling technics [11], because the purity of W foil affects Mu emission[1,2]. Recently, the surface of this W foil has been observed by laser scanning microscope (LSM), and it has been showed that there are many gaps (1-μm depth×2-10-μm wide) (Fig.2) [12]. And it has been also showed with wavelength dispersive X-ray fluorescence spectrometer (WDX) that the surface is covered with tungsten oxide in the same way as a 3N-W foil [13]. At Mu production chamber, it is expected that oxygen and carbon at the tungsten surface is removed by hydrogen and oxygen treatment, respectively.

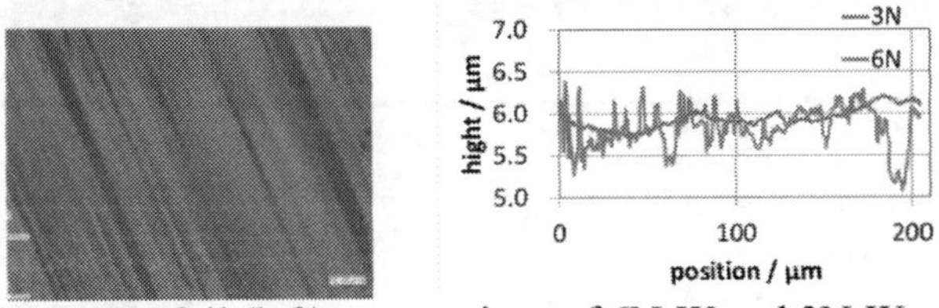

Fig. 2: Surface of the 6N-W foil (left), comparison of 6N-W and 3N-W with hot rolling (right)

The thermal Mu is ionized by laser pulses (cf. § 3), and the USM is produced. After extraction by SOA lens [7, 10], the USM beam is transported to μSR chamber [14]. The beam optics parameters have been optimized by off line beam tuning with Li^+ [10].

3. Laser system

The USM beam is produced by laser resonant ionization of thermal Mu. In this laser resonant ionization process, thermal Mu is excited (1S→2P) by coherent vacuum ultraviolet wave (VUV, 122.09 nm, Lyman-α) and ionized by 355-nm wave (required: shorter than 366 nm, UVA). This coherent VUV wave, a pulsed (25 Hz) resonance radiation, is generated in a krypton (Kr) gas cell by sum-difference frequency mixing of 212.55-nm wave (omega 1, ω_1; UVC, two photon resonant excitation) and tunable difference wave (omega 2, ω_2; near infrared, NIR, 820.65 nm for Mu). These coherent radiation sources based on a diode-pumped fiber laser and solid-state laser technologies are developed at RIKEN [15], and it will be installed in a laser cabin [16].

The Omega 1 system is an all-solid-state coherent light source, which consists of a distributed feedback laser (ω_1 seeder), three fiber amplifiers, a regenerative amplifier (regen.), an intermediate amplifier system, a power amplifier (final amp.) system, and a nonlinear frequency converter [17]. The laser pulses are planned to be amplified up to an output energy of 1 J at 1062.75 nm by use of the power amplifier. The 1062.75 nm wave is converted to the fifth harmonic wave (212.55 nm), through second harmonic generation in a LBO, fourth harmonic generation in a CLBO, and fifth harmonic generation in a CLBO in the nonlinear frequency converter.

The Omega 2 system is a tunable coherent light source, which consists of a broadband spectrum diode laser (ω_2 seeder) and a ω_2 amplifier system [18, 19]. The ω_2 amplifier contains an optical parametric oscillator (OPO) and an optical parametric amplifier (OPA) using KTP crystals as a nonlinear optical crystal.

For generation of muonium Lyman-α resonance radiation, we use a Kr cell in which argon gas is mixed to krypton to satisfy phase matching (Kr and Ar exhibit negative and positive dispersion, respectively.) [17, 20, 21] Pulsed coherent Lyman-α resonance radiation is generated in two-photon resonance four-wave mixing (sum-difference frequency mixing) using the Omega 1 and 2 according to the process as shown in Fig. 3. At present, a new way of focusing in Kr gas is extensively studied in RIKEN to adapt high intense laser pulses with the effect of plasma development and dephasing.

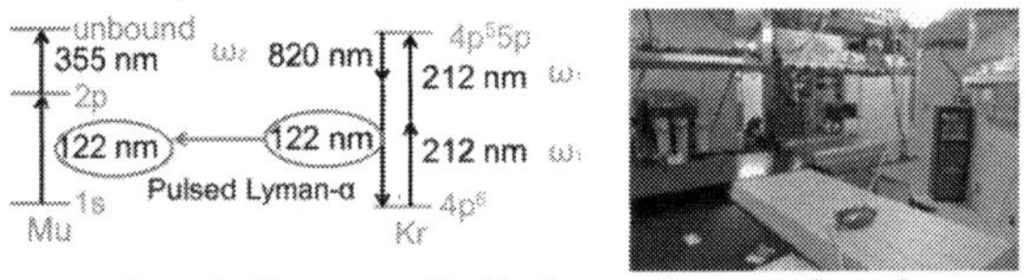

Fig. 3: Simplified energy level diagram (left), laser system in a laser cabin at J-PARC (right)

4. Transport system of laser pulses

Transport system of laser pulses [22] introduces laser pulses to the Mu production chamber (Fig. 4). The vacuum chambers for the transport system (manufactured by AVC Co., LTD. [23]) consist of a VUV steering vacuum chamber and a nitrogen monoxide (NO) gas cell. The VUV pulse generated in the Kr cell goes through the VUV steering chamber, the Mu production chamber, and the NO gas cell, sequentially.

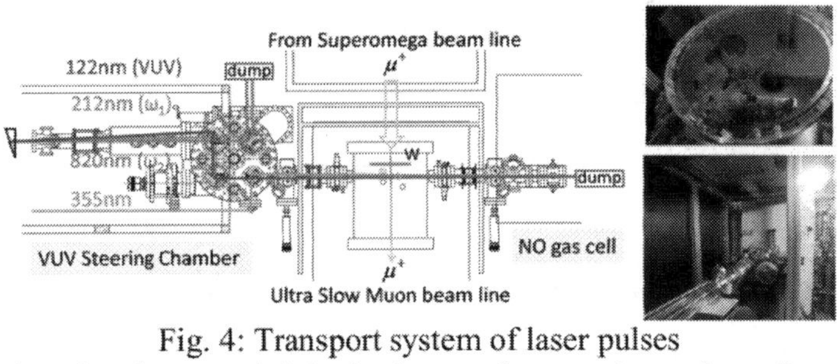

Fig. 4: Transport system of laser pulses

In the VUV steering chamber, two bright fundamental waves (ω_1, ω_2) can be separated from VUV wave. The 355-nm wave has a nearly same incident angle to the Mu production chamber as that of VUV wave. UHV ($<10^{-6}$ Pa) can be achieved in the VUV steering chamber. That is because mirrors are steered from outside of the chamber by control units with magnetic coupling mechanism, without other drive mechanisms in the chamber. Because mirror control units were prepared for each VUV pulse and 355-nm pulse, respectively, we can easily control the distance between these pulses and the W foil. By monitoring ionization rate of thermal Mu with changing this distance, it will be revealed that the velocity of thermal Mu follows a Boltzmann distribution or not. Radiation profiles and the separation of laser pulses were calculated by a ray tracing, *Optica* [22]. Although a wavelength dispersion of the refractive index at a cylindrical lens of magnesium fluoride (MgF_2) was used for pulse separation at RIKEN-RAL muon facility [2], that at a wedge plate of lithium fluoride (LiF) is planned to use for suitable horizontal pulse separation. The vertically long profile of VUV pulse is suitable for effective ionization of Mu. A cylindrical lens of LiF is used for making vertical profile. At the NO gas cell, there are parallel plates that accumulate nitrogen monoxide molecules ion for measuring intensity of the VUV pulse [24].

5. Summary

The Ultra Slow Muon Beam is supported by Muon U-line, the laser system, and the transport system of laser pulses. They are almost ready for next beam time, giving us a quarter-century waited *new* muon science, while some detailed and improved studies are performed in this shutdown term. You will investigate magnetic or electric properties of surface or subsurface of condensed or soft matter, with this new probe.

Acknowledgement

This work was supported by a Grant-in-Aid for Scientific Research on Innovative Areas "Ultra Slow Muon" (No.23108002) of the Ministry of Education, Culture, Sports, Science, and Technology, Japan.

References

[1] Miyake Y *et al.*, Nucl. Instrum. Methods Phys. Res., Sect. B 95 (1995) 265-275
[2] Bakule P *et al.*, Spectrochimica Acta Part B 58 (2003) 1019–1030;
Matsuda Y *et al.*, Physica B 326 (2003) 217–221;
Bakule P *et al.*, Nucl. Instrum. Methods Phys. Res., Sect. B : 266 (2008) 335-346
[3] Miyake Y *et al.*, Nucl. Instrum. Methods Phys. Res., Sect. A : 600 (2009) 22–24
[4] Makimura S *et al.*, Nucl. Instrum. Methods Phys. Res., Sect. A : 600 (2009) 146-149;
Makimura S *et al.*, KEK Progress Report 2012-2 KEK-MSL REPORT 2011 (2012) 5
[5] Miyake Y *et al.*, Hyperfine Interact (2013) 216:79–83
[6] Ikedo Y *et al.*, KEK Progress Report 2012-2 KEK-MSL REPORT 2011 (2012) 18;
Ikedo Y *et al.*, AIP Conf. Proc. 1382 (2011) 220
[7] Strasser P *et al.*, KEK Progress Report 2012-2 KEK-MSL REPORT 2011 (2012) 14
[8] Nishiyama K *et al.*, KEK Progress Report 2012-2 KEK-MSL REPORT 2011 (2012) 19
[9] http://legacy.kek.jp/intra-e/press/2012/112714/
[10] Nagatomo T *et al.*, 68th Annual Meeting of the Phys. Soc. of Jpn. (2013) 26pXZB-8;
Nagatomo T *et al.*, International Sympo. on Sci. Explored by Ultra Slow Muon (2013) O-5
[11] Makimura S *et al.*, International Sympo. on Sci. Explored by Ultra Slow Muon (2013) P-32
[12] Nakamura J, and Naoe T, *et al.*, private communications.
[13] Nakamura J, and Taira M, *et al.*, private communications.
[14] Higemoto W *et al.*, International Sympo. on Sci. Explored by Ultra Slow Muon (2013) P-35
[15] Saito N *et al.*, Reports on the Topical meeting of the Laser Soc. of Japan 427 (2012) 49-54
[16] Nakamura J *et al.*, KEK Progress Report 2012-2 KEK-MSL REPORT 2011 (2012) 17
[17] Oishi Y *et al.*, International Sympo. on Sci. Explored by Ultra Slow Muon (2013) P-42
[18] Miyazaki K *et al.*, International Sympo. on Sci. Explored by Ultra Slow Muon (2013) P-41
[19] Okamura K *et al.*, International Sympo. on Sci. Explored by Ultra Slow Muon (2013) P-40
[20] Louchev O A *et al.*, Phys. Rev. A 84 (2011) 033842
[21] Saito N *et al.*, International Sympo. on Sci. Explored by Ultra Slow Muon (2013) O-2
[22] Nakamura J *et al.*, 68th Annual Meeting of the Phys. Soc. of Jpn. (2013) 26pXZB-7
Nakamura J *et al.*, International Sympo. on Science Explored by Ultra Slow Muon (2013) P-43
[23] http://www.a-vc.net/
[24] Kajiwara T *et al.*, Rev. Laser Eng. 12 (1984) 367;
Samson J A R, *et al.*, Vacuum ultraviolet spectroscopy, Academic Press, London, (2000)

Neutron Brillouin Scattering Experiments with Pulsed Neutrons on High Resolution Chopper Spectrometer HRC

S Itoh[1,2], T Yokoo[1,2], D Kawana[3], Y Kaneko[4], Y Tokura[4,5], M Fujita[6], K Yoshida[7], K Saito[2,8], N Inami[2,8], Y Takeichi[2,8], K Ono[2,8], Y Endoh[1,4,6]

[1]Neutron Science Division, Institute of Materials Structure Science, High Energy Accelerator Research Organization, Tsukuba 305-0801, Japan
[2]Condensed Matter Research Center, Institute of Materials Structure Science, High Energy Accelerator Research Organization, Tsukuba 305-0801, Japan
[3]The Institute for Solid State Physics, The University of Tokyo, Tokai 319-1106, Japan
[4]RIKEN Center for Emergent Matter of Science, Wako 351-0198, Japan
[5]Institute for Materials Research, Tohoku University, Sendai 980-8577, Japan
[6]Department of Applied Physics, The University of Tokyo, Tokyo 113-8656, Japan
[7]Department of Chemistry, Fukuoka University, Fukuoka 814-0180, Japan
[8]Photon Factory, Institute of Materials Structure Science, High Energy Accelerator Research Organization, Tsukuba 305-0801, Japan

E-mail: shinichi.itoh@kek.jp

Abstract.
We improved the High Resolution Chopper Spectrometer (HRC), which is installed at MLF in J-PARC, in order to perform neutron Brillouin scattering (NBS) experiments, and successfully demonstrated a feasibility of this method. Gapless spin-wave excitations were observed in $La_{0.8}Sr_{0.2}MnO_3$, which were in good agreement with previous results using single crystals, on the other hand, a large energy gap in the ferromagnetic spin waves was found in $SrRuO_3$. Spin-wave peaks in a permanent magnet $Nd_2Fe_{14}B$ were on the dispersion curve determined previously using a single crystal. Excitations, so-called fast sound, in liquid D_2O were in a good agreement with previous inelastic neutron scattering experiments and the dispersion curve were extended to the lower Q region.

1. Introduction

Neutron Brillouin scattering (NBS) is an advantageous method for observing coherent excitations in non single-crystal materials. As a result of the powder average of the dynamical structure factor for ferromagnetic spin waves from polycrystalline samples, for instance, the scattering intensities detected in the forward direction near (000) rapidly decrease with the increase in the scattering vector Q. Owing to the kinematic constraints of neutron spectroscopy, incident neutron energy E_i in the sub-eV region is necessary for measuring scattering in the meV transferred energy E range or with a high energy resolution of $\Delta E/E_i$, further, the scattered neutrons need to be detected at very low scattering angles ϕ. Although the principles of NBS are quite old, recent efforts for realizing experimental conditions of the NBS method have been made at spectrometers, Pharos at the pulsed spallation neutron source [1] and BRISP at the steady state neutron source [2]. We also made similar approaches on the High Resolution Chopper Spectrometer (HRC) installed at the pulsed spallation neutron source of the Material and Life Science Facility (MLF) in J-PARC (Japan Proton Accelerator Research Complex) [3]. We here demonstrate the feasibility of NBS on the HRC by some experiments.

2. NBS experiment on HRC

On the HRC [4], the monochromatic neutron beam with the energy of E_i is incident upon the sample, and the energy transfer E is determined by the time-of-flight (TOF) of the detected neutron. An array of position sensitive detectors (PSD) was installed on the HRC to detect scattered neutrons with a wide energy-momentum space. For NBS measurements [3], scattered neutrons were collected by the two-dimensional bank of the PSD array at low scattering angles $\phi = 0.5 - 2.8°$. In the present experiments, the lowest angle was 0.69° due to the spread of the horizontal beam width of the direct beam. The sample was packed into a flat aluminum can. A monochromatic neutron beam was incident normal to the flat face of the sample can. The beam cross section was defined to be 3 cm × 3 cm with a cadmium plate. The accelerator was operated with approximately 200 kW of the proton beam power. The data were accumulated typically for 2 days for each condition. The intensities of the observed spectra indicated below were normalized by the accumulation time.

The reduction of background noise was essential for implementing the NBS method. A T0 chopper well cut the background noise originating from the high-energy neutron burst [4]. In order to eliminate unwanted scattering other than that from the sample, a Soller type collimator with the collimation of 0.3°, which matches the incident beam divergence, was mounted just upstream the sample for this experiment. Careful consideration of the background contamination is the most important part of the experiments, thus, we performed empty can scans.

A detector element located at ϕ scans points in the (Q, E) space is described by

$$\frac{\hbar^2 Q^2}{2m} = 2E_i - E - 2\left(E_i(E_i - E)\right)^{1/2} \cos\phi, \tag{1}$$

where $\hbar$ and m are the Planck's constant divided by 2π and the neutron mass, respectively. Low-angle detectors are essential to access the present (Q, E) space. In fact, the region above the dashed line in Fig. 1, which is $E = (\hbar^2 Q^2/2m)/\sin\phi$, or the envelope of scan loci for $\phi = 5°$ in eq. (1) with respect to E_i, can never be accessed by, for instance, a conventional spectrometer with the lowest scattering angle of $\phi = 5°$.

3. Nearly cubic perovskites $La_{0.8}Sr_{0.2}MnO_3$ and $SrRuO_3$

Spin waves were observed from two polycrystalline ferromagnets $La_{0.8}Sr_{0.2}MnO_3$ (Curie temperature, $T_C = 316$ K, and the saturation moment, $M_s = 3.6\ \mu_B$) [5, 6] and $SrRuO_3$ ($T_C =$ 165 K and $M_s = 1.6\ \mu_B$) [7]. The sample volumes were approximately 8 and 13 cm^3, respectively. Both perovskites essentially have a cubic lattice structure with a small distortion. The former is a typical example showing a colossal magnetoresistance. The latter has recently shown an enhanced anomalous Hall effect due to the spin-orbit coupling of the Ru 4d orbital [7]. Although $SrRuO_3$ has not yet been synthesized in large single-crystal form, polycrystalline samples are readily available. The measurements were performed at temperatures $T = 6$ and 254 K for $La_{0.8}Sr_{0.2}MnO_3$ and at $T = 7$ K for $SrRuO_3$, with $E_i = 102$ meV and $\Delta E = 2.0$ meV. The background-subtracted spectra showed spin-wave peaks and an elastic peak, as shown in Figs. 1 (a) and (b). The intensities of spin waves were well explained by the temperature factor as well as the magnetic form factor. Each peak was fitted to a Gaussian function with a width determined from the resolution including the dispersion slope and the peak positions (E_p) of the spin waves were determined [3], as shown in Fig. 1 (c).

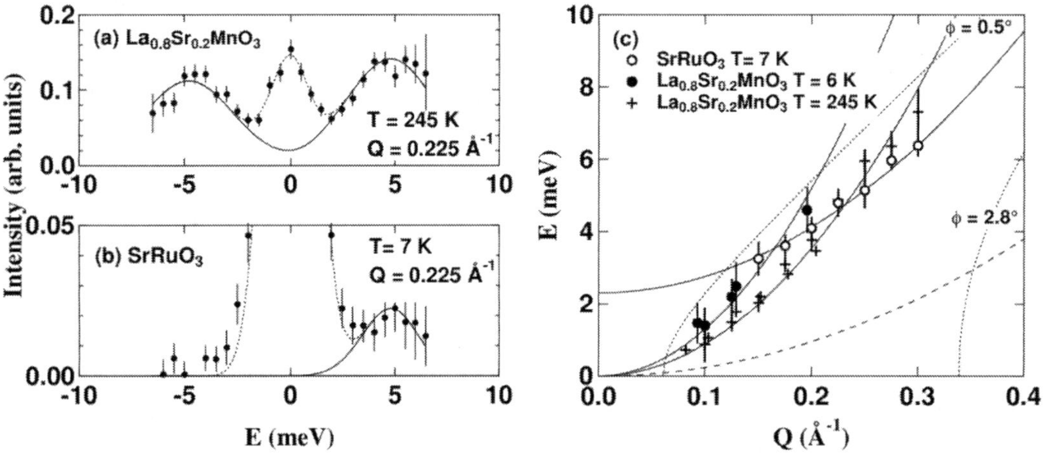

Figure 1. Typical excitation spectra for $La_{0.8}Sr_{0.2}MnO_3$ and $SrRuO_3$ (a,b). The dashed lines are fitted curves and the solid lines are the spin-wave components. Spin-wave dispersion curves (c). The solid lines are fitted dispersion curves. The dotted lines are the boundaries of the present experiment (ϕ = 0.5° and 2.8°) with E_i = 102 meV, calculated from eq. (1). The dashed line is the upper boundary accessible with a spectrometer having the lowest scattering angle of $\phi = 5°$.

The spin-wave dispersion curves in $La_{0.8}Sr_{0.2}MnO_3$ observed at T = 6 and 245 K were well fitted to $E_p(Q) = DQ^2$ with $D = 130 \pm 4$ and 88 ± 2 meVÅ^2, respectively. In previous works measuring spin waves using single crystals, an isotropic quadratic dispersion curve with D = 131 meVÅ^2 at T = 14 K and 89 meVÅ^2 at 250 K for $La_{0.8}Sr_{0.2}MnO_3$ was reported [5, 6]. At T = 14 K and $Q < 0.3$ Å^{-1} from an appropriate reciprocal point, the dispersion curve along [001] was well fitted to $E_p(Q) = DQ^2$ with D = 131 meVÅ^2 and was identical to that along [111] [6].

The spin-wave dispersion curve of $SrRuO_3$ observed at T = 7 K was well fitted to a quadratic form with an apparent energy gap: $E_p(Q) = E_0 + DQ^2$ with $E_0 = 2.2 \pm 0.3$ meV and $D = 48 \pm 5$ meVÅ^2, by using the data points for $Q = 0.15 - 0.3$ Å^{-1}. Spin waves in $La_{0.8}Sr_{0.2}MnO_3$ show the isotropic dispersion relations in smaller Q range, however the curves gradually shift with each other for $Q > 0.3$ Å^{-1} due to the crystalline effect. We applied this experimental evidence to the case of $SrRuO_3$: the dispersion relation at $Q \leq 0.3$ Å^{-1} may follow a form of $E_0 + DQ^2$.

4. Permanent magnet $Nd_2Fe_{14}{}^{11}B$

$Nd_2Fe_{14}B$ is a well-known song permanent magnet with T_C = 580 K and the saturation magnetization of 1.6 T. At room temperature, all spins are aligned along the c*-axis. Below T = 130 K the system exhibits a spin reorientation. Because the unit cell includes four chemical formulae of $Nd_2Fe_{14}B$ in a tetragonal lattice, spin-wave branches are expected to be complicated. In a previous inelastic neutron scattering experiment using a single-crystalline sample of $Nd_2Fe_{14}{}^{11}B$, where ^{11}B was used for minimizing the neutron absorption, a spin-wave branch was detected only along the c*-axis around (002) at T = 6 and 295 K [8]. We performed NBS experiments in a $Nd_2Fe_{14}{}^{11}B$ polycrystalline sample at T = 6 and 300 K with E_i = 257 meV and ΔE = 5.7 meV. The sample volume was approximately 5 cm^3. The spin-wave peaks in the observed energy spectra at T = 300 K were well fitted to Gaussian scattering functions with widths determined from the resolution including the dispersion slope, and the peak positions were determined in Fig. 2. The observed peak positions were on the dispersion curve along the

c^*-axis reported in the previous experiment. No magnetic peaks were observed at $T = 6$ K, this result is also consistent with the dispersion curve along the c^*-axis at $T = 6$ K reported in the previous experiment.

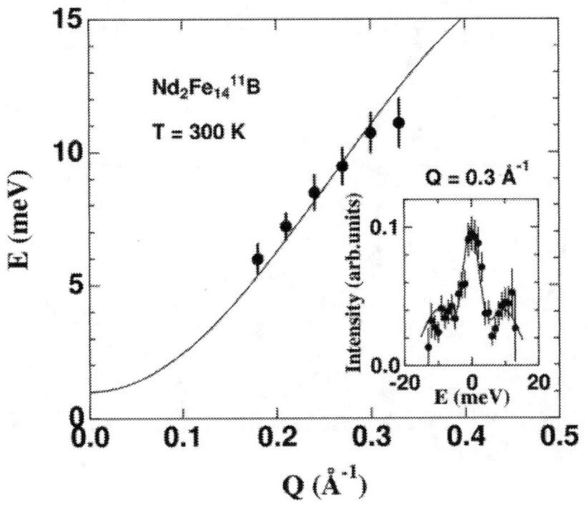

Figure 2. Spin-wave peak positions for $Nd_2Fe_{14}{}^{11}B$. The solid line is a dispersion curve along the c^*-axis determined using a single crystal in ref. [8]. The inset shows a typical excitation spectrum with a fitted curve.

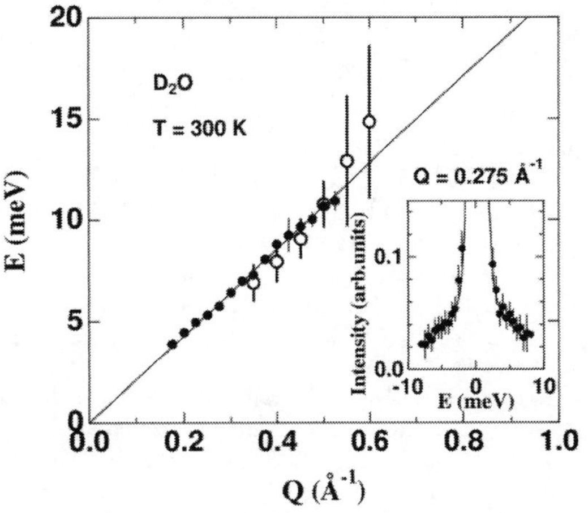

Figure 3. Dispersion relation for phononic excitations, the fast sound, in D_2O (closed circles). The open circles are previous results in ref. [9]. The solid line is a fitted line. The inset shows a typical excitation spectrum with a fitted curve.

5. Liquid D_2O

Phononic excitations in a liquid D_2O were measured at $T = 300$ K with $E_i = 102$ meV and $\Delta E = 2.0$ meV. The sample volume was approximately 15 cm^3. The background-subtracted spectra showed excitation peaks and a resolution-limited elastic peak. The excitation peaks were fitted with a damped harmonic oscillator scattering function convoluted with the resolution width multiplied by a temperature factor and the peak positions $E_p(Q)$ were determined. The present analysis was identical to that for the previous inelastic neutron scattering experiment [9]. As shown in Fig. 3, the Q dependence of the observed peak positions down to $Q = 0.02$ Å^{-1} was well fitted to $E_p(Q) = cQ$ with $c = 21.4 \pm 0.2$ meVÅ, which is equivalent to $c = 3250 \pm 30$ m/s. The observed sound velocity c was in good agreement with that for the fast sound observed in the previous experiment with $E_i = 80$ meV and $\Delta E = 4.8$ meV down to $Q = 0.035$ Å^{-1} [9].

6. Summary

We demonstrated the feasibility of NBS experiments on the HRC by observing coherent excitations in $La_{0.8}Sr_{0.2}MnO_3$, $SrRuO_3$, $Nd_2Fe_{14}{}^{11}B$ polycrystals and a liquid D_2O. In the current research in material sciences, most of the newly synthesized materials are chemically and structurally complicated such that large single crystals are not always possible to be synthesized. NBS experiments on the HRC bring about new opportunities to such material development using polycrystalline and powder samples, because the HRC at the pulsed neutron source facilitates the possibility of extending the (Q, E) space using the sub-eV range of incident neutron energies.

Acknowledgements
This neutron scattering experiment was approved by the Neutron Scattering Program Advisory Committee of the Institute of Materials Structure Science, High Energy Accelerator Research Organization (Proposal Nos. 2011S01 and 2012S01). This work was partially supported by a Grant-in-Aids for Scientific Research from the Japanese Ministry of Education, Culture, Sports, Science and Technology (MEXT), and by the Elements Strategy Initiative Center for Magnetic Materials (ESICMM) under the outsourcing project of MEXT.

References
[1] Robinson R A , McQueeney R J, and Kelley T M 1998 *Physica* B **241-243** 161
[2] Aisa D, Aisa S, Babucci E, Barocchi F, Cunsolo A, De Francesco A, Formisano F, Gahl T, Guarini E, Laloni A, Mutka H, Orecchini A, Petrillo C, Pilgrim W -C, Piluso A, Sacchetti F, Suck J -B, and Venturi G 2006 *J. Non-Cryst. Solids* **352** 5130
[3] Itoh S, Endoh Y, Yokoo T, Kawana D, Kaneko Y, Tokura Y, and Fujita M 2013 *J. Phys. Soc. Jpn.* **82** 043001
[4] Itoh S, Yokoo T, Satoh S, Yano S, Kawana D, Suzuki J, and Sato T J 2011 *Nucl. Instr. Meth. Phys. Res.* A **631** 90
[5] Endoh Y and Hirota K 1997 *J. Phys. Soc. Jpn.* **66** 2264
[6] Moussa F, Hennion M, Kober-Lehouelleur P, Reznik D, Petit S, Moudden H, Ivanov A, Mukovskii Y M, Privezentsev R, and Albenque-Rullier F 2007 *Phys. Rev.* B **76** 064403
[7] Fang Z, Nagaosa N, Takahashi K S, Asamitsu A, Mathieu R, Ogasawara T, Yamada H, Kawasaki M, Tokura Y, and Terakura K 2003 *Science* **302** 92
[8] Mayer H M, Steiner M, Stüßer N, Weinfurter H, Kakurai K, Doner B, Lindgård P A, Clausen K N, Hock S, Rodewald W 1991 *J. Magn. Magn. Matter.* **97** 210
[9] Teixeira J, Bellissent-Funel M C, Chen S H, Dorner B 1985 *Phys. Rev. Lett.* **54** 2681

Spin Waves in Ferromagnetic Phase of MnP

S Itoh[1,2], S Yano[3,5], T Yokoo[1,2], S Satoh[1], D Kawana[2,6], Y Kousaka[3], J Akimitsu[3], Y Endoh[1,4]

[1]Neutron Science Division, Institute of Materials Structure Science, High Energy Accelerator Research Organization, Tsukuba 305-0801, Japan
[2]Condensed Matter Research Center, Institute of Materials Structure Science, High Energy Accelerator Research Organization, Tsukuba 305-0801, Japan
[3]Department of Physics and Mathematics, Aoyama Gakuin University, Sagamihara 229-8558, Japan
[4]Institute for Materials Research, Tohoku University, Sendai 980-8577, Japan

E-mail: shinichi.itoh@kek.jp

Abstract. Inelastic neutron scattering experiments on an intermetallic compound, MnP, were performed by using a chopper spectrometer as well as triple axis spectrometers. Spin waves were observed in the ferromagnetic phase in the entire Brillouin zone along the a^*- and b^*-axes. The observed dispersion relations of spin waves were well described by an isotropic Heisenberg interaction adding a single ion anisotropy with two sub-lattices.

An intermetallic compound, MnP, is ferromagnetic below T_C, and transforms into a proper screw spiral phase at T^* [1, 2]. The transition temperatures were measured to be $T_C = 292.6$ K and $T^* = 46.2$ K for the present sample [3]. The crystal structure of MnP is the distorted NiAs-type structure with the lattice constants, $a = 5.917$ Å, $b = 5.260$ Å, and $c = 3.173$ Å at room temperature [1]. In the ferromagnetic phase, the easy axis of the magnetization is the c^*-axis and the hardest one is the a^*-axis. In the proper screw spiral phase, the spin rotates in the b^*-c^* plane with a propagation vector of $0.117a^*$ [2, 4]. Very recently, a canted antiferromangetic structure with weak ferromagnetic magnetization along the b^*-axis in the ferromagnetic phase and an alternatively tilted helimagnetic structure below T^* were found [5].

In order to understand the origin of these complicated properties of MnP, the microscopic interactions should be determined by observing spin waves [6, 7]. However, these early studies are limited to observations on the low energy (E) and low momentum (q) region. It was suggested that the emergence of the proper screw spiral phase is caused by a delicate balance between the nearest and the next-nearest neighbor exchange interactions [6]. Subsequently, it was reported that the spin-wave dispersion relation in the ferromagnetic phase exhibits an unusual q dependence along the a^*-axis, while the conventional q^2 law was observed along the b^*- and c^*-axes [6, 7].

It is essential to determine the spin dynamics in the entire Brillouin zone, in order to elucidate the microscopic origin of magnetism in MnP. As a first step of this study, we have determined the spin wave dispersion relations in the entire Brillouin zone along the a^*- and b^*-axes in the

[5] Present address: Department of Physics, University of Virginia, Charlottesville, VA 22904-4714, USA
[6] Present address: The Institute for Solid State Physics, The University of Tokyo, Tokai 319-1160, Japan

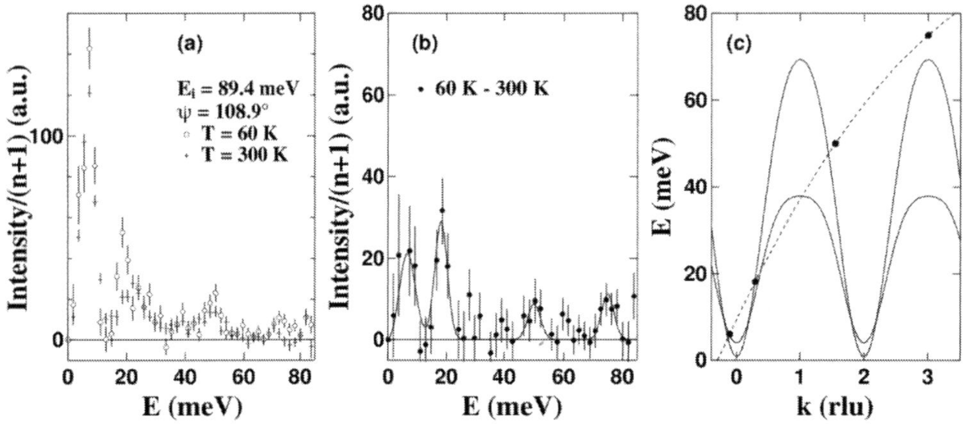

Figure 1. Typical excitation spectra in MnP along the b^*-axis, after background subtraction (a). Spin-wave spectrum with a fitted curve (solid line) (b). The observed peak positions are located at the crossing points between the scan locus (dashed line) and the spin-wave dispersion curves (solid lines, from Fig. 3) within the resolution (c).

ferromagnetic phase of MnP, by using a chopper spectrometer installed at a pulsed neutron source as well as triple axis spectrometers at a steady state neutron source [3].

Inelastic neutron scattering experiments were performed on the High Resolution Chopper Spectrometer (HRC), which is installed at the Materials and Life Science Experimental Facility (MLF) of the Japan Proton Accelerator Research Complex (J-PARC) [8], also on the triple axis spectrometers, LTAS and TAS-1, which are installed at the JRR-3M reactor in the Tokai Institute of the Japan Atomic Energy Agency.

On the HRC, the monochromatic neutron beam with the energy of E_i is incident upon the sample, and the energy transfer E is determined by the time-of-flight (TOF) of the detected neutron. The single-crystal sample of MnP was mounted with the c^*-axis vertical, and the experiments were performed at temperatures $T = 60$ and 300 K. Empty scans were also performed by removing the sample crystals, in order to estimate the background. Figure 1(a) shows the observed energy spectra at $\phi = 18.9°$ with $E_i = 89.4$ meV and $\psi = 108.9°$ on the HRC, where the background was subtracted and the intensities were divided by the temperature factor, $n + 1 = [1 - exp(-E/k_BT)]^{-1}$, with k_B the Boltzmann constant. ϕ and ψ are the scattering angle and the crystal angle between the a^*-axis and the incident neutron beam, respectively. In this setup, the detector element at $\phi = 18.9°$ scans $(2k0)$ [3]. The energy resolution was determined to be $\Delta E = 5.0$ meV by measuring the energy width (full width at half maximum) of the incoherent elastic scattering. Although excitation peaks were observed at $T = 60$ K, some peaks remained even at $T = 300$ K, which might be due to phonons. The energy spectrum subtracting the spectrum at $T = 300$ K from that at $T = 60$ K was assumed to be a spin-wave spectrum, and the peaks were fitted with Gaussians with a resolution width of $\Delta E = 5.0$ meV, as shown in Fig. 1 (b). The peak positions are located at the crossing points between the scan locus and the dispersion curve, as shown in Fig. 1 (c). To obtain the dispersion relations of the spin waves in the entire Brillouin zone for the a^*- and b^*-axes, we chose scans with various sets of E_i and ψ [3]. The peak positions determined in this way are plotted in Fig. 3 in the reduced zone scheme.

Because anomalous properties along the a^*-axis were reported [6, 7], we also measured excitations along the a^*-axis using triple axis spectrometers. The LTAS spectrometer was used for the energy region below 2 meV with $\Delta E = 0.19$ meV, and the TAS-1 was used up to 35 meV with $\Delta E = 1.5$ meV, both measurements were performed with a constant-q scan. Spin-wave peak positions obtained on these spectrometers at $T = 60$ K are plotted in Fig. 3. The peak positions at $T = 77$ K, which were previously reported [6, 7], were also plotted in Fig. 3.

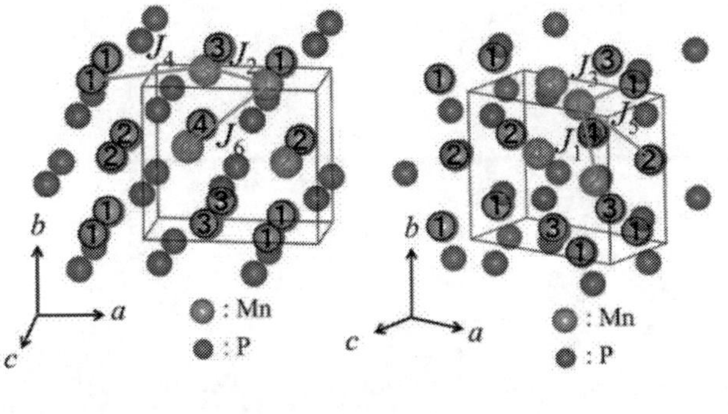

Figure 2. Crystal structure of MnP and definition of the exchange constants, J_1 - J_6, in the order of the distance between Mn pairs. The distances between the Mn pairs for J_1, ... J_6 are 2.70, 2.82, 3.17, 3.93, 4.17, and 4.24 Å, respectively.

As shown in Fig. 2, the unit cell in MnP includes four Mn sites: Mn1 at $(-x, -y, -z)$, Mn2 at $(x + 1/2, -y + 1/2, -z)$, Mn3 at (x, y, z), and Mn4 at $(-x + 1/2, y + 1/2, z)$, where $(x, y, z) = (0.303410, 0.004520, 0.25)$ [1, 9]. The coordinate system is defined as $x//a$, $y//b$, $z//c$ with the units of (a, b, c). We see that, approximately, $(x, y, z) = (0.25 + u, 0 + v, 0.25)$ with $u = 0.05$ and $v = 0.005$, as the Mn lattice deviated from a primitive orthorhombic lattice. Since v is negligibly small, the Mn lattice can be approximately divided into two sub-lattices. We consider an isotropic Heisenberg interaction adding a single ion anisotropy [9], as follows:

$$H = -\sum_{nn'jj'} J_{nn'jj'} \mathbf{S}_{nj} \mathbf{S}_{n'j'} - D \sum_{nj} (S^z_{nj})^2, \quad (1)$$

where $\mathbf{S}_{nj}$ is the spin of a Mn atom located at the nth unit cell and the jth sub-lattice in a unit cell ($j = 1, 2$). We took into account of exchange interactions between Mn pairs up to the 6th neighbors, as discussed in a theory presented before [10]. The eigenvalues of the Hamiltonian in Eq. (1) lead the two dispersion relations of spin waves: acoustic and optical modes.

The dispersion curves were well fitted to the calculated curves with the parameters of $J_1 = 0.37 \pm 0.14$, $J_2 = -0.65 \pm 0.10$, $J_4 = -0.06 \pm 0.13$, $J_5 = 0.26 \pm 0.10$, $J_6 = 0.64 \pm 0.08$, and $D = 0.10 \pm 0.02$ meV, as shown in Fig. 3. Note that J_3 will be determined by experiments along the c^*-axis. Along the a^*-axis, the acoustic and optical modes are separated from each other, but, along the b^*-axis, these two modes resonate with each other at around $k = 0.2$ rlu (reciprocal lattice unit) and switch roles, but the two modes are not crossed. As shown in Fig. 3 (c), the low energy part in the a^*-axis for the presently-improved dispersion relation was also well fitted to the present model, where the dispersion curve along the a^*-axis shows a minimum at around $h = 0.14$ rlu. This complicated q dependence is caused by the displacement parameter u, because the minimum disappeared for $u = 0$. Note that such minimum was previously observed at $T = 77$ K [6], as shown in Fig. 3 (c), and that the dispersion relation along the a^*-axis shows the q^3 dependence at $T = 150$ K [7]. The fitted curve for the acoustic mode at $T = 60$ K was lower than the data at $T = 77$ K, this is consistent with the reported T dependence [6, 7]. In the previous study, the dispersion relation along the b^*-axis was determined to be $E(k) = D_b k^2 + E_0$ with $D_b = 145$ meV at $T = 150$ K almost independently on T [7]. In fact, the spin-wave positions at $T = 77$ K are on the low enery part of the fitted acoustic mode for the data at $T = 60$ K at present, as shown in Fig. 3 (a). By using the present parameters, the stiffness constant was estimated to be $D_b = 170 \pm 31$ meV [3]. The new value is in approximate agreement with the previous value. Observed spin-wave peak positions are well explained by the two modes. In fact, two spin-wave peaks were clearly observed at $h = 0.5$ rlu on the TAS-1.

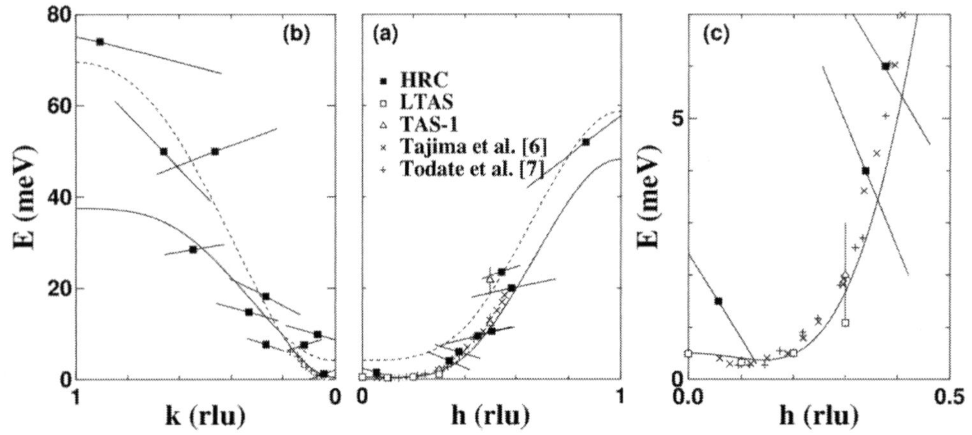

Figure 3. Dispersion relation of spin waves in the ferromagnetic phase of MnP along the a^*- (a) and b^*-axes (b) at $T = 60$ K, with calculated curves, acoustic (solid line) and optical (dashed line) modes. The bars on the marks represent statistical errors along the scan directions. Low energy part along the a^*-axis (c). The spin-wave peaks at $T = 77$ K, which were reported previously [6, 7], are also plotted.

In summary, we successfully observed spin waves in the ferromagnetic phase of MnP along the a^*- and b^*-axes in the entire Brillouin zone up to the zone boundary, by using the HRC, LTAS and TAS-1. The dispersion relations were well described by an isotropic Heisenberg model adding a single ion anisotropy with two sub-lattices, and the anomalous dispersion relation along the a^*-axis was found to be caused by the atomic displacement from the primitive lattice. The previous studies suggested that it is caused by an exchange competition [6], or attributed to itinerant nature of d electrons [7, 11]. The recent structural study suggested the existence of the Dzyaloshinsky-Moria interaction [5]. Further investigations are required to resolve this problem.

The authors are grateful to Y. Todate for supplying the specialist knowledge required for this study, and to N. Metoki, K. Kakurai and S. Wakimoto for detailed directions on the neutron scattering experiments at JRR-3M. The experiments at J-PARC were approved by the Neutron Scattering Advisory Committee of the Institute of Materials Structure Science, of the High Energy Accelerator Research Organization.

References

[1] Huber E E Jr and D. H. Ridgeley D H 1964 *Phys. Rev.* **135** A1033
[2] Forsyth J B, Pickart S J, and Brown P.J 1966 *Proc. Phys. Soc.* **88** 333
[3] Yano S, Itoh S, Yokoo T, Satoh S, Kawana D, Kousaka Y, Akimitsu J, and EndohY 2013 *J. Magn. Magn. Mater.* doi:10.1016/j.jmmm.2013.07.019
[4] Obara H, Endoh Y, Ishikawa Y, and Komatsubara T 1980 *J. Phys. Soc. Jpn.* **49** 928
[5] Yamazaki T, Tabata Y, Waki T, Sato T J, Matsuura M, Ohoyama K, Yokoyama M, and Nakamura H 2011 *arXiv:1106.4599v2*
[6] Tajima K, Ishikawa Y, and Obara H 1980 *J. Magn. Magn. Mater.* **15-18** 373
[7] Todate Y, Yamada K, Endoh Y, and Ishikawa Y 1987 *J. Phys. Soc. Jpn.* **56** 36
[8] Itoh S, Yokoo T, Satoh S, Yano S, Kawana D, Suzuki J, and Sato T J 2011 *Nucl. Instr. Meth. Phys. Res.* A **631** 90; Yano S, Itoh S, Satoh S, Yokoo T, Kawana D, and Sato T J 2011 *ibid* **654** 421
[9] Todate Y 1988 *Doctor Thesis* (Tohoku University)
[10] Takeuchi S and Motizuki K 1967 *J. Phys. Soc. Jpn.* **24** 742
[11] Taase A and Kasuya T 1979 *J. Phys. Soc. Jpn.* **47** 491

Dynamical Properties of Spins and Holes in Carrier Doped Quantum Haldane Chain

T Yokoo[1,2], S Itoh[1,2], D Kawana[3], H Yoshizawa[3] and J Akimitsu[4]

[1] IMSS, High Energy Accelerator Research Organization (KEK), Tsukuba, Ibaraki 305-0801, Japan
[2] MLF, J-PARC Center, Tokai, Ibaraki 319-1195, Japan
[3] ISSP, the University of Tokyo, Tokai, Ibaraki 319-1160, Japan
[4] Department of Physics and Mathematics, Aoyama-Gakuin University, Sagamihara, Kanagawa 229-8558, Japan

E-mail: tetsuya.yokoo@kek.jp

Abstract. Quantum spins in one-dimensional (1D) chains exhibit characteristic features such as the Haldane gap (S=1) and the spin-Peierls state (S=1/2). Recently multi degrees of freedom in cooperation with quantum spins are focused. We present here spin dynamics in the lightly hole doped $Nd_{2-x}Ca_xBaNiO_5$ (x=0.035), 1D Haldane system by means of pulsed neutron inelastic scattering method. One-magnon band with a spin gap (Haldane gap) was visible with the decrease of the zone boundary energy to 60 meV. In contrast, the spin gap slightly increases. Dynamical structures inside spin gap upon carrier doping showing incommensurate structure centered at the magnetic zone center was newly observed which is possibly originating from the doped holes in Haldane chains.

1. Introduction

Despite the three decades that have passed since Haldane's prediction on a one-dimensional (1D) Heisenberg chain with integer spin, much attentions are still paid on its research [1, 2]. Inorganic nickelate Y_2BaNiO_5 was found to be a good candidate of Haldane system with a spin gap of 10 meV at the magnetic zone center (MZC) [3]-[6]. In contrast, R_2BaNiO_5 (where R is a magnetic rare earth) has been known as a classical antiferromagnet (AF) showing three-dimensional (3D) long range magnetic ordering at low temperature [7]. However, the saturation moment of the sublattice magnetization of Ni^{2+} spins is rather suppressed indicating a remaining quantum spin fluctuation. Indeed, inelastic neutron scattering experiments revealed the existence of spin gap at the MZC even in the R_2BaNiO_5 with magnetic rare earth ions [8]-[11].

Another interest in transition metal chain oxides is the doping of carriers, which yields phenomena ranging from strongly renormalized Fermi liquid behavior to high-temperature superconductivity. In view of the enhanced quantum fluctuations in 1D systems, doping of transition metal chains could well lead to equally surprising discoveries. R_2BaNiO_5 is a charge transfer insulator [15, 16] and some amount of hole carriers are successfully introduced primarily on oxygen sites by replacing a part of the off-chain R^{3+} by Ca^{2+} [4]. The spin dynamics of hole doped Ni 1D chains have also been addressed by using a theoretical approach [17]-[19] and through neutron experiments [20, 21]. In this article, we present the spin and hole dynamics of hole doped Nd_2BaNiO_5 measured by means of a pulsed neutron scattering technique to survey the entire area in momentum and energy space of the Brillouin zone.

2. Experimental procedure

The carrier content can be controlled by substituting a portion of the Nd^{3+} ions by Ca^{2+}. Slightly doped x=0.035 in the chemical formula $Nd_{2-x}Ca_xBaNiO_5$ was chosen for neutron experiments. T_N is evaluated at 38 K. The measurement temperatures were 5 K (7.8 K for E_i=102.4 meV), 50 K and 150 K. Inelastic neutron scattering experiments were carried out on the High-Resolution Chopper spectrometer (HRC) installed at Materials and Life Science Experimental Facility in J-PARC [22, 23]. Two sets of incident neutron energies E_i=102.4 and 30.6 meV were selected to measure the wide Q- ω space and the detailed structure around the MZC, respectively. The energy resolution of our experimental condition is $\frac{dE}{E_i}$ ~7% (dE ~7 meV for E_i=102.4 meV and dE ~2 meV for E_i=30.6 meV).

3. Results and discussion

In the neutron scattering experiment, the $[H\ 0\ L]$ zone was into the horizontal scattering plane, where H corresponds to the chain direction (crystallographic a-axis). Figure 1 shows the obtained dynamical structures $S(Q,\omega)$ at T=7.8 K (well below T_N=38 K) and T=50 K (above T_N) are shown. Both spectra are normalized by the measurement time (proton current measured just before the neutron target) and also normalized by the Bose population factor as temperature correction. Intense crystal field excitations of Nd^{3+} ions lie at E=24 meV and 38 meV. The site symmetry of Nd^{3+} in Nd_2BaNiO_5 is sufficiently low to split the tenfold degenerate J=9/2 multiplet into five Kramers doublets. The stripes in 2D maps appearing as moire patterns viewed particularly at low-Q region are spurious and result from data-binning processes. We are currently working on resolving this problem.

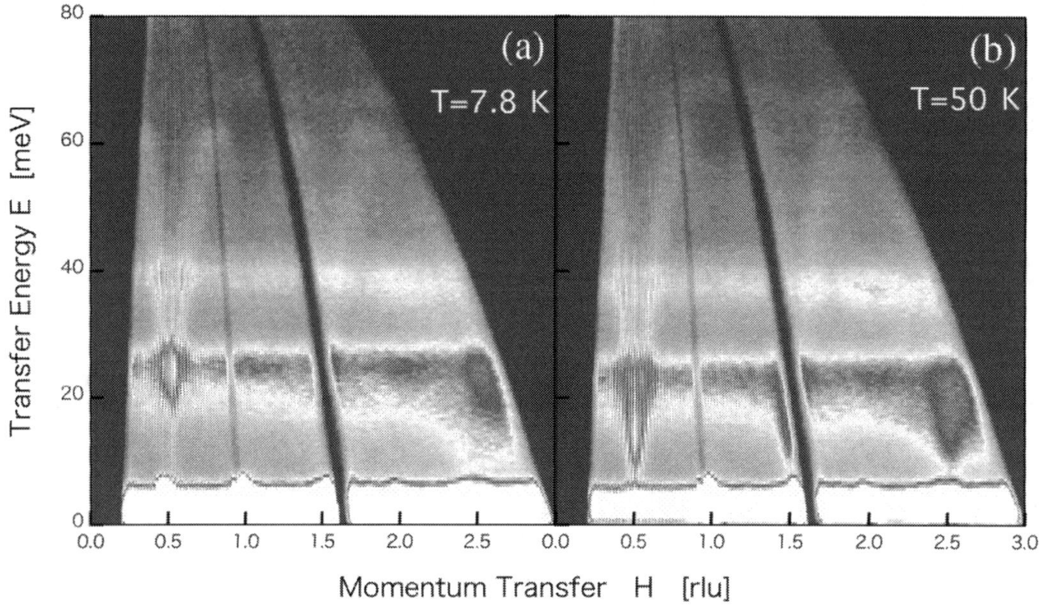

Figure 1. Observed dispersion relation of x=0.035 sample at (a) T=7.8 K $< T_N$, and (b) T=50 K $> T_N$.

Excellent dispersion relations for the Ni 1D chain are observed at both temperatures. The minima of the dispersion observed at the MZC, H=0.5, 1.5 and even 2.5, occur where magnetic scattering intensity gets weak because of the magnetic form factor and phonon scattering is dominant in this high-Q region. At low temperature, the magnon band width (energy at the zone boundary) reaches at 60 meV with a somewhat large gap energy of $\Delta \sim$20 meV. Also, a flat band lying at E=65 meV corresponding to a phonon or magnon optical branch was observed. At T=50 K, the gap is closed and the spectral weight at the zone boundary is slightly shifted with respect to the MZC. The boundary energy is almost similar both temperatures.

To survey in the vicinity of the gap, low incident energy experiments were performed [24]. Figure 2 shows a Q-ω map measured at three temperatures. At T= 50 K, in addition to the Haldane gap at Δ=10 meV, a weak excitation at the MZC (H=0.5) was appeared as a result of hole doping. Since the hole content is low, the structure below the gap still follows the commensurate respecting to 1D H direction. At well above T_N, T=150 K, the $S(Q,\omega)$ observed at T=50 K seems to be identical except for its intensity, the change of which can be interpreted by thermal smearing. However, marked changes are taking place at T=5 K. The 1D chain dispersion with the Haldane gap almost completely disappears around the MZC. Still weak signals at around 10 meV and below 10 meV remain. Also, a flat localized excitation newly appears at 4 meV. In this mixed-spin system, the Haldane gap is known to rapidly increase in energy under the effective staggered magnetic field [13, 14]. The 3D magnetic correlation of Ni^{2+} ions via rare earth (Nd^{3+}) ions is well developed at low temperature. Therefore, the gap is effectively diminishing. Note that the intense CEF excitations are located above 20 meV. The somewhat broad feature at around 10 meV corresponds to the structure newly found by Sasaki and co-workers [21]. It shows localized but broad in Q-ω space centered at the MZC. The origin of this excitation is unclear at present and further experiments are necessary. The weak feature below 10 meV vertically spreading at the MZC (H=0.5) can be interpreted as a dynamical properties of doped holes. Holes are not well conductive in this system but still dynamically moved in the 1D chains. At high hole content, an incommensurate structure along the Q direction is observed [20, 21]. The incommensurability, however, does not drop to zero even at x=0. The

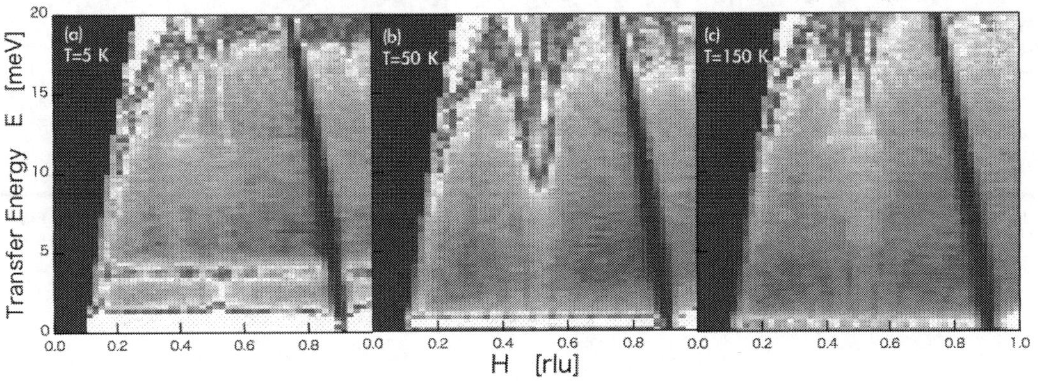

Figure 2. Low energy region with E_i=30 meV. All the temperature scans are normalized by the proton current. Also, the temperature factor is corrected to compare each spectrum directly.

excitation at E=4 meV is similar to that was reported in Refs. [8, 9]. Completely flat dispersion is found in this experiment. This appears only below the 3D magnetic ordering temperature, and developed as a magnetization order parameter possibly related to the Nd spins. Thus, the entire dynamical set of properties of a hole doped 1D Ni chain was successfully measured by pulsed neutron inelastic scattering. At T_N, the dynamical structure drastically changes presenting new features of a local excitation at $E \sim 10$ meV, doped hole dynamics below the gap energy, and 4 meV flat excitation well below T_N in addition to the whole picture of the 1D chain dispersion. We plan to further experiments with various hole content and temperature dependence in order to identify all the properties found.

Acknowledgments
One of the authors T.Y. is partly supported by a Grant-in-Aid for Scientific Research of MEXT (Scientific Research (B) No. 25287081). T.Y. and S.I. are supported by a Grant-in-Aid for Scientific Research of MEXT (Scientific Research (C) No. 24540352). The neutron scattering experiment was approved by the Neutron Scattering Program Advisory Committee of IMSS, KEK (Proposal No. 2013S01).

References

[1] Haldane F D M 1983 *Phys. Lett.* **93** 464
[2] Haldane F D M 1983 *Phys. Rev. Lett.* **50** 1153
[3] Darriet J and Regnault L P 1993 *Solid State Commun.* **86** 409
[4] DiTusa J F, Cheong S W, Park J H, Aeppli G, Broholm C and Chen C T 1994 *Phys. Rev. Lett.* **73** 1857
[5] Yokoo T, Sakaguchi T, Kakurai K and Akimitsu J 1995 *J. Phys. Soc. Jpn.* **64** 3651
[6] Sakaguchi T, Kakurai K, Yokoo T and Akimitsu J 1996 *J. Phys. Soc. Jpn.* **65** 3025
[7] Sachan V, Buttrey D J, Tranquada J M and Shirane G 1994 *Phys. Rev. B* **49** 9658
[8] Zheludev A, Tranquada J M, Vogt T and Buttrey D J 1996 *Phys. Rev. B* **54** 7210
[9] Zheludev A, Tranquada J M, Vogt T and Buttrey D J 1996 *Phys. Rev. B* **54** 6437
[10] Yokoo T, Zheludev A, Nakamura M and Akimitsu J 1997 *Phys. Rev. B* **55** 11516
[11] Yokoo T, Raymond S, Zheludev A, Maslov S, Ressouche E, Zaliznyak I, Erwin R, Nakamura M and Akimitsu J 1998 *Phys. Rev. B* **58** 14424
[12] Raymond S, Yokoo T, Zheludev A, Nagler S E, Wildes A and Akimitsu J 1999 *Phys. Rev. Lett.* **82** 2382
[13] Maslov S and Zheludev A 1998 *Phys. Rev. B* **57** 68
[14] Maslov S and Zheludev A 1998 *Phys. Rev. Lett.* **80** 5786
[15] Mattheiss L F 1993 *Phys. Rev. B* **48** 4532
[16] Eisaki H and Uchida S 1995 *J. Phys. Chem. Solids* **56** 1811
[17] Dagotto E, Riera J, Sandvik A and Moreo A 1996 *Phys. Rev. Lett.* **76** 1731
[18] Batista C D, Aligia A A and Eroles J 1998 cond-mat/9810387
[19] Penc K and Shiba H 1995 *Phys. Rev. B* **52** R715
[20] Xu G, Aeppli G, Bisher M E, Broholm C, DiTusa J F, Frost C D, Ito T, Oka K, Paul R L, Takagi H and Treacy M M J 2000 *Science* **289** 419
[21] Sasaki T, Yokoo T, Katano S and Akimitsu J 2005 *J. Phys. Soc. Jpn.* **74** 267
[22] Itoh S, Yokoo T, Satoh S, Yano S, Kawana D, Suzuki J and Sato T J 2011 *Nucl. Instrum. Methods A* **631** 90
[23] Itoh S, Ueno K and Yokoo T 2012 *Nucl. Instrum. Methods A* **661** 58
[24] By using the neutron energy coming through the Fermi chopper with different rotational phasing $2\pi n$(where n is an integer), inelastic spectra can be measured simultaneously in a multi-incident energy measurement known as a repetition rate multiplication (RRM) method.

Construction of Polarized Inelastic Neutron Spectrometer in J-PARC

T Yokoo[1,2], K Ohoyama[3], S Itoh[1,2], J Suzuki[4], M Nanbu[5], N Kaneko[1,2], K Iwasa[6], T J Sato[7], H Kimura[7] and M Ohkawara[3]

[1] IMSS, High Energy Accelerator Research Organization (KEK), Tsukuba, Ibaraki 305-0801, Japan
[2] J-PARC Center, Tokai, Ibaraki 319-1195, Japan
[3] IMR, Tohoku University, Sendai 980-8577, Japan
[4] MEC, High Energy Accelerator Research Organization (KEK), Tsukuba, Ibaraki 305-0801, Japan
[5] Center of Neutron Science for Advanced Materials, IMR, Tohoku University, Sendai 980-8577, Japan
[6] Department of Physics, Tohoku University, Sendai 980-8578, Japan
[7] IMRAM, Tohoku University, Sendai, 980-8577, Japan

E-mail: tetsuya.yokoo@kek.jp

Abstract. The instrumental design of the Polarization Analysis Neutron Spectrometer with Correlation Method (POLANO) is almost ready for its construction, although some discussion on the design still remains. POLANO is a new inelastic neutron spectrometer in the Japan Proton Accelerator Research Complex (J-PARC) utilizing polarized neutrons for comprehensive materials research, focusing on the use of quasi-elastic and inelastic scattering techniques. For instrumental construction, the basic shield design is complete and the shielding capability against radiation has been assessed. Additionally, the designs of the beam transport section using 4 Q_c supermirror guide tube are almost complete. Also, detecting section and a large vacuum chamber is now under designing. The development of polarization and chopper devices are the key to the success for the very first application of polarization analysis to inelastic scattering at spallation neutron source. High performance Fermi and T0 choppers are now under manufacturing.

1. Introduction

The success of the chopper spectrometer MARI applied to materials science at the ISIS neutron scattering facility [1], has highlighted the importance of the so-called chopper spectrometers to neutron beam line instruments. At the Japan Proton Accelerator Research Complex (J-PARC), the Materials and Life Science Experimental Facility (MLF) is a newly constructed world leading pulse neutron experimental facility, where three chopper-type spectrometers are now part of the research program. One of the remaining technical issues that should be realized in MLF is the use of a polarization neutron technique with a pulsed neutron beam. Some of elastic neutron instruments such as a small angle scattering instrument, a reflectometer, a single crystal diffractometer and an imaging instrument plan to use polarized neutron technique option but is not planned for inelastic scattering instruments yet. Polarized neutron experiments are quite significant for material science particularly for research on magnetism, hydrogen materials, and strongly correlated electron systems entangled with multiple physical degrees of freedom. Therefore new spectrometers that enable us to realize polarization analysis are very much expected as forthcoming instruments.

A joint project between the Institute of Materials Structure Science (IMSS), KEK, and the Institute of Material Research (IMR) of Tohoku University was initiated with the aim of future construction of the spectrometer and research program for functional magnetic materials in 2009. Now, the POLANO (POLarization Analysis Neutron spectrometer with cOrrelation method) project has begun. Our principal concept is to achieve higher-energy polarization analysis of inelastic scattering beyond a reactor- based neutron source. The suite of inelastic spectrometers involving an inverted geometry spectrometer and a spin-echo instrument in MLF can cover a relatively wide area in momentum (Q) and energy (E) space with various Q and E resolutions. Slow dynamics in soft matter to electron band excitation are targeted in the dynamical range of 10^{-9} eV $\leq E \leq 2\times10^{0}$ eV. For detailed dynamical studies of advanced material science, polarization analysis is an inevitable technique in addition to the wide Q-E coverage. Hence, we propose a chopper-type polarized neutron spectrometer which enables us to carry out inelastic experiments with reasonable intensity for research of dynamical properties in materials.

2. Geometrical Condition

Because the instrumental parameters of POLANO were modified from what was reported last year in Ref. [2], the revised parameter list is shown in Table 1. E_i and k_i denote incident neutron energy and wave number of incident neutron, respectively in the Table. The design features of POLANO are compactness with reasonably wide sample space design, higher neutron flux, higher neutron energy polarization up to 100~150 meV, and medium resolution. The geometrical lengths were chosen as L_1 = 17.5 m (moderator-sample distance), L_2 = 2.0 m (sample-detector distance), and L_3 = 1.85 m (Fermi chopper-sample distance). We used a slightly shorter distance in L_2 from Ref. [2], but the energy and momentum resolution are not significantly affected.

Table 1. Basic instrumental parameters in POLANO

Parameters	
beam port	beam line 23 (BL23)
moderator	H_2 decoupled
L_1 (moderator-sample distance)	17.5 m
L_2 (sample-detector distance)	2.0 m
L_3 (Fermi chopper-sample distance)	1.85 m
energy resolution $\Delta E/E_i$	3~5% at elastic position
momentum resolution $\Delta Q/k_i$	1~2%
sample size	20 × 20 mm
detector angle (horizontal)	3° ~120°
detector angle (vertical)	-7° ~7°

3. Beam Transport

A 3D representation of POLANO is shown in Fig. 1. The 23rd beam line, where POLANO will be installed, is viewing decoupled moderator with a known pulse width of $\Delta t_m = a/\sqrt{E_i}$ with a=2.5 for decoupled moderator. After moderator neutrons are transported by $m = 4$ supermirror guide tubes optimized for 110~120 meV neutron intensity with a McStas simulation [3, 4]. An elliptical guide optimized section by section (section length = 50 cm) can yield a neutron flux of 3.9×10^{5} [n/(sec·meV·cm^{2}·MW)], which is almost comparable with a coupled moderator beam

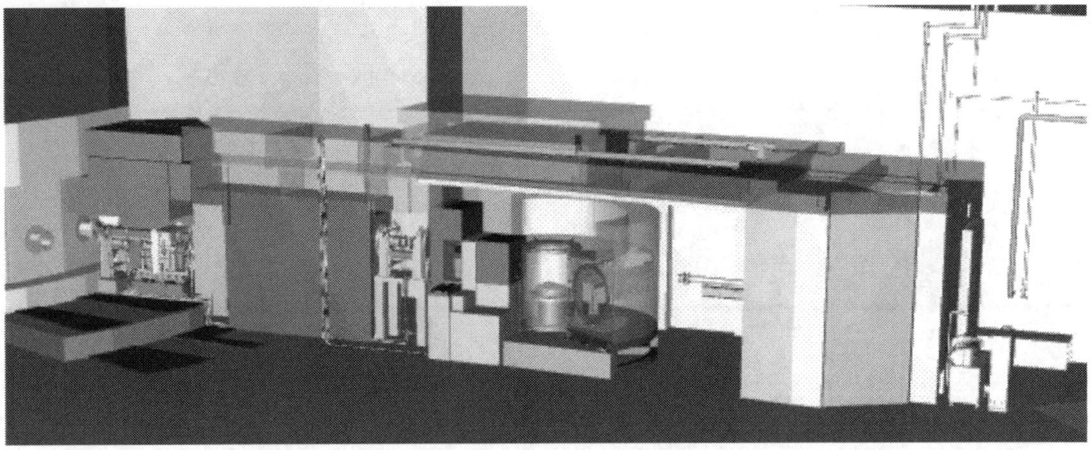

Figure 1. Schematic view of POLANO spectrometer

line at E ~100 meV. The finally designed view of the guide tube is illustrated in Fig. 2. We tested several neutron energies to be optimized, namely, the range of 20-30 meV, 60-70 meV and 110-120 meV. Additionally, the evaluated neutron flux is plotted in Fig 3 as a function of neutron energy. As mentioned above, our target energy range is higher than that of the reactor-based

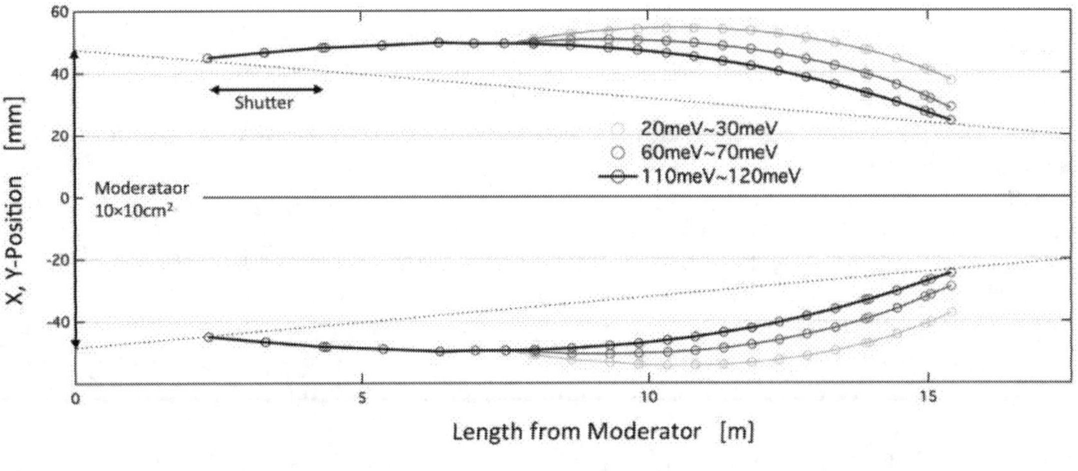

Figure 2. Evaluated guide tube cross section optimized for neutron energies 20-30 meV (light blue symbol), for 60-70 meV (blue symbol) and 110-120 meV (black symbol).

polarization instrument. Therefore, the flux of much higher neutron energy will be significant to realize an effective polarization experiment. Several energy ranges were evaluated to see the flux distribution in energy. For the calculation at 110~120 meV, we slightly lose intensity at energies less than 50 meV, but gain considerably at energies higher than 50 meV.

4. Mechanical Choppers

Below the exit of a biological shield, several types of choppers are placed. T0 chopper is the first chopper to eliminate fast and epithermal neutrons with a massive metal blade rotating with a

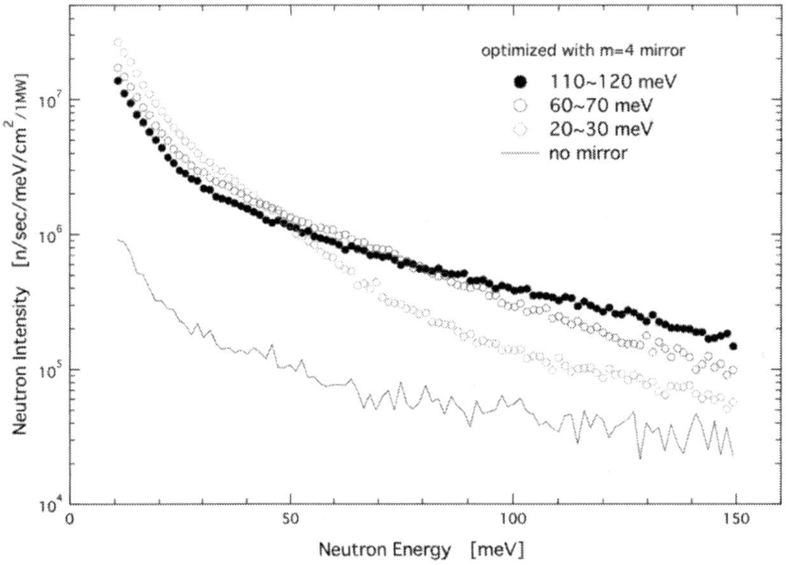

Figure 3. Simulated results of the effect of guide tubes by McStas. The $m = 4$ mirror is used to optimize for several neutron energies.

maximum speed of 100 Hz [5]. The blade is made of Inconel X750, which is known for having high strength without using radiological elements with long lives such as cobalt. Though the weight of the blade is over 120 kg, at 100 Hz of rotating speed, less than ± 5 μs of jitter (phase control accuracy) was achieved. In order to avoid the frame overlap of a 25 Hz proton (neutron) pulse in MLF, a slow rotating disk chopper will be installed in the beam line. Two disk choppers synchronized with neutron pulse are placed at $L = 7.5$ and 13.9 m. A Fermi chopper will be used for monochromatizing the incoming neutron energy. It is the same type of Fermi chopper developed for High Resolution Chopper spectrometer in MLF [6], which has a 600 Hz maximum speed with 0.3 μs jitter enabling high resolution and transmission experiments.

For more effective measurements, we adopt the cross correlation method that utilizes multiple wavelengths of neutrons for one inelastic measurement. Hence, we have intensity gains of a factor of ten or higher in principle. A special type of disk chopper (correlation chopper) is required to realize this correlation method. The disk diameter is ideally over 1000 mm with a rotational speed of 350 Hz. Additionally, 255 (2^N-1) open and close sequences are required to be on the disk [7]. These conditions are mechanically quite difficult to realize because the maximum stress evaluated is almost 4000 MPa. Totally new designing of the disks is now under way with carbon fiber reinforced plastic (CFRP) materials to realize a mechanically rotating correlation chopper.

5. Schedule

We have almost completed the design of the spectrometer and have already started manufacturing of several components such as the T0 chopper, detectors, guide tubes and so forth. The construction of a large beam line shielding will start in November of this year. Also we have started contracts for building the vacuum chamber, correlation chopper, a part of the supermirror analyzer, and other utilities. Most part of the instrument will be completed by March 2015. On the other hand, some crucial polarization devices still require R&D, and will take considerable time to be completed. Polarization devices in POLANO are described in a separate article [8].

Acknowledgments
One of the authors T.Y. is partly supported by a Grant-in-Aid for Scientific Research of MEXT (Scientific Research (B) No. 25287081). T.Y. and S.I. are supported by a Grant-in-Aid for Scientific Research of MEXT (Scientific Research (C) No. 24540352). K.I., K.O., T.J.S. and S.I. are supported by a Grant-in-Aid for Scientific Research of MEXT (Scientific Research (A) No. 23244068) The neutron scattering experiment was approved by the Neutron Scattering Program Advisory Committee of IMSS, KEK (Proposal No. 2013S01).

References

[1] Taylor A D, Arai M, Bennington S M, Bowden Z A, Osborn R, Andersen K, Stirling W G, Nakane T, Yamada K and Welz D 1991 *KEK Report (Proc. 11th Meet. Int. Collab. Advanced Neutron Sources, ICANS-XI)* **90-12** 705
[2] Yokoo T, Ohoyama K, Itoh S, Suzuki J, Iwasa K, Sato T J, Kira H, Sakaguchi Y, Ino T, Oku T, Tomiyasu K, Matsuura M, Hiraka H, Fujita M, Kimura H, Sato T, Suzuki J, Takeda M, Kaneko K, Hino M and Muto S 2013 *J. Phys. Soc. Jpn.* **82** SA035
[3] Lefmann K and Nielsen K 1999 *Neutron News* **10/3** 20
[4] Willendrup P, Farhi E and Lefmann K 2004 *Physica B* **350** 735
[5] Itoh S, Ueno K, Ohkubo R, Sagehashi H, Funahashi Y and Yokoo T 2011 *Nuclear Instruments and Methods in Physics Research A* **654** 527
[6] Itoh S, Ueno K and Yokoo T 2012 *Nuclear Instruments and Methods in Physics Research A* **661** 58
[7] Rosenkranz S and Osborn R 2008 *Pramana Journal of Physics* **71** 705
[8] Ohoyama K, Yokoo T, Itoh S, Ino T, Ohkawara M, Oku T, Tasaki S, Iwasa K, Sato TJ, Ishimoto S, Taketani K, Kira H, Sakaguch Y, Nanbu M, Hiraka H, Shimizu HM, Takeda M, Hino M, Hayashi K, Kenzelmann M, Fliges U and Hautle P 2014 *Journal of Physics Conference Series* this issue

Journal of Physics: Conference Series **502** (2014) 012047 doi:10.1088/1742-6596/502/1/012047

High Intensity Neutron Beams for Small Samples

Peter Böni

Physik-Department E21, Technische Universität München, James-Franck-Str. 1, D-85748 Garching, Germany.

E-mail: peter.boeni@frm2.tum.de

Abstract. As novel materials of excellent homogeneity can often only be grown in small quantities it is important to optimize the transport of neutrons from the moderator to the sample while keeping the background low. Using elliptically or parabolically tapered guides the losses can be strongly reduced such that 50% - 90% of the useful neutrons arrive at the sample. If not properly designed, however, the divergence at the sample becomes inhomogeneous. In contrast, pairs of nested Kirkpatrick-Baez mirrors in Montel geometry yield well focused beams with a compact phase space. The mirrors extract only the useful neutrons from the moderator and effectively interrupt the line of sight leading to a very low background. As the focal distances are typically several meters, the extraction of the neutrons and the installation of bulky sample environment is facilitated.

1. Introduction

An increasing number of metal and oxide samples for investigating effects of strong correlations are grown by optical float zoning. The resulting high quality single crystals have typically a volume of 1 cm^3. If investigated under extreme conditions the sample size shrinks further to a few 0.01 mm^3 - 1 mm^3 [1]. In the field of macromolecular crystallography and in biophysics the sample volumes are very small too. Therefore, one of the recent trends in neutron scattering is directed towards optimizing beam lines for the investigation of small samples. Indeed, the recent years have witnessed a rapid development of simple devices to focus neutron beams to small areas [2] and to transport neutrons with a high efficiency [3].

With the invention of neutron guides by Maier-Leibnitz [4] it became possible to transport neutrons over large distances by Ni-coated guide tubes. However, due to the small critical angle of total reflection given by $\theta(^0) = 0.099m\lambda(\text{Å})$ where $m = 1$ for Ni, the transport was only efficient for cold neutrons. Using supermirror coatings [5], the index m can be increased up to $m = 7$ [6] thus allowing to even transport epithermal neutrons with reasonable divergence to instruments at spallation sources.

To reduce the reflection losses inherent to straight guides, truly curved elliptic guides [7] were proposed thus reducing the number of reflections of neutrons to approximately 2 while keeping the phase space of the neutrons compact [8]. For HRPD at ISIS, a gain of up to two orders of magnitude was reported [9]. There is no doubt that elliptic guides will improve the performance of instruments at spallation sources where each guide can be individually optimized. Klenø et al. [10] have shown recently that approximately 50% to 90% of the brilliance of cold and thermal neutron beams can be transported from the moderator to the sample using parabolic or elliptic guide geometries. Instruments at existing straight neutron guides can be improved by adding focusing guides at the down stream ends.

It is instructive to estimate the maximum possible intensity I at the sample position to judge the performance of a beam line. Assuming a loss free transport of neutrons from the moderator having a brilliance Ψ to the sample the intensity is given by

$$I = A \cdot \Delta\lambda \cdot \Omega \cdot \Psi. \tag{1}$$

Here, A, $\Delta\lambda$ and Ω designate beam size, wavelength band, and solid angle at the sample position, respectively. As an example, we consider the thermal beam port H12 at the ILL. Its brilliance at $\lambda = 1.2$ Å is $\Psi = 8 \cdot 10^{13}$ cm^{-2}s^{-1}Å^{-1}sterad^{-1} [11]. Assuming $\Delta\lambda/\lambda = 1\%$ and a horizontal and vertical divergence of 1^0 and 4^0, respectively, we end up with $1.2 \cdot 10^7$ neutrons per second hitting a sample with an area $A = 1$ mm^2. This procedure of estimating the intensity is very useful to benchmark the smallest sample size for which an experiment is feasible. Due to the theorem of Liouville there is no way to improve I further using passive means of focusing. If the beam line is designed and built properly, the quoted intensity can be approached within a factor of $\simeq 2$ using state of the art neutron optics while keeping the background at a low level.

In the following we discuss some basic principles on the adaption of neutron beams and the transport and focusing of neutrons. We will use simple concepts mostly based on parabolic and elliptic neutron optics. We do not show any Monte-Carlo simulations. Of course for the final optimization of a beam line numerical techniques are mandatory [12] but will not be discussed here. Due to the restriction in length not all details can be completely worked out in depth.

2. Adaption of Neutron Beams to the Sample

Often, beam lines are arranged along powerful guides, which extract as many neutrons as possible from the moderator. Beam size and divergence are defined close to the sample by means of collimators and slits. Thus, extensive shielding close to the guide and the sample is required to reduce the background to acceptable levels. A more efficient way of designing an instrument is to define the required phase space of the neutrons (beam size, divergence, bandwidth, time structure) as close as possible to the moderator and to extract only the neutrons to be used.

For the time being we neglect effects of bandwidth and time structure of the beam. Then, according to the theorem of Liouville, the phase space density $B = A\Omega$ is a constant. Let us consider a loss-less neutron guide 6 cm $\times$ 15 cm transporting neutrons with a divergence of 1^0. Within the guide, $B_{guide} = 90$ (cm ^{0})^2. Assuming a sample area of 1×1 cm^2 and an accepted divergence of 2^0, $B_{sample} = 4$ (cm ^{0})^2. Therefore, only $\simeq 4.0\%$ of the neutrons may be useful for the experiment while 96% have to be absorbed. In addition, the selection of the bandwidth for spectrometers reduces the intensity by another factor of 100, i.e. only $\simeq 4 \cdot 10^{-4}$ of the neutrons are actually being used requiring expensive shielding.

Concluding, the efficiency of a beam line can be strongly improved if only neutrons with the required phase space are extracted from the moderator. To reduce the contribution of fast neutrons and γ radiation it is best to extract the neutrons from a small area of the moderator. If this is not possible, a beam defining slit should be placed as close as possible to the moderator.

3. Elliptic Neutron Guides

Following the ideas of the last paragraph, neutrons are ideally extracted from an area of the moderator that has a similar size as the sample and the required divergence. Neglecting gravitational effects, it is the reflection by an ellipse that performs the required point to point transport of the neutrons (Fig. 1). As shown by Schanzer et al. [7], elliptic guides yield a reasonably homogenous beam at the sample position. Most neutrons are reflected while the contribution of the non-reflected neutrons (if there is no beam stop) is very small leading to an inhomogeneous divergence ψ at the sample. There will be a dip in intensity for angles $\psi < W/L$, where W is the width of the entrance and L is the length of the guide. For example, for $L = 40$

m and $W = 40$ mm, the dip is 10^{-3} rad (0.06^0) wide, i.e. small compared to the divergence of the neutrons. In two dimensions 4 peaks develop in the divergence distribution [13, 14].

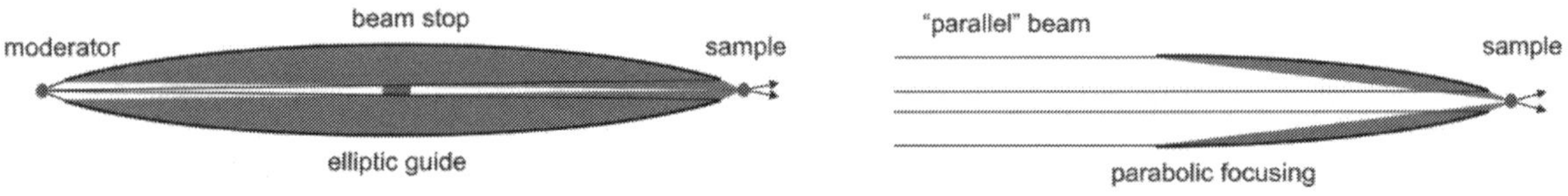

Figure 1. The transport of neutrons using an elliptic guide provides a point to point imaging of neutrons from the moderator to the sample. As the ellipse is not complete the divergence exhibits two peaks. The direct beam may be blocked by a beam stop. Parabolic focusing using a short neutron guide leads to a well-defined beam spot. However, due to the short length the divergence may become very inhomogeneous.

For an optimum design of elliptic guides, coma aberrations and halo effects should be considered [14]. Their importance scales with D/L where D is the spot size. Surprisingly, the waviness η has only rarely been taken into account although it is of similar importance. For example, $\eta = 1.0 \cdot 10^{-4}$ rad leads for a point-like object to a blurring of 1 cm for $L = 100$ m. Gravitation will further deteriorate the beam scaling with L^2. If properly considered one may use gravitation to interrupt the line of sight [15]. The experience from operating beam lines with elliptic guides, the mentioned deteriorating effects seem to be a minor issue [9].

Finally, let us make some practical remarks concerning the implementation of elliptic guides. Simulations for shielding demonstrate that the signal to noise for elliptic guides may be better than for curved guides [8]. Although, the beam stop (Fig. 1) interrupts the direct line of sight, the background at the sample position is still significant. The reason is that the guide sections close to the sample are directly illuminated by radiation from the moderator (Fig. 2) [16]. A simple remedy is dividing a single guide in two guides. Using an even number of guides has the further advantage that effects of halo, coma aberration [17], gravitation etc. are minimized.

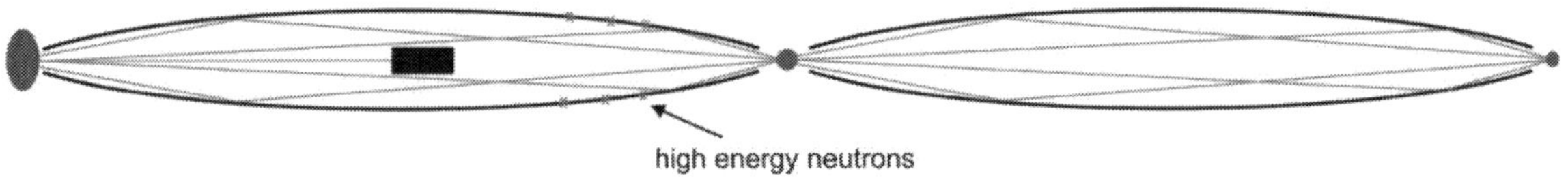

Figure 2. Guide sections close to the sample are directly illuminated by radiation from the moderatore (red stars) leading to a significant background. Dividing the guide into two sections improves the homogeneity of the beam and reduces the fast neutron background.

4. Focusing of Neutrons

Traditionally, neutron beams for small samples are configured by collimators and beam defining apertures leading to large losses in intensity and non-uniform illumination of the sample. If the last slit cannot be placed very close to the sample, the sample environment will be illuminated too leading to unwanted background. A large improvement can be made by installing truly curved elliptic or parabolic guides for focusing the neutron beam on the sample yielding gains of 25 and more [18].

Fig. 3 shows beam profiles of a focusing guide at the exit, focal position, and 200 mm away from the exit of the guide. The length of the guide is 500 mm and the exit size is 4×8 mm^2. The sides are coated with supermirror $m = 3$. The data was measured at the beam line CONRAD at HZB using neutrons with a wavelength of 3 Å and a CCD-camera [18]. Four beams evolve at the edges leading to one single spot at the sample position. Clearly, the divergence will be inhomogeneous showing four peaks by reasons explained in Fig. 1. A pronounced inhomogeneity is intrinsic to short focusing guides and becomes even more pronounced when the divergence of the incident beam is reduced as it leads to sharper peaks. Indeed, the width of the spot can be decreased to $\simeq 0.3$ mm if the divergence of the incident beam is reduced to 10'.

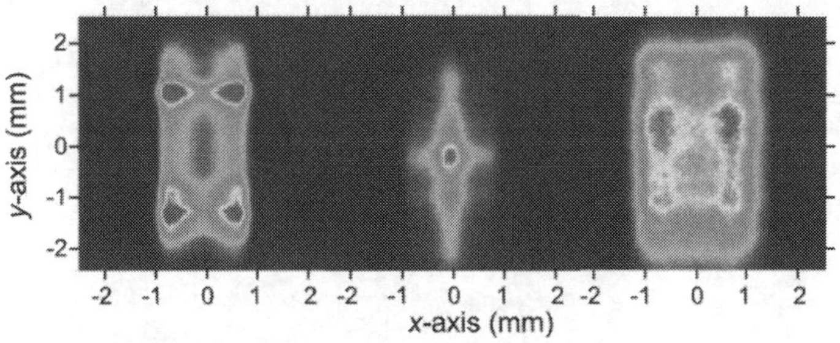

Figure 3. At the exit of the focusing guide, the neutrons leave the guide along the edges leading to a single sharp peak at the focal position $d = 80$ mm. Further away, the beam splits again into 4 beams.

To prove the usefulness of focusing neutrons with short flight tubes we have conducted diffraction experiments using the spectrometer PANDA at FRM II. Before the sample, the focusing guide from Fig. 3 was installed. Fig. 4(a) shows that the antiferromagnetic peaks in NiS_2 can be nicely resolved using a single crystal with the dimensions $1 \times 1 \times 0.1$ mm^3 [19]. Such small crystals fit into Bridgman cells for pressures up to 10 GPa. Fig. 4(c) shows inelastic data as measured at the thermal triple axis spectrometer PUMA using quartz with a volume of 0.75 mm^3. Although the signal is weak, the transverse acoustic phonon can clearly be resolved. More data on inelastic neutron scattering from small samples is reported in Ref. [20]. By adding a second (de)focusing guide after the sample the intensity can be further increased.

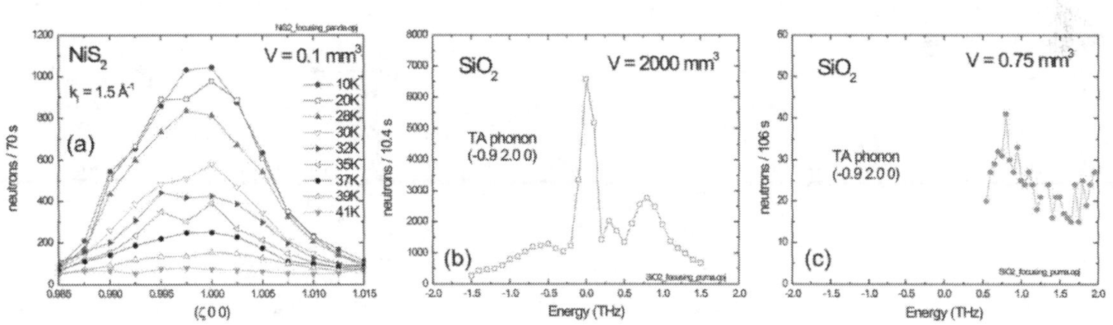

Figure 4. (a) Detection of magnetic order in NiS_2 using a focusing guide. The sample volume is only 0.1 mm^3. (b) Measurement of a transverse acoustic phonon in quartz from a sample with a volume $V = 2000$ mm^3. (c) Measurement using a focusing guide set-up with $V = 0.75$ mm^3.

5. Montel Optics

A major drawback of strong elliptic and parabolic focusing is the required small distance between the exit of the guide and the sample, $d \simeq 80$ mm, hindering the installation of bulky sample environment. A solution to this problem is the use of Montel optics [21, 22], where two elliptic (or parabolic) Kirkpatrick-Baez mirrors are arranged side by side. As shown in Fig. 3, the Montel geometry corresponds simply of using one beam emerging from one of the 4 corners of the guide leading to a focused beam with a well-defined divergence. The loss of a factor of 4 in intensity can be compensated for by doubling m from $m = 3$ to $m = 6$.

The major advantages of Montel optics are (Fig. 5): i) the beam size scales with the length L of the mirrors, ii) the distance $d \simeq L/2$ between the focal points and the mirrors is large, iii) the beam size can be tuned by an aperture at the first focal point, iv) the divergence can be adjusted by slits, and v) there is no direct line of sight between the two focal points. For an application for neutron reflectometry see Refs. [17, 23] and for a time-of-flight MIEZE spectrometer see Ref. [24]. Recently, design studies were conducted for the beam line SNAP, at SNS [25]. Using metallic substrates [26] it is anticipated that beam sizes of 30 μm or below can be obtained.

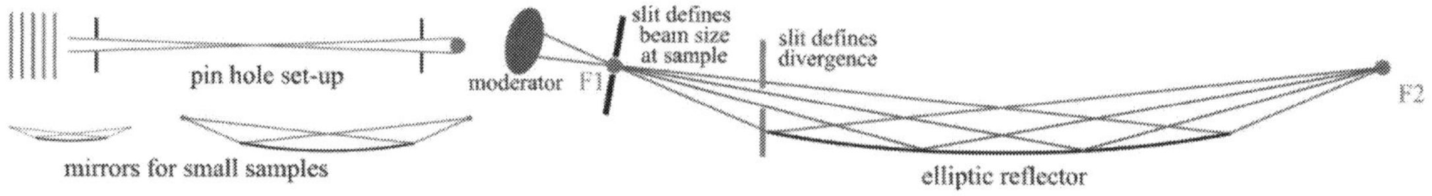

Figure 5. For a pin-hole set up [27] the beamsize is defined by means of slits close to the sample, which usually restrict the divergence leading to low intensity and high background. An elliptic mirror provides a point to point imaging. The beam size and the divergence can be adjusted by slits close to the moderator thus allowing to reduce the size and improve the performance of the moderator. Large (small) mirrors allow the illumination of large (small) samples.

Typically, a minimal guide system will consist of two or more Montel mirrors. An even number of elements is of advantage as coma aberrations, halos, and gravitational effects are reduced. In between the mirrors, beam defining elements such as choppers, apertures, monochromators, polarizers etc. can be inserted [24]. As typical lengths of Montel mirrors are of the order of 10 m - 20 m, the entrance of the first mirror can be placed approximately 5 m away from the moderator, i.e. in a regime of reduced radiation. This geometry allows a simple adaption of beam optics for new neutron scattering instruments as the optical parts are placed at the boundary of the biological shielding and can easily be modified. Moreover, as neutrons are extracted from a small spot of the moderator only, its size can be strongly reduced leading possibly to a higher brilliance. It is clear that the use of Montel optics for future beam lines will significantly improve the performance of neutron instrumentation due to the well-defined phase space and the extremely low background. Moreover, costs for shielding can be massively reduced.

6. Conclusions

Summarizing, we have shown that the significant technological advances in the development of supermirror coatings and the step from using straight guides via elliptic guides to Montel optics will lead to a significant improvement of neutron beam lines. Respecting the limits imposed by Liouville's theorem it is rather straightforward to design a beam line and predict its performance.

For beam lines requiring highest intensity such as inelastic chopper spectrometers, a combination of an even number of elliptic guides may be the optimum choice. For most other beam lines an even number of Montel mirrors may yield highest performance. This view is at variance with propositions of going away from a purely elliptic transport for neutrons [28]. Many beam lines built so far have proven that the elliptic approach is rather straightforward from the design point of view and that it works very well yielding the expected beam quality [9].

To guarantee the proper transport of the neutrons from the moderator to the sample, an active surveillance system to control the alignment of the neutron guides may be required. Such a system would also allow to actively adapt the complete guide system for the optimum transport of neutrons within a predefined wavelength band thus reducing gravitational effects. Sophisticated alignment techniques have been developed for the Compact Linear Collider CLIC at CERN, where transverse positions of components have to be aligned and stabilized to within 10 μm over distances of 200 m [29].

To take full advantage of all the possible neutron gains using advanced reflecting optics attention should also be paid to the development of radiation-hard chopper systems that can be placed as close as possible to the moderator to allow i) a precise chopping of the beam due to the small beam size [3] and ii) a reduction of frame overlap. Further, as the neutron yield from spallation sources and high flux reactors is limited intense research should be devoted to producing highly brilliant neutron beams using for example intense γ-rays from electron recovery linacs [30]. If such neutron sources can be realized, similar dramatic increases in brilliance will become feasible as realized with x-ray synchrotron sources.

Acknowledgments

Part of the work was supported by the Swiss National Science Foundation through the National Centre of Competence in Reasearch MaNEP and by the Deutsche Forschungsgemeinschaft via the Transregional Research Center TRR 80. The close collaboration with C. Schanzer, M. Schneider, U. Filges, J. Stahn, N. Kardjilov, B. Roessli, and K. Hradil is gratefully acknowledged.

References

[1] Goncharenko L N, Mireaubeau I, Molina P and Böni P 1997 *Physica* B **234-236** 1047
[2] Hils T, Böni P and Stahn J 2004 *Physica* B **350** 166
[3] Böni P 2008 *Nucl. Instr. and Meth.* A **586** 1
[4] Maier-Leibnitz H and Springer T 1963 *Nuclear Science and Technology (Journal of Nuclear Energy Parts A/B)* **17** 217
[5] Mezei F 1976 *Commun. Phys.* **1** 81; *Commun. Phys.* (1977) **2** 41
[6] SwissNeutronics, www.swissneutronics.ch/products/coatings.html
[7] Schanzer C, Böni P, Filges U, and Hils T 2004 *Nucl. Instr. and Meth.* A **529** 63
[8] Böni P, Grünauer F and Schanzer C 2010 *Nucl. Instr. and Meth.* A **624** 162
[9] Ibberson R M 2009 *Nucl. Instr. and Meth.* A **600** 47
[10] Klenø K H , Lieutenant K, Andersen K H, and Lefmann K 2012 *Nucl. Instr. and Meth.* A **696** 75
[11] Yellow Book of the ILL 1986, p 8, Fig. 4
[12] McStas webpage: http://neutron.risoe.dk/mcstas
[13] Harjo S, Kamiyama T, Torii S, Ishigaki T and Yonemura M 2006 *Physica* B **385-386** 1025
[14] Cussen L D, Nekrassov D, Zendler C and Lieutenant K 2013 *Nucl. Instr. and Meth.* A **705** 121
[15] Klenø K H, Willendrup P K, Knudsen E and Lefmann K 2011 *Nucl. Instr. and Meth.* A **634** S100
[16] Grünauer F, private communication
[17] Stahn J, Filges U and Panzner T 2012 *Eur. Phys. J. Appl. Phys.* **58** 11001
[18] Kardjilov N, Böni P, Hilger A, Strobl M and Treimer W 2005 *Nucl. Instr. and Meth.* A **542** 248
[19] Niklowitz P G, Pfleiderer C, Mühlbauer S, Böni P, Keller T, Link P, Wilson J A, Vojta M, and Mydosh J A 2009 *Physica* B **404** 2955
[20] Böni P, Roessli B and Hradil K 2011 *J. Phys.: Condens. Matter* **23** 254209
[21] Montel M 1957 *X-Ray Microscopy with Catamegonic Roof Mirrors, in X-ray Microscopy and Microradiography* (Academic Press, New York) pp 177-185
[22] http://www.x-ray-optics.de/index.php?option=com_content&view=article&id=59&Itemid=71&lang=en
[23] Stahn J, Panzner T, Filges U, Marcelot C and Böni P 2011 *Nucl. Instr. and Meth.* A **634** S12
[24] Weber T, Brandl G, Georgii R, Häußler W, Weichselbaumer S, and Böni P 2013 *Nucl. Instr. and Meth.* A **713** 71
[25] Ice G I *et al* 2009 *J. Appl. Cryst.* **42** 1004
[26] Schanzer C, Böni P and Schneider M 2010 *J. Phys.: Conf. Series* **251** 012082
[27] First pictures with a pin-hole camera were taken around 1850
[28] Bentley P, Kennedy S J, Andersen K H, Rodríguez D M and Mildner D F R 2012 *Nucl. Instr. and Meth.* A **693** 268
[29] Mainaud-Durand H 2010 *CERN-TS_note-2005-028* (Geneva, CERN) 6, http://www.psi.ch/
[30] Habs D, Gross M, Thirolf P G and Böni P 2011 *Appl. Phys.* B **103** 485

Applications of a micro-pixel chamber (μPIC) based, time-resolved neutron imaging detector at pulsed neutron beams

J D Parker[1,6], M Harada[2], K Hattori[1], S Iwaki[1], S Kabuki[1], Y Kishimoto[1], H Kubo[1,3], S Kurosawa[1], Y Matsuoka[1], K Miuchi[4], T Mizumoto[1], H Nishimura[1], T Oku[2], T Sawano[1], T Shinohara[2], J-I Suzuki[2], A Takada[1], T Tanimori[1], K Ueno[1], M Ikeno[3,5], M Tanaka[3,5] and T Uchida[3,5]

[1] Department of Physics, Kyoto University, Kyoto 606-8502, Japan

[2] Materials and Life Science Facility Division, Japan Atomic Energy Agency, Tokai, Ibaraki 319-1195, Japan

[3] Open Source Consortium of Instrumentation (Open-It), KEK, Tsukuba, Ibaraki 305-0801, Japan

[4] Department of Physics, Kobe University, Hyogo 657-8501, Japan

[5] Institute of Particle and Nuclear Studies, KEK, Tsukuba, Ibaraki 305-0801, Japan

E-mail: jparker@cr.scphys.kyoto-u.ac.jp

Abstract. The realization of high-intensity, pulsed spallation neutron sources such as J-PARC in Japan and SNS in the US has brought time-of-flight (TOF) based neutron techniques to the fore and spurred the development of new detector technologies. When combined with high-resolution imaging, TOF-based methods become powerful tools for direct imaging of material properties, including crystal structure/internal strain, isotopic/temperature distributions, and internal and external magnetic fields. To carry out such measurements in the high-intensities and high gamma backgrounds found at spallation sources, we have developed a new time-resolved neutron imaging detector employing a micro-pattern gaseous detector known as the micro-pixel chamber (μPIC) coupled with a field-programmable-gate-array-based data acquisition system. The detector combines 100μm-level (σ) spatial and sub-μs time resolutions with low gamma sensitivity of less than 10^{-12} and a rate capability on the order of Mcps (mega-counts-per-second). Here, we demonstrate the application of our detector to TOF-based techniques with examples of Bragg-edge transmission and neutron resonance transmission imaging (with computed tomography) carried out at J-PARC. We also consider the direct imaging of magnetic fields with our detector using polarized neutrons.

Neutrons, being neutral, highly penetrating particles, have found extensive use in fields ranging from materials science, engineering/manufacturing, medical, and security applications. In particular, materials science and manufacturing applications use neutrons as a probe for non-destructive testing and to study the structure and properties of bulk materials from the millimeter to nanometer scales, helping to understand and improve the materials and technologies used in our everyday lives. Presently, the realization of next-generation, high-intensity spallation neutron sources, including at J-PARC (Japan Proton Accelerator Research Complex) and the SNS (Spallation Neutron Source) in the United States, has brought time-of-flight (TOF)-based measurement techniques to the fore and spurred the development of new detector technologies to cope with the high intensities, while the ongoing development of compact accelerator-driven neutron sources in Japan, the United States, and China promises to bring these techniques and technologies to a wide audience [1].

At such pulsed sources, we can determine the energy of each neutron by measuring the time-of-flight over a known path length. This allows direct measurement of various quantities (e.g.,

[6] Author to whom any correspondence should be addressed.

transmission, scattering angle) as a function of neutron energy and requires a detector with good time resolution (<1 μs). Furthermore, a detector that combines good time resolution with good spatial resolution (100μm-order or less), low gamma sensitivity, and high rate capability opens the door to new, powerful direct imaging techniques. The present paper considers the following such techniques:

- *Bragg-edge transmission* [2] provides direct imaging of crystal structure and internal strain. It requires good time resolution to resolve small changes in Bragg-edge positions.
- *Neutron resonance transmission imaging (NRTI)* [3] provides direct imaging of isotopic/temperature distributions and requires good time resolution to resolve narrow resonance structures.
- *Magnetic imaging* [4] utilizes polarized neutrons to provide direct imaging of internal and external magnetic fields and requires good spatial resolution for imaging small details.
- *Computed tomography (CT)* allows three-dimensional image reconstruction from a set of two-dimensional images and can be combined with NRTI and magnetic imaging. A high rate capability enables shorter measurement times.

The high-intensities and large gamma backgrounds at spallation sources place strict demands on the detector system, facilitating the need for new detectors using cutting edge technologies. Traditional detectors for high-resolution neutron imaging such as imaging plates (IP) [5] and CCD (Charge-Coupled Device) cameras [6] lack adequate time resolution (IPs record no time information at all) and, being integrating-type detectors, have no means for gamma discrimination. Scintillator-based detectors [7] provide good time resolution and decent gamma discrimination, but spatial resolution is limited to the order of a millimeter. Recently developed micro-pattern detectors such as a boron-coated stacked-GEM (Gas Electron Multiplier) detector [8], a boron-doped MCP (Micro-Channel Plate) detector [9], and our μPIC (Micro-PIxel Chamber)-based detector [10,11] have the potential to satisfy most or all of the above requirements. Of these, the μPIC detector (figure 1) is the only one that provides the combination of ultra-low gamma sensitivity and 100μm-level spatial resolution (given as the Gaussian width, σ) with the ability to cover large areas (up to 40×40 cm^2) at relatively low cost. With further development, the μPIC detector is expected to reach 50~60μm spatial resolution (similar to IPs) and significantly increased counting rates. Additionally, such micro-pattern detector systems can be made quite compact, making them equally at home at large spallation sources and compact accelerator-driven sources.

Figure 1. Schematic diagram of the μPIC detector and section of μPIC surface is shown in (a), with a photograph of the prototype detector in (b). An image of a wristwatch taken with our detector at J-PARC, NOBORU/BL10 is shown in (c). The resolution is ~120 μm (σ).

Our detector was developed under the Quantum Beam Technology Program of the Japan Ministry of Education, Culture, Sports, Science and Technology (MEXT) as part of a program to develop new detectors and technologies for the high-intensity spallation neutron source at J-PARC. The work was carried out in collaboration with members from the Japan Atomic Energy Agency (JAEA) and the High-Energy Research Organization (KEK). The prototype detector (figure 1a, b) consists of a time-projection-chamber with 2.5-cm gas depth and a 10×10-cm^2 μPIC. The μPIC [12] features a 400μm pitch with two-dimensional strip readout. Neutrons are detected via absorption on ^{3}He with 18%

efficiency at 25.3 meV using a gas mixture of Ar-C_2H_6-^{3}He (63:7:30) at 2 atm. The μPIC is coupled to a fast, FPGA (Field Programmable Gate Array)-based data acquisition system that records both the energy deposition (via time-over-threshold) and three-dimensional track (two-dimensional position plus drift time) for each neutron-induced event. This detailed tracking information allows our detector to achieve a spatial resolution of 100~120 μm (σ) and an ultra-low gamma sensitivity of $<10^{-12}$. The detector also features a time resolution of 0.6 μs and is capable of neutron rates on the order of Mcps (mega-counts-per-second) over the whole detector. The maximum count rate is achieved for a mostly homogeneous beam profile, as is typical for imaging measurements, while, due to our modular data encoder system and strip readout, the whole-detector count rate for a localized beam profile (such as encountered in small-angle neutron scattering) is reduced by at most half. Using this detector, we carried out a series of experiments at NOBORU (NeutrOn Beamline for Observation and Research Use)/BL10 at the MLF (Materials and Life Science Facility), J-PARC [13] to test detector performance and demonstrate various measurement techniques.

As an example, we used Bragg-edge transmission to make a two-dimensional map, in a single measurement, of the residual strain in a welded steel sample (figure 2a). In general, Bragg-edges appear as sudden jumps in the transmission where the neutron wavelength exceeds the Bragg-scattering condition for a given set of crystal planes. From the wavelength at the Bragg-edge, one can determine the crystal spacing ($d = \lambda/2$), and by studying the shape of the spectrum in the vicinity of the edge, one can learn about crystal properties such as grain size and texture [2]. For strain mapping, the crystal is used as a microscopic strain gauge, with the strain component derived from the deviation, in a given direction, of the crystal spacing from the unstrained value. Figure 2b shows the transmission spectra, taken with a single exposure of ~2 hours at a time-averaged neutron rate of 29 kcps, for several different regions of the 1-cm thick, double-TIG-welded 316L stainless steel sample, displaying very clear Bragg-edges with some difference between weld and steel plate regions. Figure 2c shows the estimated strain distributions for the upper and lower weld regions (indicating that the weld is under tension). Further details about this measurement and the data analysis can be found in [14].

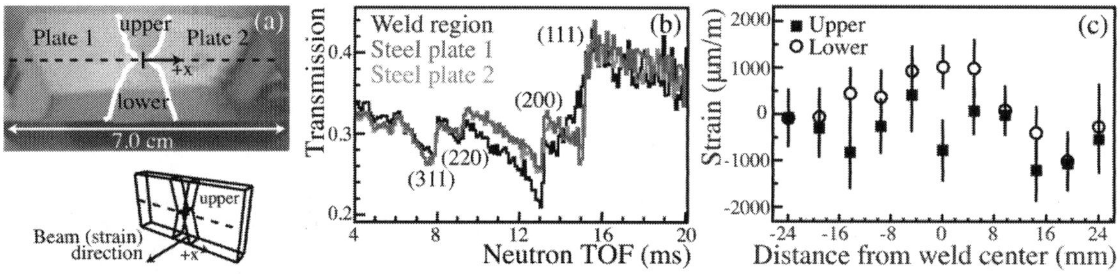

Figure 2. Strain scanning via Bragg-edge transmission. Shown here: (a) welded steel sample (white lines indicate the weld area), (b) neutron TOF spectra for weld region and steel plates (with Miller indices for some edges), and (c) longitudinal strain for upper/lower weld regions.

As another example, we combined neutron resonance transmission imaging with computed tomography. Neutron resonance techniques take advantage of the tendency of some nuclides to preferentially absorb neutrons at specific energies, unique to each nuclide, to study isotopic composition and internal temperature [3]. Resonance absorption can be directly observed by measuring the transmission, determined by taking the ratio of neutron intensities measured with and without the sample, as a function of neutron TOF. Resonances appear as sharp drops at specific energies where preferential absorption occurs. The depth of the resonance is related to the isotopic density, while Doppler-broadening of the resonance width gives information about temperature. Figure 3 shows preliminary results of a CT measurement performed at NOBORU. The sample, consisting of Cu, Mo, W, Ag, and In rods with diameters from 2 to 6 mm (figure 3a), was measured at 16 angles for ~30 minutes each (~14×10^6 neutrons per angle), with the axis of rotation aligned along the length of the rods, and images of a single CT slice were reconstructed (figures 3b—f). The indicated neutron TOF cuts, chosen to select resonances at 2039 and 579 eV (Cu), 45 eV (Mo), 19 eV (W), 5.2 eV (Ag), and 1.5 eV (In), clearly enhance the individual elements in each image. A more accurate method using fits to the known cross-sections is now under development.

Finally, we plan to apply our detector in combination with a neutron polarizing apparatus developed at J-PARC [4] to image magnetic materials with 100μm spatial resolution. Magnetic

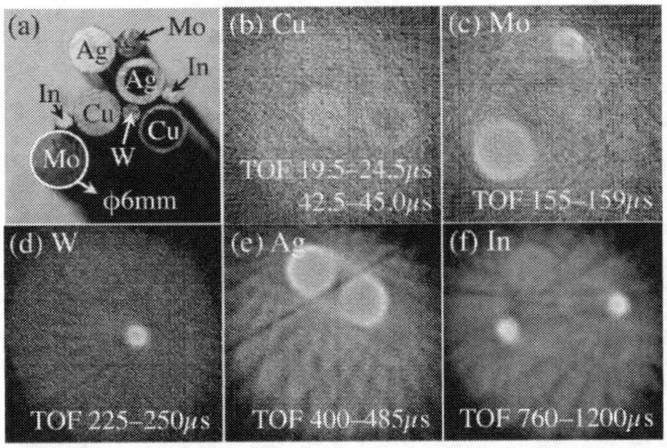

Figure 3. Resonance CT. (a) Photo of sample from above, and (b—f) computed tomography images of specific elements selected by neutron TOF (b: Cu, c: Mo, d: W, e: Ag, and f: In).

materials are at the heart of most modern technology and, in particular, emerging green technologies (e.g, electric motors for hybrid vehicles). Direct imaging of internal and external magnetic fields *in situ* with high-resolution would be an invaluable tool for the understanding and optimization of these important materials. A proposed experiment, accepted for the J-PARC/MLF 2013A run period but subsequently cancelled due to the radiation accident in May, would use this setup to image the magnetic domains in a sheet of Permalloy before and after stressing. Bragg-edge transmission would also be used to study the alterations of the crystal structure. We hope to carry out this first test experiment for magnetic imaging during in the next MLF user time.

Using high-intensity pulsed neutron sources, the combination of TOF-based measurement with high-resolution imaging produces a set of powerful techniques that allow direct imaging of bulk properties with short measurement times. Such techniques require an equally powerful and flexible detector, for which our μPIC-based detector is an excellent candidate. Additionally, although not discussed here, the application of such a high-resolution detector to small-angle neutron scattering (SANS) would provide access to small values of momentum transfer (i.e., smaller scattering angles) in a compact setup. Development and validation of our detector system, to improve spatial resolution and data rate, and of the TOF-based direct imaging techniques are ongoing.

Acknowledgments
This work was supported by the Quantum Beam Technology Program of the Japan Ministry of Education, Culture, Sports, Science, and Technology (MEXT). The neutron experiments were performed at NOBORU (BL10) of the J-PARC/MLF with the approval of the Japan Atomic Energy Agency (JAEA), Proposal No. 2009A0083.

[1] Carpenter J M 2012 *Phys. Proc.* **26** 1—7; Union for Compact Accelerator-driven Neutron Sources (UCANS) http://www.indiana.edu/~lens/UCANS/
[2] Sato H, Takada O, Iwase K, Kamiyama T and Kiyanagi Y 2010 *J. of Phys. Conf. Series* **251** 012070; Santisteban J R, Edwards L, Fizpatrick M E, Steuwer A and Withers P J 2002 *Appl. Phys.* A **74** S1433—6
[3] Sato H, Kamiyama T and Kiyanagi Y 2009 *Nucl. Instr. and Meth.* A **605** 36—9
[4] Shinohara T *et al.* 2011 *Nucl. Instr. and Meth.* A **651** 121—5
[5] Kobayashi H and Satoh M 1999 *Nucl. Instr. and Meth.* A **424** 1—8
[6] Lehmann E H, Frei G, Kühne G and Boillet P 2007 *Nucl. Instr. and Meth.* A **576** 389—96
[7] Hirota K *et al.* 2005 *Phys. Chem. Chem. Phys.* **7** 1836—8
[8] Uno S *et al.* 2012 *Phys. Proc.* **26** 142—52
[9] Tremsin A S, McPhate J B, Vallerga J V, Siegmund O H W, Feller W B, Lehmann E and Dawson M 2011 *Nucl. Instr. and Meth.* A **628** 415—8
[10] Parker J D *et al.* 2013 *Nucl. Instr. and Meth.* A **697** 23—31
[11] Parker J D *et al.* 2013 *Nucl. Instr. and Meth.* A **726** 155—61
[12] Ochi A *et al.* 2001 *Nucl. Instr. and Meth.* A **471** 264—7
[13] Maekawa F *et al.* 2009 *Nucl. Instr. and Meth.* A **600** 335—7
[14] Parker J D *et al.* 2010 *IEEE Nuclear Science Symposium Conference Record* (©2010 IEEE) pp 291—7

Journal of Physics: Conference Series **502** (2014) 012049 doi:10.1088/1742-6596/502/1/012049

Water-enhanced Adhesion at Interface in Immiscible Bilayer Film of Polystyrene and Poly(methyl methacrylate)

M Harada[1], T Koga[1], K Fukumori[1], J Sugiyama[1] and T Geue[2]

[1] Toyota Central R&D Labs., Inc., 41-1 Yokomichi, Nagakute, Aichi 480-1192, Japan
[2] Laboratory for Neutron Scattering, ETHZ & PSI, CH-5232 Villigen PSI, Switzerland

E-mail: harada@mosk.tytlabs.co.jp

Abstract. From nano-scratch tests, strong interfacial adhesion has been found for polystyrene (PS) and poly(methyl methacrylate) (PMMA) bilayer films prepared by a water floating (WF) method, while a PS layer on a PMMA film produced by a spin coating (SC) method peels off easily at the interface. Neutron reflectivity measurements demonstrated a clear difference in the interfacial width (σ) between the two bilayers; σ = 9 nm for the film obtained by the WF method, whereas σ = 5 nm for that by the SC method. Plasticization of the films by water would be responsible for broadening of the interface to enhance adhesion strength.

1. Introduction

Control of interfacial toughness between two kinds of polymers is one of the most important targets in research and development of coatings and blends. To understand the mechanical properties of the interfaces between polymers, the interfacial structure was investigated and the correlation between the width of the interfaces and the energy of adhesion was clarified for polymer bilayers [1, 2]. The origin of the interfacial strength was interpreted with an entanglement model of polymer chains that the interface energy is governed by the number density of the entanglement points per the interface area [3]. Therefore, the adhesion strength between polymer films increases by thickening of the mixed layer at the interface, and above a critical thickness cohesive failure could occur in the films.

The interfacial thickness of immiscible polymers for annealed bilayers are dominated by the Flory-Huggins interaction parameter and the Kuhn segment length. However, most of interfaces between polymers in practice are far from equilibrium. Here, we focus on the effect of coating process, where some small molecules mediate the formation of the interface. Bilayer films consisting of polystyrene (PS) and poly(methyl methacrylate) (PMMA) are employed, since the interface between PS and PMMA is well understood system in equilibrium [4].

For investigation of the interfacial width between immisible polymers, neutron reflectivity (NR) measurements have been a powerful tool. However, the samples must be thin films on very flat substrates. Although the energy of adhesion has been studied macroscopically by various experimental techniques, the mechanical properties for thin films can be analyzed by nano-scratch (NS) tests. By combining the NR measurements and the NS tests, structure and strength of interfaces are evaluated for the same bilayer films.

2. Experiments

Monodisperse polystyrene (PS) and poly(methyl methacrylate) (PMMA) with a number-average molecular weight of 115,000 and 105,000 were purchased from Polymer Source Inc. Note that a perdeuterated polymer is actually employed for PS to give enough contrast against PMMA when the polymer reflect neutrons. Silicon wafers with a 3-inch diameter were used as substrates for bilayer films.

Bilayer films were prepared by two kinds of processes. In each process, the bottom layers were commonly spin-casted from a 2 wt% toluene solution of PMMA onto a silicon wafer at 2000 rpm for 30 seconds and dried under vacuum at 120 °C for 24 hours. The thickness of the bottom films was evaluated to be around 60 nm from X-ray reflectivity measurements. The top layers were formed by a spin coating (SC) method or a water floating (WF) method, respectively. In the SC method, a PS film was spin-coated on the PMMA bottom layer from 2 wt% solution of a selective solvent for PS. We employed 1-chloropentane as a selective solvent, which is reported to dissolve PS but not to dissolve PMMA [5] . PS/PMMA bilayers with a clear interference color were obtained and dried under vacuum at room temperature for 24 hours. On the other hand, in the WF method, a floating PS film on a water surface was skimmed off by the PMMA-coated silicon wafer from the water side. Here, the PMMA film was once dipped in water and swollen by uptaking water molecules [6] . The floating PS film was provided by peeling off a spin-cast PS film from a Si wafer on a water surface. After water at the PS/PMMA interface was evaporated, the bilayers were kept in a vacuum chamber at room temperature overnight. Since room temperature is far below glass transition temperature of PS (100 °C) and PMMA (110 °C), we expect that the interface of the bilayers does not reach to its equilibrium structure. In each coating process, two bilayer samples were prepared to estimate errors.

Nanoscratch tests were performed by TI 900 TriboIndenter^{TM} (Hysitron Inc.). We used a cono-spherical diamond probe with an included angle of 60° and a radius of 1 μm. The bilayer samples were scratched in a constant slope and a speed with the probe held perpendicular to the samples. The probe was typically moved 100 nm vertically and 10,000 nm horizontally in 30 seconds (333 nm/s in speed). Normal force working on the probe was measured, while monitoring the probe position controlled by a feed-back circuit. The measurements were carried out twice for one sample at different positions.

Neutron reflectivity measurements were performed on AMOR reflectometer at SINQ facility of Paul Scherrer Institut in Switzerland [7]. Pulsed neutrons reflected specularly by the bilayer samples were counted in a time-of-flight mode. Neutron reflectivity spectra were analyzed by *Motofit* package to obtain cross-section profiles of scattering length density (ρ_{SL}) [8]. A bilayer model on a silicon substrate was employed for the fitting of the spectra. In the fitting, the parameters, such as ρ_{SL} and thickness (t) of the two layers and the roughness (σ) of the three interfaces, were optimized to minimize the chi squared, while the ρ_{SL} values of the air and the substrate were fixed.

3. Results and discussion

3.1. Nanoscratch tests

The results of the nanoscratch tests for the bilayers prepared by the SC method and the WF method are shown in figure 1. The top view of the AFM image presents a groove on the bilayers generated by the nanoscratch probe. The width of the groove has increased as the probe penetrates into the bilayer. The two normal force curves measured at different points are almost identical for each samples. The strength of the normal force has risen, as the probe goes down from the surface and the contact area between the probe and the bilayers expands.

The increase of the normal forces for the bilayer prepared by the SC method is smooth until the probe reaches to 50 nm and the normal force drops at 60 and 70 nm, as shown in figure 1(a). Since the decrease of the normal force could happen when the contact area shrinks, the drops should be attributed to the fracture of the film.

On the other hand, as shown in figure 1(b), the normal force curve of the WF method presents a stick-slip motion from 50 nm to 90 nm. The stick-slip motion could be induced by inhomogeneity in the film on a nanoscale, which may be generated in the preparing process of the bilayer by the WF method.

Since the fracture of the film should be followed by the peel-off at the interface, the adhesion energy of the bilayer by the SC method is low. Meanwhile, the interface layer of the bilayer by the WF method could be thick enough to have a similar mechanical property as the bulk of the polymer.

3.2. Neutron reflectivity measurements

Neutron reflectivity spectra and scattering length density profiles of the bilayers prepared by the two kinds of procedures are shown in figure 2. The spectra are displayed using the scattering vector q $(= (4\pi/\lambda)\sin\theta)$, where λ and θ are the wavelength and the incident angle of the neutrons. The profiles inserted on the upper right of the spectra are the cross-section profiles of $\rho_{\rm SL}$ from the surface to the substrate of the bilayers. Figure 2(a) and figure 2(b) represent the results of the SC method and the WF method, respectively. The results obtained for separately prepared bilayers with the same method are overlaid in the neutron reflectivity spectra. Since no distinct differences are observed for the spectra, the cross-section profiles showed little change as well.

The both spectra show interference fringes, whose amplitude damps down with an increase of q. The number of fringes in figure 2(b) is smaller than that in figure 2(a), corresponding to a broader interface in the bilayer of the WF method. From the cross-section profiles, the thickness of the each layer and the position of the interface have been clarified. The thickness of the bottom PMMA layer was consistent with the evaluation by the X-ray reflectivity measurements

The difference in the interfacial structures of the bilayer films are presented in the cross-section profiles of figure 2. The top layer with a high $\rho_{\rm SL}$ value and the bottom layer with a low $\rho_{\rm SL}$ value are observed between the air ($\rho_{\rm SL} = 0$) and the substrate ($\rho_{\rm SL} = 2.07$). The observed two $\rho_{\rm SL}$ values are identical with the calculated values of PS and PMMA. The transition region between the two layers is the interface layer, which should not be ideally sharp but has a width including roughness and mixing. After fitting with a hyperboric tangent function, the thickness of the interface σ was evaluated as $\sigma = 5$ nm for the bilayer obtained by the SC method, whereas $\sigma = 9$ nm for that by the WC method.

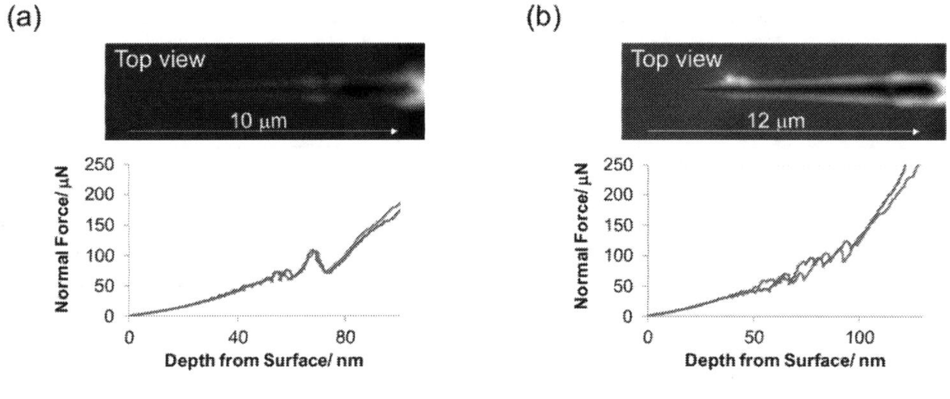

Figure 1. AFM images after nanoscratch test and normal force curves for nanoscratch test of bilayers prepared by (a) spin coating method and (b) water floating method.

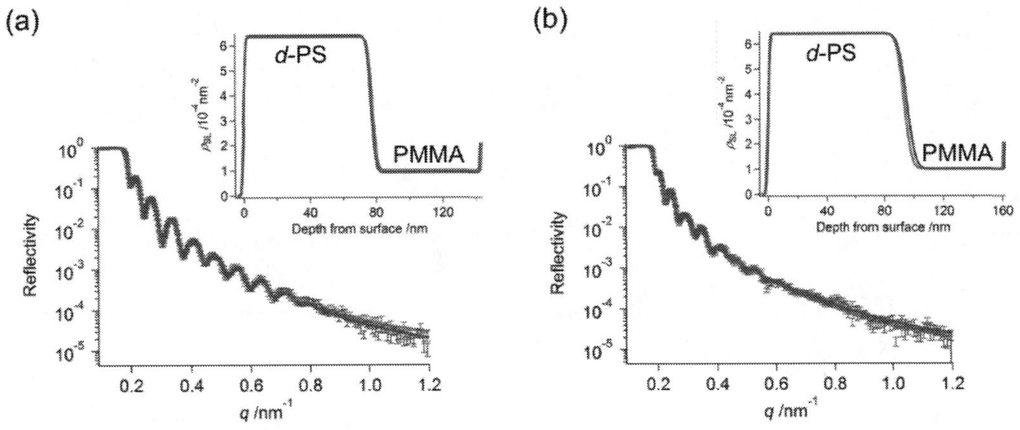

Figure 2. Neutron reflectivity spectra and scattering length density profile of bilayers prepared by (a) spin coating method and (b) water floating method.

3.3. Strength and thickness of interfaces
The interfacial width σ between a strong segregation pair of PS and PMMA is around 5 nm in equilibrium. Since the interface of the bilayer prepared by the SC method is same in thickness, the peel-off behavior observed for the PS film on the PMMA is reasonable. When a bilayer is prepared by the WF method, a thicker interface, generated by increase in the roughness and the mixing, has enhanced the strength of adhesion. Thus, the normal force curve indicates that the mechanical property of the bilayer could be similar to that of a PS film. A comprehensive understanding has not given for the factors to determine the interfacial thickness in various kinds of selective solvents or coating processes. However, we believe that water uptake in the bottom PMMA layer would be responsible for the broadening on the interface.

4. Conclusion
Two kinds of polymer bilayers of PS and PMMA were prepared by the spin coating method and the water floating method. The strength of the interface was enhanced by the water floating method, while the spin-coated PS layer is not adhesive to the PMMA layer. Although the interfacial structure of the spin-coating bilayer was similar to that of the bilayers in equilibrium, the broadening at the interface was observed for the PS/PMMA bilayer produced by the water floating method. Water should have worked as a plasticizer in the coating process, which may be associated with stickers.

Acknowledgments
This work is based on experiments performed at the Swiss spallation neutron source SINQ, Paul Scherrer Institute, Villigen, Switzerland.

References
[1] Schnell R, Stamm M and Creton C 1998 *Macromolecules* **31** 2284–2294
[2] Schnell R, Stamm M and Creton C 1999 *Macromolecules* **32** 3420–3425
[3] Brown H R 1991 *Macromolecules* **24** 2752
[4] Sferrazza M, Xiao C, Jones R A L, Bucknail D G, Webster J and Penfold J 1997 *Phys. Rev. Lett.* **78** 3693–3696
[5] Ennis D, Betz H and Ade H 2006 *J. Polym. Sci. Part B: Polym. Phys.* **44** 3234–3244
[6] Tanaka K, Fujii Y, Atarashi H, Akabori K, Hino M and Nagamura T 2008 *Langmuir* **24** 296–301
[7] Clemens D, Gross P, Keller P, Schlumpf N and Koennecke M 2000 *Physica B* **140** 276–278
[8] Nelson A 2006 *J. Appl. Cryst.* **39** 273–276

Detector development for a high-flux neutron reflectometer

S Satoh, N L Yamada and S Muto

High Energy Accelerator Research Organization, Oho 1-1 Tsukuba, Ibaraki 305-0801, Japan

E-mail: setsuo.satoh@kek.jp

Abstract Neutron reflectometry is a powerful tool for investigating the surface and interfacial structures of various materials on the nanometer to submicrometer spatial range. We have been developing a horizontal-type neutron reflectometer, named SOFIA, at BL16 in J-PARC/MLF for investigating such interfacial structures, including those of free liquid surfaces. Utilizing the high-flux neutron beam at J-PARC, we have, until now, efficiently performed reflectivity measurements with SOFIA. However, the maximum count rate of the two-dimensional (2-D) detector currently in use will be insufficient for the full-power operation of J-PARC (1 MW) in the near future. In order to increase the count rate, we developed a new detector comprising 21 one-dimensional (1-D) detectors. 1-D detectors were used because neutron counts for a two-dimensional (2-D) detector can be reduced by dividing it into an array of 1-D detectors. In this study, we evaluated the performance of the newly developed detector and successfully carried out neutron reflectivity measurements using it.

1. Introduction

Neutron reflectometry is widely used for investigating surface structures and buried soft material interfaces. Beam line 16 (BL16) at Materials and Life Science Facility (MLF) in Japan Proton Accelerator Research Complex (J-PARC) is dedicated to a horizontal-type neutron reflectometer in which two downward neutron beams (2.22° and 5.71°) are transported from a coupled hydrogen moderator to irradiate a free surface such as the air–water interface. In order to measure neutron reflectivity by using the high-flux neutron beam at J-PARC/MLF as soon as possible, we utilized an old reflectometer, named ARISA [1], which was relocated from KENS facility to MLF in 2008. However, the range of movement of the slit, sample, and detector stages was limited and only the 2.22° beam line at BL16 was available for neutron reflectivity measurements, because the distance between the two downward beams (2.22° and 5.71°) was too large for ARISA. This is a serious disadvantage that prevented the use of the 5.71° beam line for measuring neutron reflectivity at air–liquid interfaces measurement with a high incident angle. To overcome this disadvantage, the components of ARISA were replaced by a new reflectometer—SOFt-Interface Analyzer (SOFIA)—in 2011 through a joint collaboration between JST/ERATO and KEK [2-4]. The slit, sample, and detector stages of SOFIA allowed us to utilize both the 5.71° and 2.22° beam lines. This, in turn, enabled us to measure neutron reflectivity over a wide momentum-transfer region on free liquid surfaces.

We employed a two-dimensional (2-D) photomultiplier tube equipped with a $^{6}LiF/ZnS$ scintillator as the detector for SOFIA [5]. Using this detector, we were able to measure specular reflection, off-specular reflection, and background at a time. Neutron signals were recorded by a NeuNET module [6] and stored as event data with time stamps inserted by a GateNET module. This recording system enabled us to change time slicing bins for kinetics measurement even when the original measurement had been completed. In addition, SOFIA provided the double frame mode (12.5 Hz operation) that extended the wavelength band up to 1.76 nm; this is significantly higher than the maximum wavelength available in the single frame mode (0.88 nm at 25 Hz operation). This setup of SOFIA reflectometer enabled us to efficiently perform kinetics measurements, and thus, numerous kinetics studies were carried out using this setup [7, 8].

However, this detector started showing pulse pile-up at around 10,000 cps (counts per second) because of scintillator afterglow. This count rate would have been insufficient for the full-power operation of J-PARC (1 MW) in the near future. In order to increase the count rate, we developed a new photon detector consisting of an opto-semiconductor device, multi-pixel photon counter (MPPC). Figure 1 (a) schematically illustrates the new detector comprising 21 one-dimensional (1-D) detectors and a readout module. The sensitive area of each detector was 5 mm in width and 128 mm in length, and the total detector area, including the areas of all 21 1-D detectors, was 105 mm × 128 mm. Since the neutron beam used for reflectivity measurements is collimated along one dimension but is very wide along the other dimension, neutron counts for a detector can be reduced by dividing a 2-D detector into an array of 1-D detectors. Figure 1 (b) shows the side view of a single 1-D detector. The neutron position along the length direction was evaluated by a charge division method, and the spatial resolution was about 1 mm. Detailed information is provided in our previous report [9].

In this study, we evaluated the performance of the new detector and carried out neutron reflectivity measurements using this detector with the SOFIA reflectometer.

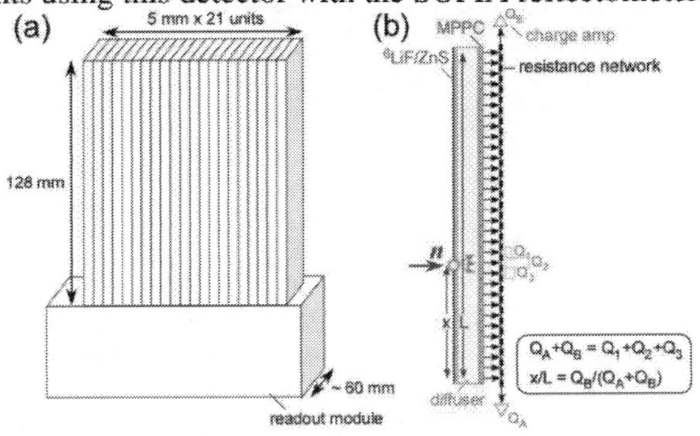

Figure 1. Schematic illustration of the new detector. (a) Dimensions of the 1-D detector array. (b) Side view of a single 1-D detector.

2. Performance and reflectivity measurements

To determine the position linearity and spatial resolution of the new detector, we performed a position scan with a beam width of 0.1 mm and a step of 4 mm. Figure 2 (a) shows the obtained beam profile for a single detector. The peak position and full width at half maximum (FWHM) were evaluated by multi-peak fitting performed using a Gaussian function. The position linearity translated very well, as shown in Figure 2 (b), and the slope was evaluated to be 0.183 mm/channel (total number of channels: 1024). Using this slope, we also evaluated the FWHM depending on the beam position, as shown in Figure 2 (c). The FWHM was about 1.4 and 2.0 mm at the center and edge of the detector, respectively; this resolution is sufficient for neutron reflectivity measurements.

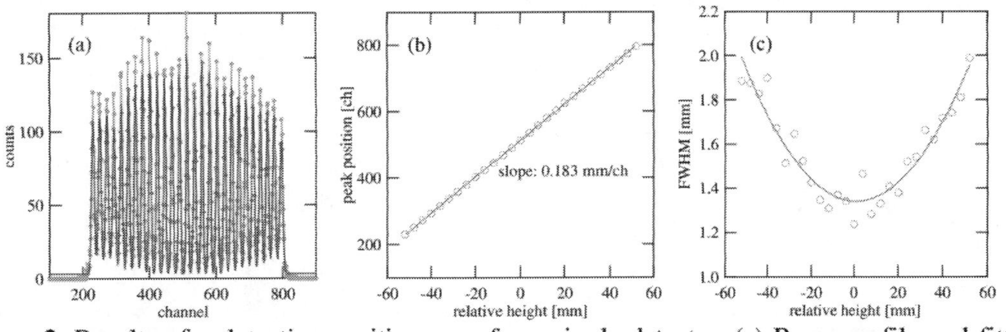

Figure 2. Results of a detection position scan for a single detector. (a) Beam profile and fitting

results. (b) Peak position depending on the beam position. (c) FWHM depending on the beam position.

Next, we evaluated the homogeneity of the neutron detection sensitivity by directing a homogeneous beam on the detector. For this purpose, the neutron beam was first scattered by a poly(methyl methacrylate) plate and incoherent scattering from the plate was measured by the detector at a distance of 1.8 m from the plate. Figures 3 (a) and (b) show the obtained intensity map obtained for various detection positions and frequency distribution of the neutron count, respectively. The intensity map appears flat. Further, considering the standard deviation of the frequency distribution, it was concluded that about 10% of the average frequency was sufficient for use in reflectivity measurements.

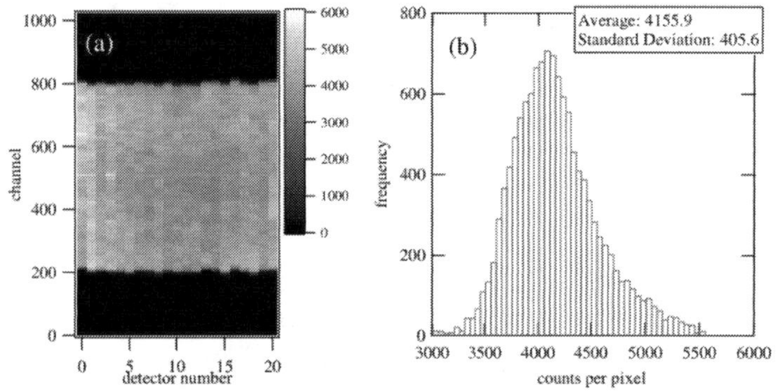

Figure 3. Results of homogeneity evaluation for neutron detection sensitivity. (a) Intensity map based on the detection position. (b) Frequency distribution of the neutron count.

Furthermore, we evaluated the count rate of the detector. A neutron beam was collimated by double slits and neutrons directly counted by the detector. The horizontal width of the slits was adjusted according to the beam width at a sample position of 40 mm (width at the detector was 55 mm). This means that in the case of the new detector, eleven 1-D detectors should work because the width of each 1-D detector was 5 mm. The vertical width of the downstream slit was fixed at 0.2 mm and that of the upstream slit was varied between 0.2 mm and 8.0 mm. Figures 4 (a) and (b) show the time-of-flight profiles for various beam collimations obtained using the new and old detector, respectively. All the profiles in Figure 4 (a) show a Maxwellian peak around 11 ms; however, the peak height was suppressed at 3.2 mm and completely disappeared at 6.4 mm in Figure 4 (b). This indicates that the maximum count rate of the new detector was considerably higher than that of the old one. Figure 4 (c) shows the count rate of the new detector at the Maxwellian peak as a function of the slit width. For low slit widths, the count rate is proportional to the slit width. With increasing slit width, however, the count rate lessens than that indicated by the fit line around 40,000 cps because of counting loss. Although the eleven detectors worked, this count rate was only four times that of the old detector. The region at the detector that exhibited high flux was 20 mm in width. This is the reason for the maximum count rate being limited to 40,000 cps, and the count rate in reflectivity measurements should be maintained at a value less than this.

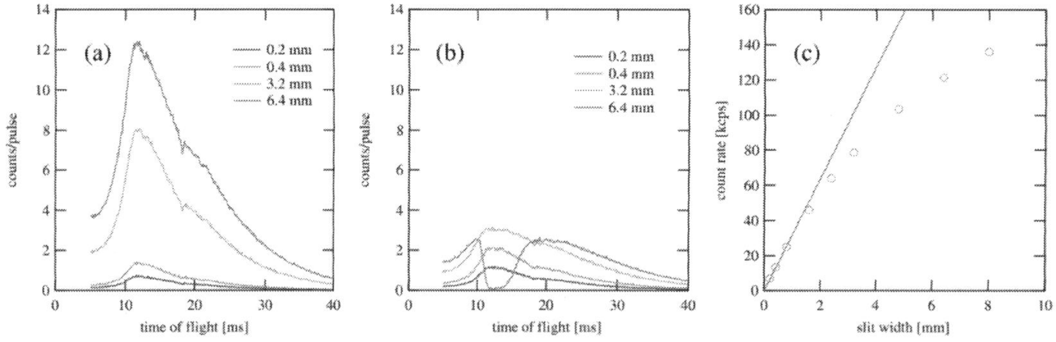

Figure 4. Results of count rate evaluation. (a) Time-of-flight (TOF) profile obtained using the new detector. (b) TOF profile obtained using the old detector. (c) Count rate of the new detector at the Maxwellian peak with varying slit width.

Finally, we performed reflectivity measurements using the new detector on a 65-nm-thick deuterated polystyrene thin film coated on a silicon wafer. Figure 5 (a) shows the TOF profiles for a direct beam and the reflection at an incident angle of 0.6°. The reflection intensity modulates for up to 30 ms owing to interference from the polystyrene film, and overlaps with that of the direct beam above 30 ms because of total reflection at the film surface. To compare the obtained reflectivity data with the measured data obtained using the old detector, the reflection intensity was normalized by the direct beam profile. Figure 5 (b) shows the reflectivity profile depending on the momentum transfer normal to the substrate, Q_z. The profile obtained using the new detector agreed well with that obtained using the old detector. This is direct evidence that the new detector is suitable for neutron reflectivity measurements.

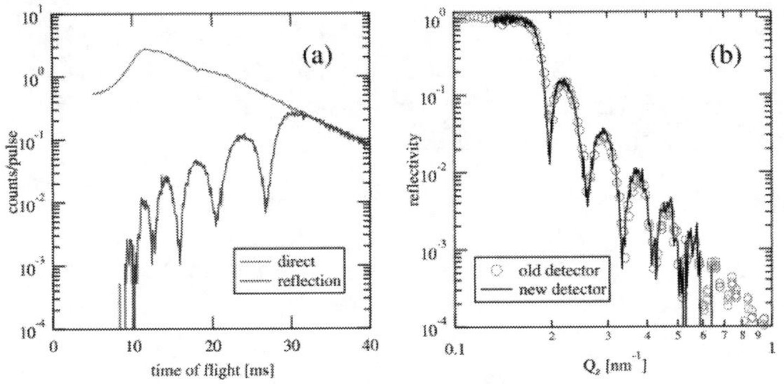

Figure 5. Reflectivity measurements for a deuterated polystyrene thin film performed using the new detector. (a) TOF profiles for a direct beam and reflection at an incident angle of 0.6°. (b) Reflectivity profile depending on momentum transfer normal to the substrate, Q_z.

3. Conclusion

We developed an array of one-dimensional, position-sensitive detectors for use with a neutron reflectometer, SOFIA, at J-PARC. The FWHM of the detector was about 1.4 mm at the center, and the homogeneity of neutron sensitivity was about 10%. These values are sufficient for using the detector for neutron reflectivity measurements. Remarkably, the maximum count rate of the new detector was four times that of the old detector used with SOFIA. This improved count rate is insufficient to accept a direct beam with a wide slit width, but it is enough for time-slicing reflectivity measurements with a short accumulation time performed using the high-flux neutron beam operating at 1 MW in J-PARC. Furthermore, we used the new detector for performing reflectivity measurements and confirmed that the detector is suitable for this purpose. Installation of the new detector in SOFIA has been planned in the near future.

The neutron scattering experiment was approved by the Neutron Scattering Program Advisory Committee of IMSS, KEK (Proposal No. 2009S08).

References

[1] Torikai N *et al.*2002 *Appl. Phys.* A **74** S264.
[2] Mitamura K *et al.*2011 *J. Phys.: Conf. Ser.* **272** 012017.
[3] Yamada N L *et al.*2011 *Euro. Phys. J. Plus* **126** 108.
[4] Mitamura K *et al.*2013 *Polymer J.* **45** 100.
[5] Hirota K *et al.*2005 *Phys. Chem. Chem. Phys.* **7** 1836.
[6] Satoh S *et al.*2009 *Nucl. Inst. Meth.* A **600** 103.
[7] Kawaguchi D *et al.*2011 *Macromolecules* **44** 9424.
[8] Xia T *et al.*2013 *Macromolecules* **46** 4540.
[9] Satoh S *et al. Nucl. Inst. Meth. A* (in press; DOI: nima.2013.04.067).

Concepts of Neutron Polarisation Analysis Devices for a New Neutron Chopper Spectrometer, POLANO, in J-PARC

K Ohoyama[1], T Yokoo[2,3], S Itoh[2,3], T Ino[2,3], M Ohkawara[1], T Oku[4], S Tasaki[5], K Iwasa[6], T J Sato[7], S Ishimoto[2], K Taketani[2], H Kira[8], Y Sakaguch[8], M Nanbu[1], H Hiraka[2], H M Shimizu[9], M Takeda[4], M Hino[10], K Hayashi[1], U Fliges[11], and P Hautle[11]

[1]Center of Neutron Science for Advanced Materials, Institute for Materials Research, Tohoku University, Sendai 980-8577, Japan
[2]Neutron Science Division, High Energy Accelerator Research Organization, Tsukuba, 305-0801, Japan
[3]J-PARC Center, Tokai, 319-1195, Japan
[4]Japan Atomic Energy Agency, Tokai, 319-1195, Japan,
[5]Department of Nuclear Engineering, Kyoto University, Kyoto, 615-8530, Japan,
[6]Department of Physics, Tohoku University, Sendai 980-8578, Japan
[7]Institute of Multidisciplinary Research for Advanced Materials, Tohoku University, Sendai, 980-8577, Japan
[8]Comprehensive Research Organization for Science and Society, Tokai, 319-1106, Japan
[9]Department of Physics, Nagoya University, Nagoya 464-8602,Japan
[10]Kyoto University Research Reactor Institute, Kyoto University, Kumatori, 590-0494, Japan
[11]Paul Scherrer Institut, 5232 Villigen PSI, Switzerland

E-mail: `ohoyama@imr.tohoku.ac.jp`

Abstract. The basic concept of polarisation analysers with bender type supermirrors on the polarised neutron spectrometer, POLANO, which is now under construction in J-PARC/MLF, is described.

1. Introduction

The construction of a new polarisation analysis neutron chopper spectrometer, named POLANO [?] is now in progress in J-PARC/MLF, Tokai, Japan, as a collaboration project between High Energy Accelerator Research Organization (KEK) and Tohoku University. POLANO is a compact chopper spectrometer ($L_1 = 17.5$ m, $L_2 = 2$ m) with polarisation analysis ability, installed at BL23 (decoupled H_2 moderator) in the second experimental hall of MLF. The present status of this project is described by Yokoo et al. in this conference [?].

As already reported elsewhere [?, ?], a ^{3}He spin lter with a continuous polarising spin exchange optical pumping (SEOP) system and a bender type supermirror device will be used as the polariser and analyser in the rst stage of the project for investigations in the region of $\epsilon \leq 30$ meV. Discussions on the development and design of polarising devices are in progress under international collaboration.

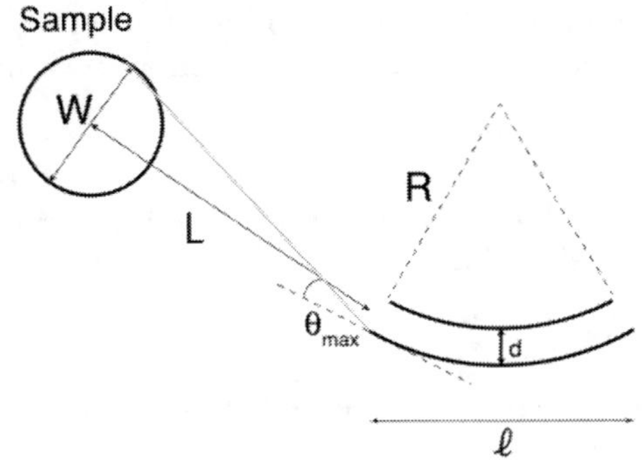

Figure 1. De nition of parameters based on Ref. 3. ℓ is the length of direct sight, the length of the supermirrors required to avoid the direct line of sight and to ensure at least one reflection on a mirror. The green line indicates a neutron flight pass from the edge of the sample.

In this short article, we estimate the parameters of the bender type supermirror analyser of POLANO based on Stewart's calculations in Ref. 3. Figure **??** shows the de nitions of the parameters of a bender type supermirror analyser in this text. For such analysers, the sample size is limited because of an acceptable angular divergency as shown in Fig. **??**. The neutrons with an incident angle on the supermirror that is larger than the critical angle, θ_c, cannot be reflected; that is, they cannot be detected at a detector. Thus, L and W must satisfy the following condition [**?**].

$$\theta_c \geq \theta_{max} = W/2L \tag{1}$$

This directly means that samples with a large size within the scattering plane are not e ective for instruments with supermirror-type analysers. Figure **??** shows the acceptable diameter of a sample for several m values obtained from eq (1). Because of the structure of the sample vacuum chamber, the shortest L for POLANO will be ~55 cm; this covers the solid angle of the scattered beams as much as possible. For m=5, which is almost the maximum value of commercial products at present, samples with a diameter of 20 mm can be used for ~20 meV, and 15 mm for 40 meV; these sample diameters are comparable to an incident beam size of 20×20 mm^2 at the sample position, which is given by the focusing guide tubes at POLANO [**?**, **?**]. This means that m=5, (or 5.5 if possible) is the best choice for bender type supermirror analysers. Triple axis spectrometers with a polariser and analyser that show advantages in a few 10 meV region; thus, large m supermirrors should be installed on POLANO to access a much wider energy region. We should point out that, for a cylindrical sample with a diameter of 10 mm, even scattered neutrons with E_f ~ 60 meV can be analysed with m=5 mirrors, in principle.

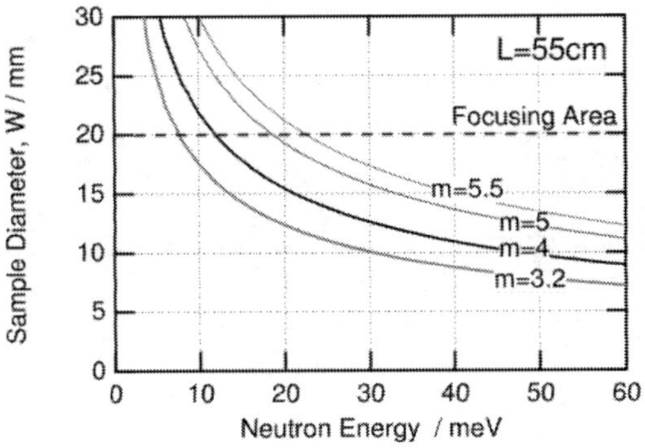

Figure 2. The maximum acceptable sample size for several m values. For POLANO, L will be ~55 cm.

Next, we estimate the gap d between adjacent mirrors and the minimum length of curved mirrors, ℓ, the length of direct sight. To avoid a direct pass with no reflection on a mirror, d and ℓ should satisfy,

$$\ell = (8Rd)^{1/2} \tag{2}$$

for $d \ll R$ [?]. On the other hand, for the reflection that does not touch the upper mirror, the garland reflection, d and R satisfy the following relation [?]:

$$\sin\theta_c = (2d/R)^{1/2} \tag{3}$$

Figure **??** shows ℓ for several d for m=5 and 3.2 obtained from eq (2) and (3). Smaller d values naturally give a shorter ℓ; for instance, one can use a compact analyser system with $\ell \leq$ 100 mm for d=0.32 mm for m=5 to analyser neutrons with energies up to 40 meV. However, because the substrate of borosilicate glass has a nite thickness (a typical value is ~0.3 mm), a smaller d reduces the transmissions. For POLANO, we choose d=0.7 mm, for which the minimum mirror length ℓ can be approximately 230 mm for m=5 at E=40 meV; this length satis es the space limitation. For a practical device, 1.3 times longer supermirrors (ℓ ~300 mm) will be used to avoid the pass inside the substrate glass plates. Typical borosilicate glass includes approximately 10-13 wt% of boron; the transmitted neutrons through the substrate glass which degrade polarisation are negligible for ℓ ~300 mm.

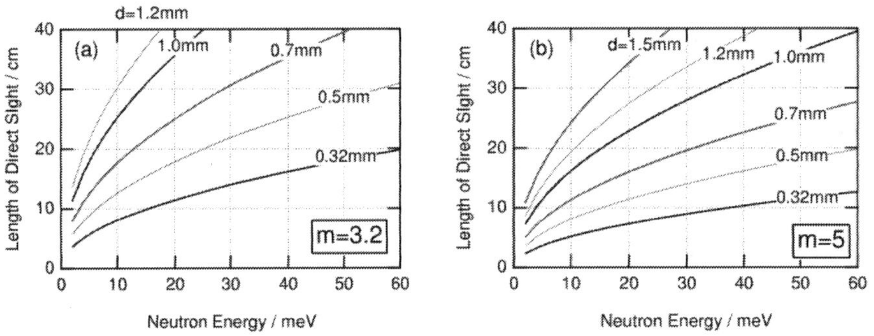

Figure 3. Mirror length of direct sight ℓ for several d for (a) m=3.2 and (b)5.0.

Typical expected values of the parameters of the supermirror analyser in the present design are summarised in Table 1. The vertical width is determined from the conditions of the magnetic elds which saturate the magnetisation of the supermirrors. One should be careful about the relation between d and the total area of supermirrors; for a larger d, the number of supermirrors necessary to cover a particular scattering angle can be reduced, but the total area of the necessary supermirrors becomes larger; that is, the cost will increase.

The solid angles of the analyser are quite limited because of the size of the supermirror. Thus, we will use position sensitive detectors (PSDs) with a length of 600 mm centred at the equatorial plane, which is much shorter than the length of standard PSDs in J-PARC/MLF. Supermirrors with the parameters in Table 1 cover 76% of the length of a PSD. Because one PSD package of POLANO which consists of 32 PSDs covers the scattering angle 2θ over a range of 18 °, the analyser device with ~180 mirrors in Table 1 corresponds well to one PSD package. The coverage of 2θ is much smaller than the coverage of other polarisation analysis spectrometers in pulsed neutron sources, such as HYSPEC in SNS. Thus, the analyser device will be installed on a rotating table take measurements at di erent scattering angle positions up to 2θ=120°. The number of supermirrors will be increased in stages in the future so as to cover larger solid angles.

Table 1. Summary of expected parameters of the supermirrors of the present design

m value		5.0 or 5.5 (concave surface) 1.5 (convex surface)
Optimised Energy		41 meV
Length, ℓ		300 mm
Vertical width	(sample side)	∼140 mm
	(PSD side)	∼180 mm
Radius of curvature, R		10.1 m
Gap, d	(sample side)	0.7 mm
	(PSD side)	1.0 mm
Thickness of substrate		0.3 mm
Distance from the sample, L		550 mm
No. of mirrors		∼180
Covered 2θ range		∼16°

Though the supermirror analyser must be the best for the $\epsilon \leq 30$ meV region, POLANO should be available in higher energy regions too. SEOP ^{3}He spin lter analysers are now under development for the $E_f \sim 100$ meV region in collaboration with KEK, JAEA, and Tohoku University. In particular, GE180 glass cells with a diameter of 10 cm have already been prepared in Tohoku University [?]. We will develop banana-shaped analyser cells for POLANO in this project as well. We are also developing the dynamic nuclear polarisation (DNP) technique in KEK with international collaboration, which is important for a higher energy polariser as well as for sample environment equipment for investigations of hydrides. For the further development of the DNP technique, as well as supermirror benders, discussions with PSI are in progress.

Acknowledgments

This work was partly supported by the S-type project of KEK (2009S09), and a Grant-in-Aid for Challenging Exploratory Research (24656004), (A)(23244068) and (S)(23224009) from MEXT, Japan.

References

[1] K. Ohoyama, T. Yokoo, S. Itoh, J. Suzuki, K. Iwasa, T.J. Sato, H. Kira, Y. Sakaguchi, T. Ino, T. Oku, K. Tomiyasu, M. Matsuura, H. Hiraka, M.Fujita, H. Kimura, T. Sato, J. Suzuki, H.M. Shimizu, T. Arima, M. Takeda, K. Kaneko, M. Hino, S. Muto, H. Nojiri, C.H. Lee, J.G. Park, and S. Choi, Proceedings of 5th Workshop on Inelastic Neutron Spectrometers (WINS2012), in print.

[2] T. Yokoo, K. Ohoyama, S. Itoh, J. Suzuki, K. Iwasa, T. J. Sato, H. Kimura, N. Kaneko, and M. Ohkawara, Proceedings of LPBMS2013.

[3] J. R. Stewart, P. P. Deen, K. H. Andersen, H. Schober, J.-F. Barthélémy, J. M. Hillier, A. P. Murani, T. Hayes and B. Lindenau, J. Appl. Cryst. (2009). 42, 69.

Journal of Physics: Conference Series **502** (2014) 012052 doi:10.1088/1742-6596/502/1/012052

Improvement of Instrument Devices for Super High Resolution Powder Diffractometer at J-PARC

S Torii[1]*, M Yonemura[1], Y Ishikawa[1], P Miao[1], R Tomiyasu[1], S Satoh[1], Y Noda[2] and T Kamiyama[1]
[1]High Energy Accelerator Research Organization, Tokai, Ibaraki 319-1106, JAPAN
[2]Tohoku University, Aoba-ku, Sendai 980-8577, JAPAN

E-mail: torii@post.kek.jp

Abstract. Super High Resolution Powder Diffractometer SuperHRPD [1] at J-PARC have suffered damage by the Great East Japan Earthquake on 11 March 2011. After temporarily restoration work, the general user program has been started since April 2012. Full-fledged restoration of the temporarily restored devices is being carried out in 2013. The absorber materials for two disk choppers were replaced to $^{10}B_4C$ which improved quality of incident spectrum. The new detector system of 8 mm diameter position sensitive detectors was developed, and the test experiments showed excellent results.

1. Introduction

SuperHRPD accepted the first neutron beam from the pulsed source at Materials and Life Science Experimental Facility (MLF) of Japan Proton Accelerator Research Complex (J-PARC) in May 2008 and achieved the world best resolution of $\Delta d/d$ = 0.035 % in June 2008. The beam power was operated around 4 kW at the beginning, but it is operated stably around 300 kW up to now. Several upgrading works to improve the performance of instrument such as replacement of vacuum scattering chamber, updating of the DAQ system, enhancement of detector system, *etc.* have also been successfully done. We have evolved the instrument with a very high *S/N* ratio. Coupled with the increase of the beam power, the high resolution and the high *S/N* ratio make it possible to carry out detailed analysis of complex crystal structures. More than 50 research proposals as general user program have been conducted at SuperHRPD and some research results have been published in various academic papers [2, 3, 4].

The Great East Japan Earthquake that occurred in March 2011 has caused very heavy damages to SuperHRPD instrument: vertical subsidence and horizontal shift of the annex building for SuperHRPD beam line, damage of the super mirror guide tube, deviation of the beam alignment, tilt of the vacuum scattering chamber, shifting of the sample center from its normal position, *etc.* After temporarily restoration work that took mostly over one period of the 2011 fiscal year, SuperHRPD became capable to accept the first neutron beam in April 2012.

Full-fledged restoration of the temporarily restored devices is being carried out during shut down in 2013; many of beam line devices, vacuum chamber base and shielding, *etc.* are modified and improved by considering that the slight subsidence is continuing and the risk of future aftershock is not past. We will report them in future, and herewith report on other developments and improvements.

* To whom any correspondence should be addressed.

2. Instrument Upgrading

2.1. Change of absorber materials for the disk chopper

SuperHRPD has about 100 m length beam line in order to achieve super high resolution. Since accelerator cycle of the MLF is 25 Hz, the wavelength window of the instrument with such long beam line would be extremely narrow. Therefore, installation of disk choppers painted with absorber materials were required in order to achieve the wide wavelength window and to suppress the frame overlap. In order to ensure the wide wavelength window of SuperHRPD, the disk choppers are usually set for the measurement at 5 Hz (1/5) and 8.33 Hz (1/3) in operation mode.

Furthermore, in order to avoid the effects of the burst and high-energy neutron, the curved guide section of 32 m in length was provided. The characteristic wavelength λ^* calculated from equation (1) is 0.61 Å.

$$\lambda^* = \sqrt{\frac{2a}{\rho}} \cdot \sqrt{\frac{\pi}{Nb_{coh}}} \qquad (1)$$

where a is the guide width (=25 mm), ρ is the radius of curvature (=5 km), N is the atomic density and b_{coh} is the mean coherent scattering length. The value of Nb_{coh} is called as the scattering length density, which denotes the value of specific materials. The supermirror guide used at SuperHRPD has three-times larger critical angle than nickel guide ($\sqrt{Nb_{coh}/\pi}$=1.73×10^{-3} Å^{-1}). Although λ^* is in the region where the total cross section of gadolinium decreases rapidly (absorption end), the cross section is still large value (see Figure 1). Therefore, we initially chose gadolinium oxide as absorber materials for the disk choppers. However, we learned the leakage of the short-wavelength components was large due to the excellent reflectivity of the supermirror. Then we changed the absorber materials from gadolinium oxide to boron carbide $^{10}B_4C$ (^{10}B > 95 at%). The incident intensity spectra obtained using measurement of the V-Ni alloy with $^{10}B_4C$ and Gd_2O_3 disk choppers are compared in Figure 2 (left). It is evident that the extra humps were removed by the adoption of $^{10}B_4C$.

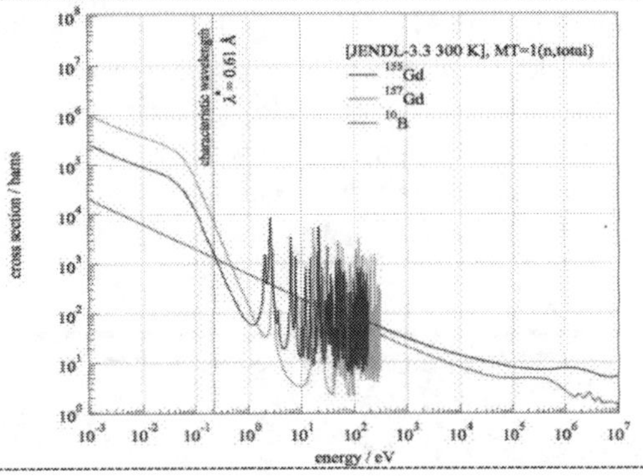

Figure 1. Total neutron cross section of gadolinium (^{155}Gd, ^{157}Gd) and boron (^{10}B) calculated from JENDL-3.3 code program [5]. The vertical dotted line in figure shows the energy value that converts the characteristic wavelength. A large absorption end exists in gadolinium near the characteristic wavelength. Boron shows smooth energy dependence in the same region.

Although any trace of hump has been detected in the background measurements without samples, weak humps were detected when strong scattering samples are placed at the sample positions; for example, diffraction patterns of a silicon with both choppers are shown in log scale in Figure 2 (right); the extra humps hardly influenced for crystal structure analyses.

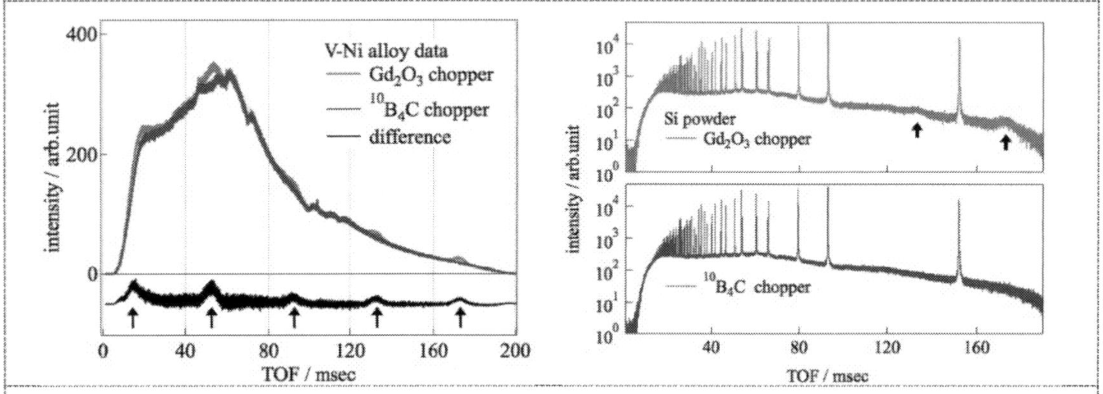

Figure 2. (left) The incident intensity spectra with Gd_2O_3 and $^{10}B_4C$ disk chopper absorber materials at 5 Hz operation. The extra humps were removed. (right) The diffraction patterns of a NIST silicon powder sample by both choppers. The extra humps existed shown by black arrows.

2.2. New detector system with 8 mm diameter Position Sensitive Detector

Because of shortage of initial buget and delay of development in a high spatial resolution detector system, we initially installed the 1/2 inch diameter position sensitive detectors (PSDs) that had ever used at KENS facility of KEK. This imposed limit on resolution compared with the designed one. Then we developed a new detector system based on 8 mm diameter PSDs and could improve resolution as shown below. The 16 PSDs of 8 mm in diameter were composed together into one unit and an amplifier circuit was connected to each PSD. In the new detector system, 8 amplifier circuits were built into one ASIC chip and this contributes to the more compact circuit (Figure 3).

The test measurement using the new detector system has been performed in May 2013. For the measurement, four new detector units have been arranged at the backward detector bank to cover the area about $2\theta = 157° \sim 172°$. A diamond powder sample enclosed into a sample holder of 6 mm in diameter was used for the measurement. We were able to obtain a very sharp diffraction peak shown in Figure 4 with $\Delta d/d = 0.0637$ %. Since the resolution of the normal sample with the 1/2 inch PSDs system is about $\Delta d/d = 0.1$ %, the improvement has been achieved by the 8 mm PSDs system. Based on the test measurement, all backward bank detectors are being replaced to the 8 mm PSDs in 2013.

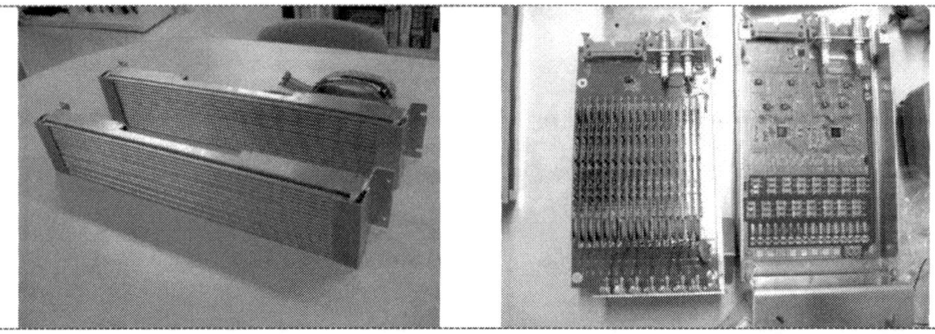

Figure 3. (left photograph) A new detector unit consists of 16 PSDs with 8 mm diameter. (right photograph) The conventional amplifier circuit (left) and the advanced one with an ASIC chip (right); the 16 amplifier circuits are accumulated in two ASIC chips.

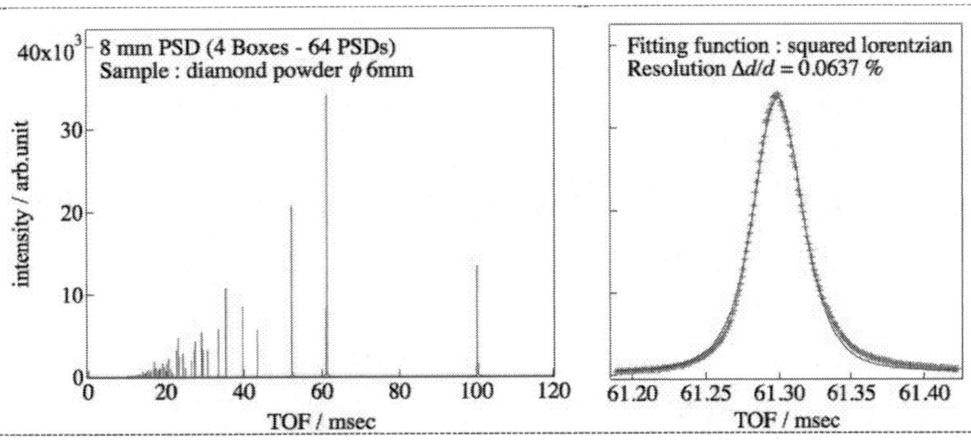

Figure 4. (left) Neutron diffraction pattern of diamond powder sample data using new detector system. (right) The peak fitting result using squared Lorentzian function.

3. Conclusion

SuperHRPD has been upgraded repeatedly and the high performance with the best resolution has been successfully obtained. During long-term shut down in 2013, the full-scale restoration works of re-alignment of the beam line would be carried out. The absorber materials for two disk choppers were replaced to $^{10}B_4C$ after the earthquake. This improvement rather reduced leakage of the short-wavelength components and no extra hump was not observed in the sample data. The test experiments of the new detector system with 8 mm diameter position sensitive detectors showed excellent results, and all backward bank detectors are being replaced.

Acknowledgements

We would like to thank Mr. Takashi Muroya, Mr. Masahiro Shioya and Mr. Masataka Sakaguchi for their technical supports. The neutron scattering experiment was approved by the Neutron Scattering Program Advisory Committee of IMSS, KEK (Proposal No. 2009S05).

References:

[1] Torii S, Yonemura M, Surya Panca Putra Y T, Zhang J, Miao P, Muroya T, Tomiyasu R, Morishima T, Satoh S, Sagehashi H, Noda Y and Kamiyama T, 2011 *J. Phys. Soc. Jpn.* **80** SB020.

[2] Kamaya N, Homma K, Yamakawa Y, Hirayama M, Kanno R, Yonemura M, Kamiyama T, Kato Y, Hama S, Kawamoto K and Mitsui A, 2011 *Nature Materials* **10**, 682-686.

[3] Ohshima Y, Noguchi Y, Oguchi T, Kitanaka Y, Miyayama M, Torii S and Kamiyama T, 2012 *ECS Trans.* **45** 195-207.

[4] Takai S, Doi Y, Torii S, Zhang J, Surya Panca Putra Y T, Miao P, Kamiyama T and Esaka T, 2013 *Solid State Ionics* **238** 36-43.

[5] Shibata K et al., 2002 *J.Nucl.Sci.Technol.* **39** 1125-1136.

Development of SPICA, New Dedicated Neutron Powder Diffractometer for Battery Studies

M Yonemura[1], K Mori[2], T Kamiyama[1], T Fukunaga[2], S Torii[1], M Nagao[1], Y Ishikawa[1], Y Onodera[2], D S Adipranoto[3], H Arai[3], Y Uchimoto[4] and Z Ogumi[3]

[1] Institute of Materials Structure Science, High Energy Accelerator Research Organization, 203-1 Shirakata, Tokai, Ibaraki 319-1107, Japan
[2] Research Reactor Institute, Kyoto University, Kumatori-cho, Sennan-gun, Osaka 590-0494, Japan
[3] Office of Society-Academia Collaboration for Innovation, Kyoto Univiersity, Uji-shi, Kyoto 611-0011, Japan
[4] Graduate School of Human and Environmental Studies, Kyoto University, Sakyo-ku, Kyoto 606-8501, Japan

masao.yonemura@kek.jp

Abstract. SPICA, a new special environment powder neutron diffractometer was built at BL09 in the Material and Life science Facility (MLF) of the Japan Proton Accelerator Research Complex (J-PARC). This is the first instrument dedicated solely to the study of next-generation batteries in J-PARC and is optimized for *in situ* measurements to clarify structural changes of materials in batteries. The basic design and instrumentation of SPICA have been completed. The highest $\Delta d/d$ resolution achieved at the commissioning stage was 0.09% at the back scattering bank of SPICA. The reliability of the diffraction data has achieved a sufficiently high level for the structural analysis of materials using the Rietveld method. The air scattering banks with the blades made of B_4C for *in situ* measurements also function very well.

1. Introduction

Mankind's technology has been largely based on the use of energy since the industrial revolution. In modern society, batteries are the key technology for energy storage because most mobile devices such as miniaturized laptop computers and cellular phones are powered only by batteries with high power, large capacity and long cycle characteristics. [1,2] Such high performance power units are required not only for small devices, but also for electric vehicles (EVs) and hybrid electric vehicles (HEVs). The present lithium ion battery still has potential applications for these applications; however, it is gradually becoming insufficient to meet requirement. Therefore, it is necessary to push the envelope in the next-generation battery systems.

The reactions in typical batteries are not so complicated and can be described as charged ions that move between positive and negative electrodes through an electrolyte; the ions transfer in the electrodes and at the interfaces of the electrodes and electrolyte. However the relationship between the

[1] To whom any correspondence should be addressed.

structure changes, ionic distributions in the electrodes and battery fading are still not clear. To improve present batteries toward the development of next-generation battery systems, it is necessary to make every effort to investigate the structural changes that occur at the electrodes and interfaces in an operational battery system. The neutron is a powerful probe to determine structural information regarding the locations of light elements such as hydrogen and lithium, especially in comparison with X-rays. Thus, an advantage of the neutron diffraction technique is the acquisition of more information on the location of light ions or the changes of atomic distribution in both electrodes and electrolyte materials.

The SPICA diffractometer has been newly installed in MLF/J-PARC. This diffractometer is optimized for *in situ* measurements to clarify the structural changes of battery materials at the atomic level. Neutrons have deep penetration, so that they enter deeply into a battery case and provide diffraction patterns of the electrode materials. Our approach with this diffractometer is to reveal the reactions in batteries and to determine factors of safety and degradation over long periods in practical battery systems.

In the present work, we have conducted design and installment of the diffractometer. Several neutron diffraction data were successfully collected using *in situ* techniques. Here, we discuss the basic characteristics of this new instrument and present a quick overview of its potential impact based on recent experimental results.

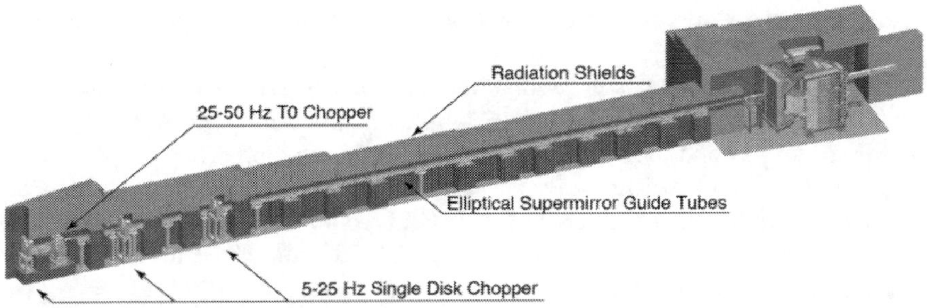

Figure 1. Overview of SPICA instrument. Neutrons are transported by elliptical supermirror guides. The first 3-section of Fe collimators that are defined the maximum beam divergences start 2.3 m from decoupled poisoned moderator and end at 7.2 m in biological shielding before the 1st chopper. The supermirror guide is installed from 7.3 to 48 m positions. Three single disk choppers are installed at 7.25, 12.8, 18 m.

2. Instrument design for SPICA

To make *in situ* measurements of real batteries more fruitful, we need high $\Delta d/d$ resolution with wider d ranges to detect many phases during chemical reaction, high neutron intensity to know the specific reaction process in high speed charge/discharge, low background and large sample area to install big sample environment and a dedicated chemistry area to carry out long-term scheduled experiments with many sets of *on-beam* measurements and off-beam charge-discharge measurements. The instrumental design is started with the assumption that (1) the sample size is 20 × 40 mm^2 and (2) the $\Delta d/d$ resolution is under 0.08% at the back scattering bank.

SPICA is installed at BL09 in 1st experimental hall of MLF. The sample position is 52 m from the poisoned and decoupled moderator. The thin side of the poisoned moderator was selected. The neuron intensity on the thin side is weaker than that on the thick side; however, the thin side has an advantage of a symmetrical peak profile, which can be extracted the strain information from samples.

A focusing supermirror guide was required to increase the number of neutrons that reach the sample position. The elliptical guide with off-focusing in the vertical and horizontal planes is one of the candidates for focusing. After several elliptical guides designs were simulated using McStas [3], the final design of the supermirror guide, which has a gradually changed mirror coating from m=3 to 6 at the posterior half of the guide, was adopted.

The basic layout of the SPICA instrument is shown in Figure 1. The flight path is 52 m from the moderator to the sample position and 2 m from the sample position to the detectors. The T0 chopper rotates with a frequency of 25 Hz, which is the same as the repetition rate of the proton beam injection. The natural bandwidth of the instrument is 2.9 Å. Three single disc choppers are used for bandwidth selection to prevent frame overlaps and operated with 25/4 to 25/1 Hz repetition to select bandwidths. The minimum bandwidth is from 0.4 to 2.5 Å at 25 Hz repetition, and the maximum bandwidth is from 1.6 to 10.1 Å at 25/4 Hz repetition.

The detector system has been successfully installed (Figure 2). SPICA utilizes 1568 ^{3}He gas position sensitive detectors (PSDs) with 1/2 inch diameters and 0.60 m active length of a total 0.67 m. Eight detectors were assembled in one detector-component box. The detector-banks are formally grouped according to scattering angle; back scattering bank (175-150°), multi purpose bank including high angle (150-120°), 90° bank (120-60°), low angle bank (60-10°) and small angle bank (15-5°). The horizontal detector crates are arranged on a cylindrical locus from 2θ =175 to 5° and are located approximately 2 m from the sample position. The back scattering bank and multi purpose bank have two other up and down vertical crates (totally 3 crates) that cover the maximum elevation angle (+/- 35°), because these banks are the main banks for high resolution and high intensity *in situ* measurements. At the low angle bank, there is just one extra down vertical crate that has a maximum elevation angle (-35°) to quickly collect low-Q data for long-range periodic structures. This detector-formation the collection of several diffraction data in a shorter time shot pulse by using multi-angle data with different wavelengths.

SPICA has a "sample vacuum chamber" with a diameter of 1m. The back scattering bank also has a "scattering vacuum chamber" located between the sample chamber and detectors. These chambers are basically filled with argon gas and evacuated for precise measurements. The sample vacuum chamber can be removed to set large special sample environments. The multi purpose bank has air scattering chambers installed with aluminum blades coated with B_4C resin as course collimators. The blades are spaced in view of each detector box. In the previous simulation study, the loss of neutrons in the air scattering chamber was approximately 6 - 11% in the wavelength range of 0.2 to 6.0 Å [4]. Among all detected neutrons, the ratio of air-scattered neutrons was suppressed to less than 1% with the blades. Therefore, the air scattering chamber is very reliable for *in situ* measurements because the SPICA instrument is capable of adapting to a large space for sample environments, in addition, the use of the removable sample vacuum chamber with very small loss of scattered neutrons.

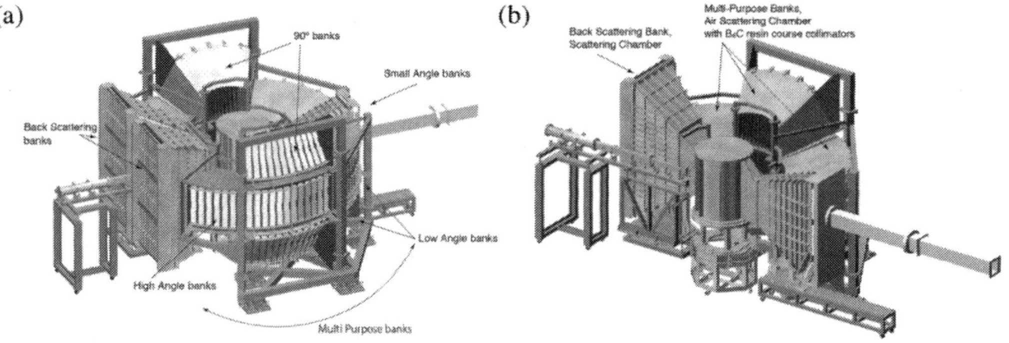

Figure 2. Design illustrations of SPICA instrument. (a) ^{3}He PSDs encircled equidistant from the sample position. The back scattering, 90° and low angle banks have not only the horizontal crate but also extra vertical crates. (b) The back scattering banks and incident neutron beam path can be evacuated. The multi-purpose banks have air scattering chamber with B_4C resin collimators.

3. Performance and recent data under commissioning

Several reference materials such as Si, CeO_2, TiO_2 provided by National Institute of Standards and Technology (NIST) were measured to confirm the reliability of the SPICA instrument. All diffraction profiles indicated low background and good signal to noise ratios. A typical measurement time was 0.5hr for 0.5g of NIST Si powder at 25/3 Hz repetition. The best $\Delta d/d$ resolution using implosion diamond powder was recognized to be less than 0.09% at the stage of this commissioning. Figure 3 shows the Rietveld refinement of CeO_2 collected from the back scattering and 90° banks. The observed intensities are consistent with those calculated from the crystal model. Figure 4 shows a typical *in situ* measurement for a commercialized Li-ion battery using the air scattering chamber. The structural change of the material, which is dependent on the lithium content, was clearly observed. The lattice parameters for the anode and cathode materials as a function of the lithium content were extracted from the diffraction patterns. The structure refinements are in progress. These results promise that SPICA has the high potential of structure refinement for *in situ* measurement even using air scattering bank.

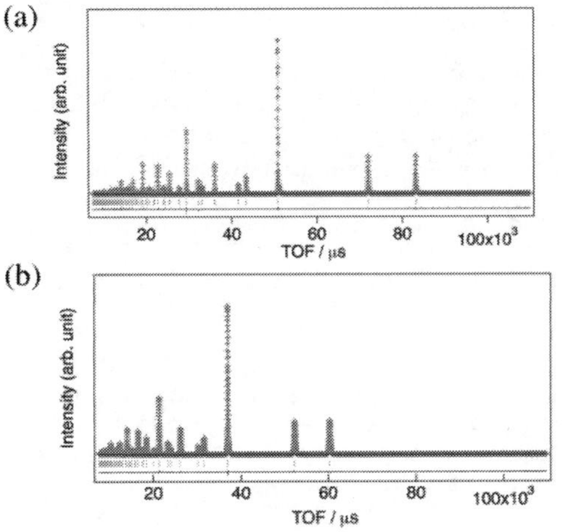

Figure 3. Rietveld refinement pattern of NIST CeO_2 on SPICA. Observed (red +), calculated (solid light blue line) and difference (dark blue line) data are plotted. (a) Back scattering banks and (b) 90° banks (air scattering chambers).

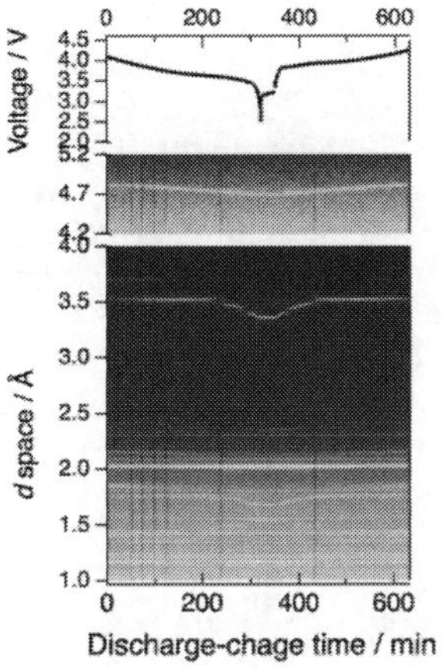

Figure 4. Relationship between the discharge/charge time and the diffraction profile changes of materials in a 18650-type lithium battery.

4. Conclusion

The SPICA instrument has been built to perform *in situ* measurements for battery studies. All components for experimental implementation using SPICA have been also been installed. After the basic commissioning, the structure refinements of NIST standard reference materials were successfully completed. Initial *in situ* experiments have also been conducted. The SPICA instrument was confirmed to provide good performance, low background, high reliability and high resolution.

Acknowledgements

This work was supported by the Research and Development Initiative for Scientific Innovation of New Generation Batteries (RISING) project of the New Energy and Industrial Technology Development Organization (NEDO). The neutron scattering experiment was approved by the Neutron Scattering Program Advisory Committee of IMSS, KEK (Proposal No. 2009S10).

References

[1] Tarascon J-M and Armand M 2001 *Nature* **414** 359-367
[2] Armand M and Tarascon J-M 2008 *Nature* 451 652
[3] Nielsen K, Lefmann K 2000 *Physica B* 283 (2000) 426
[4] Kino K, Mori K, Yonemura M, Torii S, Kawai M, Fukunaga T and Kamiyama T 2011 *J. Phys. Soc. Jpn.* **80** SB001

Development of Spectroelectrochemical Cells for *in situ* Neutron Reflectometry

M Yonemura[1], M Hirayama[2], K Suzuki[2], R Kanno[2], N Torikai[3] and N L Yamada[1]

[1] Neutron Science Division, Institute of Materials Structure Science, High Energy Accelerator Research Organization, 203-1 Shirakata, Tokai, Ibaraki 319-1107, Japan
[2] Department of Electronic Chemistry, Interdisciplinary Graduate School of Science and Engineering, Tokyo Institute of Technology, 4259 Nagatsuda, Midori-ku, Yokohama 226-8502, Japan
[3] Department of Chemistory for Materials, Graduate school of Engineering, Mie University, 1577 Kurimamachiya-cho, Tsu 514-8507, Japan

masao.yonemura@kek.jp

Abstract. A new spectroelectrochemical cell for *in situ* neutron reflectometry was developed. Electrochemical reactions were determined using this *in situ* cell with a $LiMn_2O_4/SrRuO_3$ multi layer thin film. Changes in interfacial structures on the surface of the thin film are also observed by *in situ* neutron reflectivity measurements. The performances of this *in situ* cell were expected the determination of changes on the interface between the electrodes and electrolytes.

1. Introduction

Rechargeable lithium-ion batteries have been studied to improve their performance even after they were commercialized in 1990s. Applications of the batteries such as hybrid electric vehicles (HEV) and pure electric vehicles (EV) require fast charge and discharge performance. In order to achieve high current and power, various studies have examined electrochemical reactions in bulk structures. [1-6] Studies of crystal structural changes have resulted in a better understanding of the reaction mechanism in the bulk electrodes. However, electrochemical reactions occur not only in the bulk electrode but also at the electrode/electrolyte interface of lithium batteries. A comprehensive understanding of interface reactions provides a new viewpoint for the designing of electrode materials that can operate at high powers and have long-life properties. The surface layers of electrodes have been widely examined using *ex situ* methods such as transmission electron microscopy, spectroscopic and electrochemical measurements. [7-13] However, the surface structural changes during the battery reactions are still unclear. High-resolution data of the interface is absolutely essential for gaining an atomistic understanding to control the kinetics in lithium battery reactions.

In order to investigate the surface structure, an ideal electrode with very flat and highly orientated model electrodes is required because poly crystal electrodes have surface morphology, grain boundaries and surface detecting which affect surface reactions. Epitaxial thin film is a possible one of the candidates for the ideal electrode that can be analyzed. Epitaxial thin film electrodes have advantages for mechanistic studies on lithium (de)intercalation and surface structural changes because the two dimensional interface restricts their reaction fields, which could enables the lattice plane dependence of the reaction to be examined directly. The surface structural changes have already been observed using *in situ* x-ray scattering techniques. [14-20] It is well known that neutrons are powerful probes for determining structural information about light elements such as hydrogen and lithium, especially in comparison with X-ray. Furthermore, the neutron scattering technique will allow us to

[1] To whom any correspondence should be addressed.

Journal of Physics: Conference Series **502** (2014) 012054 doi:10.1088/1742-6596/502/1/012054

elucidate the behavior of lithium ions on the surfaces between electrodes and electrolytes. In the present study, we developed a new spectroelectrochemical *in situ* cell to carry out *in situ* neutron reflectivity measurements.

2. *In situ* cell design

An original *in situ* spectroelectrochemical cell was designed for neutron reflectivity measurements. It is necessary to irradiate a wide area of the thin film because the neutron beam intensity is much weaker than synchrotron X-rays. Therefore, the fabricated thin films should be wider than those used in X-ray reflectivity measurements. In this study, an epitaxial thin film with a very flat surface was successfully synthesized on a 20 × 20 mm^2 square $SrTiO_3$ (STO) single crystal substrate with a height of 5 mm. To measure reflectivity over a wide q range, the incoherent scattering of hydrogen, ^{1}H in the electrolyte should be considered. In general, to reduce the background noise from the incoherent scattering of ^{1}H, the cells should be designed to favor neutron beams incident on the substrate side of the film and reflected at the interface between the electrolyte and thin film electrode. (Figure.1(a))

A basic layout of the new *in situ* cell is shown in Figure 1(b). The top and bottom flange of the cell was made from aluminum metal. The substrate put on the cell body made from poly-chloro-trifluoro-ethylene (PCTFE). The substrates and cell body structure was put between the top and bottom flanges and tightened by the screw pillars. The window hole in the top flange was used to make an electric contact to connect between the substrate and a potentiostat/galvanostat for carrying out electrochemical measurements. The in/outlets on the both sides of the cell body supply the electrolyte to the surface of the thin film electrode. The big tapered screw in Figure 1(d) is an anodic electrode. This screw has trench shape so that it can be wrapped with Li metal.

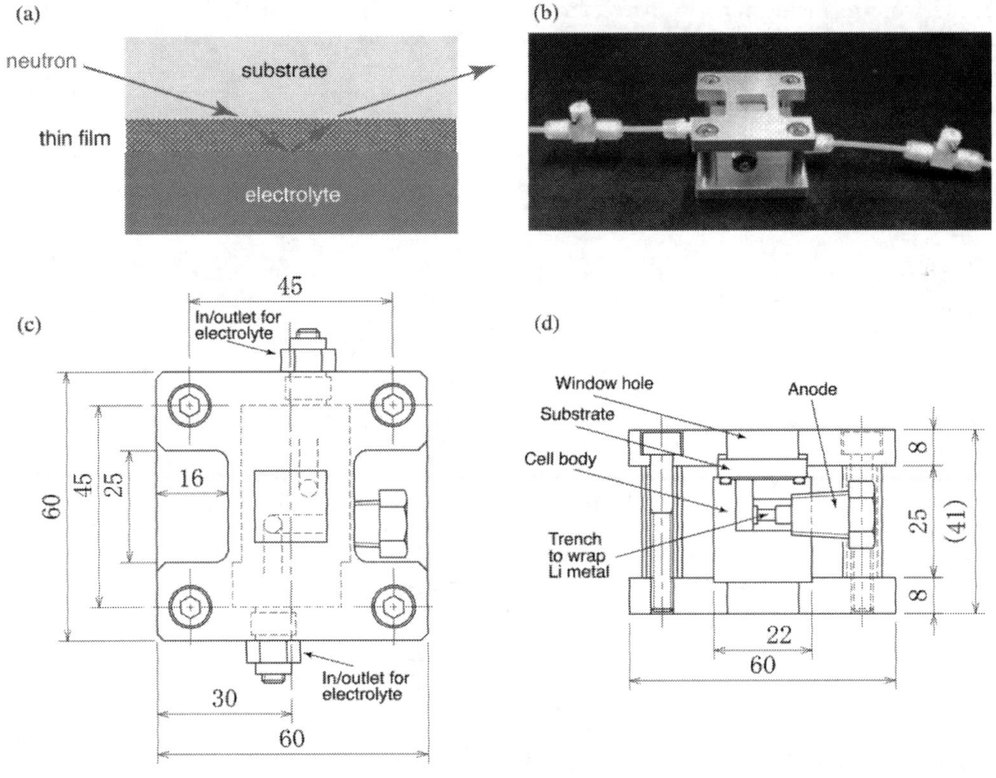

Figure 1. Design illustrations of a new *in situ* spectroelectrochemical cell for neutron reflectivity measurements. (a) Neutron beam passes through the substrate before reflection at the interface between the electrode and electrolyte. (b) Photo of the assembled *in situ* cell. (c) Design schematic from above. Foot print size is 16 ×16 mm^2. The depth of the electrolyte reservoir on the surface of the thin film is 0.5 mm. (d) Design schematic side-on view. The anode is located under the thin films and connected through the duct filled with electrolyte.

Journal of Physics: Conference Series **502** (2014) 012054 doi:10.1088/1742-6596/502/1/012054

3. Performance and recent data

Electrochemical and reflectivity measurements have been demonstrated using the developed *in situ* spectroelectrochemical cell. A $LiMn_2O_4/SrRuO_3$ multi-layer thin film was fabricated on an Nb-doped STO (111) substrate (0.5%Nb) using a KrF excimer laser with a wavelength of 248 nm and PLD apparatuses (PLAD131, AOV inc. and PLD 3000, PVD Products, Inc.), to serve as a test model electrode. [21] The *in situ* cell was assembled in an argon-substituted glove box with lithium metal as the counter electrode and the $LiMn_2O_4/SrRuO_3$ thin film as the working electrode. Deuterated-propylene carbonate (d-PC) was employed as a solvent and a supporting electrolyte of 1M-$LiPF_6$ was used. Cyclic voltammetry (CV) was carried out using a potentiostat/galvanostat (Ivium Tech., CompactStat). Neutron reflectivity measurements were performed using the BL16, Soft Interface Analyzer (SOFIA) at the Materials and Life Science Experimental Facility (MLF) in the Japan Proton Accelerator Research Complex (J-PARC), which is a time-of-flight (TOF) reflectometer. [22]

Figure 2. shows the 2nd - 5th CV cycles of the model electrode, which were recorded with a scan rate of 5 mV/s in the potential range of 3.7 to 4.5 V. Two pairs of well-separated oxidation and reduction peaks were observed. These redox peaks were occurred in bulk electrode and are attributed to the lithium insertion and extraction. [23,24]

The *in situ* neutron reflectivity spectra and fitting curves of the model electrode are shown in Figure 3(a). They were obtained during the *in situ* charge-discharge process; specifically, after setting in the thin film/d-PC/Li cell (OCV), during charging (4.1 V, 4.5 V) and during discharging (4.0 V, 3.7 V) in first cycle. Figure 3(b) shows the voltage dependence of the scattering length density profiles, which provides information on the density changes in the depth direction. The scattering length density of $LiMn_2O_4$ increased and decreased in accordance with the delithiation and lithiation processes, respectively. An interfacial layer (approx. 40 nm) was observed between the $LiMn_2O_4$ electrodes and the electrolytes. This interfacial layer could be related to the concentration gradients that results from diffusing ions in the electrolyte when an electric double layer forms. These results indicate that this *in situ* cell functions successfully in electrochemical and neutron reflectivity measurements.

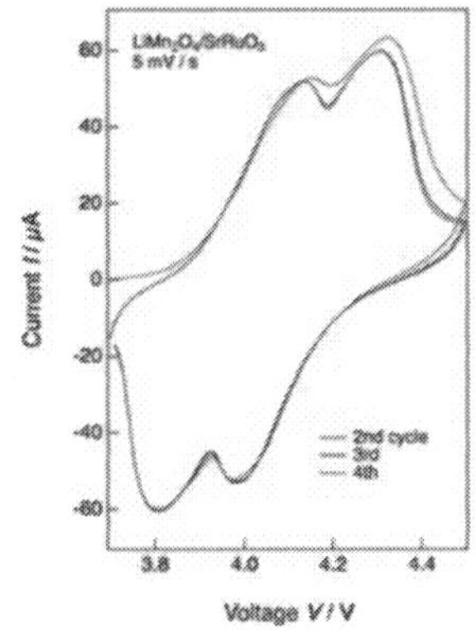

Figure 2. CV curves of the model electrode, $LiMn_2O_4$ /$SrRuO_3$.

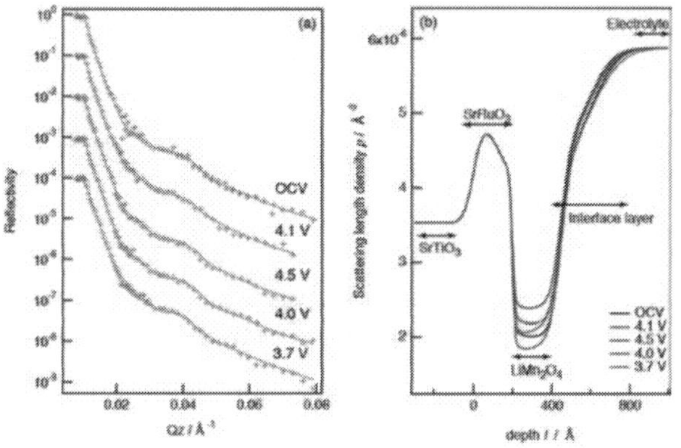

Figure 3. (a) in situ neutron reflectivity spectra and (b) scattering length density profiles of $LiMn_2O_4/SrRuO_3$ multi-layer thin film in d-PC during the first charging and discharging conditions.

4. Conclusion

A new spectroelectrochemical *in situ* cell has been developed for neutron reflectivity measurements. Initial *in situ* experiments have been conducted. The redox reactions of a multi layer thin film electrode during CV cycles were determined. They were consistent with those for bulk electrodes. Changes in the interfacial reaction between the electrode and the electrolyte were adequately observed by neutron reflectivity measurements. This cell was confirmed to provide good performance for *in situ* neutron reflectivity measurements and electrochemical measurements.

5. Acknowledgements

The neutron scattering experiments were approved by the Neutron Science Proposal Review Committee of J-PARC/MLF (Proposal No. 2010B0040, 2012A0108, 2012A0109, 2013A0113) and supported by the Inter-University Research Program on Neutron Scattering of IMSS, KEK.

Journal of Physics: Conference Series **502** (2014) 012054 doi:10.1088/1742-6596/502/1/012054

References

[1] Mizushima K, Jones P.C, Wiseman P.J and Goodenough J.B. 1980 *Mater. Res. Bull.* **15** 783-789
[2] Kanno R, Takeda Y, Tajima Y, Nakanishi and K, Yamamoto O, 1989 *J. Power Sources* **26** 535-543
[3] Mohri M, Yanagisawa N., Tajima Y, Tanaka H, Mitate T, Nakajima S, Yoshida M, Yoshimoto Y, Suzuki T and Wada H 1989 *J. Power Sources* **26** 545-551
[4] Thackeray M.M, David W.I.F, Bruce P.G and Goodenough J.B 1983 *Mater. Res. Bull.* 18 461-472
[5] Thackerray M.M, 1997 *Prog. Solid State Chem.* **25** 1-71
[6] Tarascon J-M and Armand M 2001 *Nature* **414** 359-367
[7] Aurbach D, Levi M D, Levi E, Teller H, Markovsky B and Salitra G 1998 *J. Electrochem. Soc.* **145** 3024-3034
[8] Aurbach D, Gamolsky K, Markovsky B, Salitra G, Gofer Y, Heider U, Oesten R and Schmidt M 2000 *J. Electrochem. Soc.* **147** 1322-31
[9] Araki K and Sato N 2003 *Journal of Power Sources* **124** 124-32
[10] Wang Y, Guo X, Greenbaum S, Liu J and Amine K 2001 *Electrochem. Solid-State Lett.* **4** A68-A70
[11] Menetrier M, Vaysse C, Croguennec L, Delmas C, Jordy C, Bonhomme F and Biensanb P 2001 *Electrochem. Solid-State Lett.* **7** A140-A3
[12] Abraham D P, Twesten R D, Balasubramanian M, Petrov I, McBreen J and Amine K 2002 *Electrochem. Commun.* **4** 620-5
[13] Abe T, Sagane F, Ohtsuka M, Iriyama Y and Ogumi Z 2005 *J. Electrochem. Soc.* **152** A2151-A4
[14] Hirayama M, Sakamoto K, Hiraide T, Mori D, Yamada A, Kanno R, Sonoyama N, Tamura K, Mizuki J, 2007 *Electrochimica Acta* **53** 871-881
[15] Hirayama M, Sonoyama N, Abe T, Minoura M, Ito M, Mori D, Yamada A, Kanno R, Terashima T, Takano M, Tamura K and Mizuki J 2007 *J. Power Sources* **168** 493-500
[16] Sakamoto K, Konishi H, Sonoyama N, Yamada A, Tamura K, Mizuki J and Kanno R 2007 *J. Power Sources* **174** 678-82
[17] Sakamoto K, Hirayama M, Sonoyama N, Mori D, Yamada A, Tamura K, Mizuki J and Kanno R 2009 *Chem. Mater.* **21** 2632
[18] Hirayama M, Ido H, Kim K, Cho W, Tamura K, Mizuki J and Kanno R 2010 *J. Am. Chem. Soc.* **132** 15268-76
[19] Hirayama M, Yonemura M, Suzuki K, Torikai N, Smith H, Watkinsand E, Majewski J and Kanno R 2010 *Electrochemistry* **78** 413-415
[20] Sakamoto K, Hirayama M, Konishi H, Sonoyama N, Dupre N, Guyomard D, Tamura K, Mizuki J and Kanno R 2010 *Phys. Chem. Chem. Phys.* **12** 3815-3823
[21] Suzuki K, Kim K, Taminato S, Hirayama M and Kanno R 2013 *J. Power Sources* **226** 340-5
[22] Yamada N L, Torikai N, Mitamura K, Sagehashi H, Sato S, Seto H, Sugita T, Goko S, Furusaka M, Oda T, Hino M, Fujiwara T, Takahashi H and Takahara A 2011 *Eur. Phys. J. Plus* **126** 108
[23] Gummow R J, Kock A d and Thackeray M M 1994 *Solid State Ionics* **69** 59-67
[24] Sun X, Yang X Q, Balasubramanian M, McBreen J, Xia Y and Sakai T 2002 *J. Electrochem. Soc.* **149** A842-A8

Journal of Physics: Conference Series **502** (2014) 012055 doi:10.1088/1742-6596/502/1/012055

Structure analysis on small molecular crystal by high resolution neutron powder diffraction

P. Miao[1,2], S. Torii[2], M. Yonemura[2], Y. Ishikawa[2], J. Zhang[2] and T. Kamiyama[1,2]

[1] The Graduate University for Advanced Studies, Hayama, Kanagawa 240-0115, Japan

[2] High Energy Accelerator Research Organization, Tokai, Ibaraki 319-1106, Japan

Email: miao@post.kek.jp

Abstract. Deuterated α–glycine was studied by high resolution neutron powder diffraction (NPD). We performed a careful data analysis by Rietveld refinement and compared the result with those of the previous neutron single crystal (NSD) studies. The comparison demonstrates that the deviation of NPD study is generally within 1.2% from the NSD studies, indicating high resolution NPD is by no means inferior to NSD in studying small organic molecular crystals.

1. Introduction

For structure analysis of a crystal, single crystal diffraction has been proved to be the most powerful approach due to its high precision and accuracy. However, the requirement of large size of single crystal imposes tremendous difficulties on the application of this technique to the compound which is hard to grow into adequate size. Under the increasing demand of material science with complicated structures such as inorganic-organic composite materials, super molecules, pharmaceuticals, biomaterials *etc*., giant effort has been made to develop the capability of neutron powder diffraction (NPD) on studying small molecular crystals. As molecular crystals usually possess large unit cells and low symmetries, many overlapped peaks appear in the powder diffraction pattern, making it more challenging to analyze the structure. High resolution powder diffraction seems to be the only way to overcome the overlap problem. Thus, using the high resolution powder diffractometer, we attempted to push the limit of neutron powder diffraction (NPD) forward by performing a careful data analysis on small molecular crystals of amino acids and peptides such as α-glycine ($C_2NO_2H_5$), L-alanine ($C_3NO_2H_5$), L-serine($C_3NO_3H_7$) and glycyl-glycine-hydrochloride ($C_4N_2O_4H_{11}Cl$) *etc*.. Here, we are mainly reporting the result of α–glycine.

2. Experiment

The powder α -glycine specimen, D_2NCD_2COOD, was purchased from Cambridge Isotope Laboratories, with deuteration higher than 98%. Neutron powder diffraction for α-glycine was carried out at room temperature, using SuperHRPD, the high resolution neutron power diffractometer installed at the pulsed spallation neutron source of Japan Proton Accelerator Research Complex (J-PARC) in Tokai, Japan. The data were collected by detectors of the back scattering bank in order to achieve a high resolution. The back scattering bank covers a 2θ range of 150 - 175°, corresponding to d-spacing range of 0.3 - 4.0 Å. The best resolution ($\Delta d/d$) of 0.035% can be achieved at 2θ of 175°. In this study all the signals of the bank are focused to 172° by the time focusing method. We repeated the

Journal of Physics: Conference Series **502** (2014) 012055 doi:10.1088/1742-6596/502/1/012055

experiment twice in order to check the reproducibility. The two data are referred to as SuperHRPD1 and SuperHRPD2 hereafter. The diffraction data were analyzed by Rietveld refinement with the Z-Rietveld program in a software suit, Z-Code. A new type of peak shape function optimized for the diffractometer, so called Type 0 profile function, was adopted in the refinement.

3. Result and Discussion

The refinement on SuperHRPD1 was performed at the time-of-flight range of 40 – 190 ms, corresponding to *d*-spacing range of 0.8 – 4 Å, and the initial structural model was taken from the NSD study [1]. It was found that some weak second-phase peaks appear in the diffraction pattern, the typical intensity of which is approximately 3% of the strongest main-phase peak, and they were all excluded from the refinement. Hydrogen as an impurity of deuterium was taken into account by assigning hydrogen to the corresponding deuterium position and constraining the sum of the occupancy of the two atoms to 1. The occupancies of deuterium atoms are demonstrated in Table 1, the average value of which agrees well with total 98% deuteration of the specimen. As shown in Figure 1, the fitting itself looks good, with R_{wp} value achieving 5.21%. The final atomic coordinates and anisotropic displacement parameters are listed in Table 2.

Table 1. Occupancies of deuterium at different atomic positions

D(1)	D(2)	D(3)	D(4)	D(5)
0.975(2)	0.985(2)	1.000(2)	0.996(2)	0.972(2)

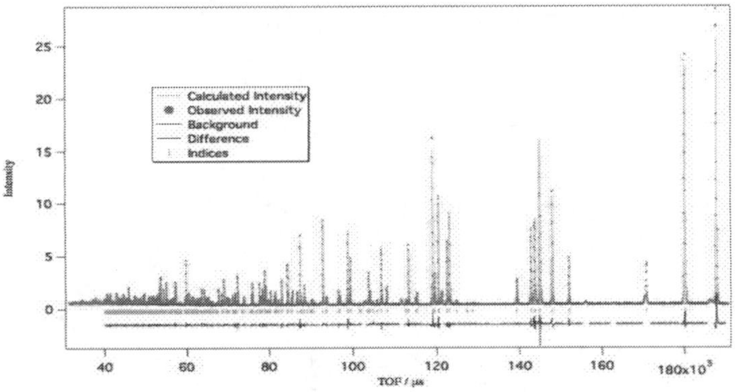

Figure 1. The observed (solid circle) and calculated (light blue line) profiles of α-glycine. Black line indicates the background intensity and dark blue line indicates the difference between observed and calculated intensities.

Table 2. Atomic coordinates and anisotropic displacement parameters with space group P2₁/n [a = 5.103494(5) Å, b = 11.948140(11) Å, c = 5.466384(5) Å, β = 111.73497(5)°]

	x	y	z	B11	B22	B33	B12	B13	B23
C(1)	0.07463(10)	0.12552(4)	0.06636(8)	1.29(2)	1.68(2)	0.94(2)	-0.07(2)	0.59(2)	0.04(2)
C(2)	0.06512(11)	0.14505(5)	-0.21214(4)	1.36(3)	1.85(3)	0.99(2)	0.57(2)	0.72(2)	0.32(2)
O(1)	0.30675(11)	0.09358(4)	0.23592(10)	1.60(2)	3.16(3)	0.85(2)	0.46(2)	0.86(2)	0.35(2)
O(2)	-0.14769(11)	0.14193(5)	0.10768(9)	1.78(3)	3.85(4)	1.78(3)	0.29(2)	1.49(3)	-0.20(2)
N	0.30056(9)	0.08989(4)	-0.25836(8)	1.49(2)	2.41(2)	0.86(2)	0.59(2)	0.55(2)	0.16(2)
D(1)	0.29119(10)	0.10129(4)	-0.45054(10)	2.17(3)	3.70(3)	1.16(2)	0.20(2)	0.83(2)	0.42(2)
D(2)	0.49414(12)	0.11978(5)	-0.13061(11)	1.82(3)	4.79(4)	1.54(3)	-0.23(3)	0.72(2)	-0.29(2)
D(3)	0.29909(12)	0.00638(6)	-0.22645(11)	4.88(4)	2.43(3)	2.39(3)	1.72(3)	1.55(3)	0.02(2)
D(4)	0.07932(11)	0.23362(5)	-0.24202(11)	4.30(3)	2.33(3)	3.61(3)	1.28(2)	2.08(3)	1.32(2)
D(5)	-0.13255(11)	0.11458(5)	-0.35675(10)	1.66(3)	5.31(4)	1.68(3)	-0.02(3)	0.56(2)	-0.75(2)

Table 3. Intramolecular bond lengths (Å) and angles (degree) of α-glycine from SuperHRPD1, SuperHRPD2, Ref. [1] and Ref. [2]. Discrepancies between them are calculated by "|P-J|/J", "|S1-J|/J" and "|S2-J|/J", where J is short for Jönsson and Kvick [1], P for Power *et al*. [2], S1 for SuperHRPD1 and S2 for SuperHRPD2.

	SuperHRPD1	SuperHRPD2	Jönsson	Power	|P-J|/J	|S1-J|/J	|S2-J|/J
C(1)-O(1)	1.263(1)	1.261(1)	1.250(1)	1.251(1)	0.08%	1.04%	0.88%
C(1)-O(2)	1.251(1)	1.253(1)	1.251(1)	1.252(1)	0.08%	0.00%	0.16%
C(2)-N	1.472(1)	1.473(1)	1.476(1)	1.475(1)	0.07%	0.27%	0.20%
C(1)-C(2)	1.523(1)	1.522(1)	1.526(1)	1.525(1)	0.07%	0.20%	0.26%
N-H(1)	1.043(0)	1.044(1)	1.054(2)	1.053(2)	0.09%	1.04%	0.95%
N-H(2)	1.039(1)	1.038(1)	1.037(2)	1.036(2)	0.10%	0.19%	0.10%
N-H(3)	1.014(1)	1.015(1)	1.025(2)	1.029(2)	0.39%	1.07%	0.98%
C(2)-H(4)	1.077(1)	1.079(1)	1.090(2)	1.094(2)	0.37%	1.19%	1.01%
C(2)-H(5)	1.088(1)	1.088(1)	1.089(2)	1.087(2)	0.18%	0.09%	0.09%
C(2)-C(1)-O(1)	116.99(3)	117.21(3)	117.46(6)	117.38(7)	0.07%	0.40%	0.21%
C(2)-C(1)-O(2)	117.46(3)	117.39(3)	117.09(7)	117.08(7)	0.01%	0.32%	0.26%
O(1)-C(1)-O(2)	125.55(3)	125.4(3)	125.45(8)	125.53(8)	0.06%	0.08%	0.04%
C(1)-C(2)-N	112.17(5)	112.05(4)	111.85(5)	111.79(6)	0.05%	0.29%	0.18%
C(2)-N-H(1)	112.27(5)	112.07(5)	112.09(10)	112.08(11)	0.01%	0.16%	0.02%
C(2)-N-H(2)	111.36(13)	111.1(12)	111.73(12)	112.00(12)	0.24%	0.33%	0.56%
C(2)-N-H(3)	110.65(7)	110.79(8)	110.37(12)	110.22(14)	0.14%	0.25%	0.38%
H(1)-N-H(2)	108.03(8)	107.89(8)	108.71(15)	109.05(16)	0.31%	0.63%	0.75%
H(1)-N-H(3)	107.54(15)	107.9(14)	107.13(16)	107.02(17)	0.10%	0.38%	0.72%
H(2)-N-H(3)	106.75(6)	106.88(6)	106.56(17)	106.20(19)	0.34%	0.18%	0.30%
C(1)-C(2)-H(4)	108.84(13)	108.78(13)	108.81(12)	108.55(13)	0.24%	0.03%	0.03%
C(1)-C(2)-H(5)	110.65(8)	110.55(7)	110.50(11)	110.38(13)	0.11%	0.14%	0.05%
N-C(2)-H(4)	107.97(7)	108.07(8)	108.51(12)	108.41(13)	0.09%	0.50%	0.41%
N-C(2)-H(5)	108.77(13)	108.75(12)	109.05(12)	109.47(13)	0.38%	0.26%	0.28%
H(4)-C(2)-H(5)	108.32(7)	108.55(7)	108.03(18)	108.14(21)	0.10%	0.27%	0.48%

To evaluate the accuracy of our result, two neutron single crystal diffraction studies were referred [1, 2]. Since α-glycine is a molecular crystal, we look straight into the molecular structure. Table 3 illustrates the intramolecular bond lengths and angles of α-glycine from SuperHRPD1, SuperHRPD2, Jönsson and Kvick [1] and Power *et al.* [2], respectively. It should be noted that the specimens studied in both references [1, 2] were not deuterated. As we know, geometric isotope effect often takes place in hydrogen bonds [3]. If this also exists in α-glycine crystal, which is bound by the hydrogen bonds as well, the two results from NSD cannot be taken as references. So before comparison, it is necessary to check whether or not geometric isotope effect occurs in this α-glycine crystal. Actually, isotope effect in O-H•••O hydrogen bond has been widely studied in the past 30 years. Ikeda *et al.* [4] proposed a model which successfully explains both geometric and dynamic isotope effects. According to their model, geometric isotope effect was predicted to occur in the limited range of $2.43 < R < 2.6$ Å, where R is hydrogen bond distance (O•••O distance). And the accumulated experimental data on geometric isotope effect of various hydrogen bonds [3] agree well with the Ikeda's prediction. As for the α -glycine crystal, molecules are connected by two roughly linear N-H•••O hydrogen bonds forming individual layers parallel to the *ac* plane. Two neighboring layers are first linked by bifurcated N-H•••O hydrogen bonds to become anti-parallel layers and then connected by bifurcated C-H•••O hydrogen bonds along the *b* axis, becoming the three dimensional crystal structure. We found that all the hydrogen bond distances are over 2.7 Å, thus no geometric isotope effect is anticipated in α-glycine. As a result, the two NSD results can be adopted to check the accuracy of the NPD result, even though the NSD specimen is not deuterated as in our NPD experiment.

We evaluated the discrepancy by calculating the percentage that one deviates from the other. First, the discrepancy between the two single crystal studies [1, 2] is demonstrated in the "|P-J|/J" column of Table 3. It is found that the differences are within 0.4%, so either is qualified to be set as the criterion for comparison. The discrepancies between the result from SuperHRPD1 and that from Jönsson and Kvick [2] were calculated and listed in the "|S1-J|/J" column of Table 3. And the deviations are all within 1.2%. Compared with the 0.4% discrepancy of two single crystal studies, we can draw the conclusion that the molecular structure obtained from SuperHRPD maintains a high accuracy.

To make sure that the high resolution NPD can be applied to the small molecular crystal, reproducibility is also a very crucial criterion to check. By comparison between "|S1-J|/J" column and "|S2-J|/J" column of Table 3, It is unambiguous that the results from both SuperHRPD1 and SuperHRPD2 achieve the same level of accuracy, around 1%, indicating a good reproducibility.

4. Conclusion

It is proved that the application of high resolution neutron powder diffraction can be extended into small molecular crystals. This would be quite helpful for the modern materials science when coming across the problem of growing large single crystal.

5. Acknowledgement

The authors are grateful to Prof. S. Ikeda for useful discussions.

References

[1] P.-G. Jönsson and Å. Kvick, Acta Cryst. B28 (1972) 1827-1833
[2] L. F. Power, K. E. Turner and F. H. Moore, Acta Cryst. B32 (1976) 11-16
[3] M. Ichikawa, J. Mol. Struct. 552 (2000) 63-70
[4] S. Ikeda, H. Sugimoto and Y. Yamada, Phys. Rev. Lett. 81 (1998) 5449-5452

Journal of Physics: Conference Series **502** (2014) 012056 doi:10.1088/1742-6596/502/1/012056

Inelastic neutron scattering study of phonon anomalies in $La_{1.5}Sr_{0.5}NiO_4$

R Kajimoto[1], **M Fujita**[2], **K Nakajima**[1], **K Ikeuchi**[3], **Y Inamura**[1], **M Nakamura**[1] **and T Imasato**[4]

[1] J-PARC Center, Japan Atomic Energy Agency, Tokai, Ibaraki 319-1195, Japan
[2] Institute for Materials Research, Tohoku University, Sendai, Miyagi 980-8577, Japan
[3] Research Center for Neutron Science and Technology, Comprehensive Research Organization for Science and Society, Tokai, Ibaraki 319-1106, Japan
[4] Department of Physics, Tohoku University, Sendai, Miyagi 980-8578, Japan

E-mail: `ryoichi.kajimoto@j-parc.jp`

Abstract. The high-energy phonons in $La_{1.5}Sr_{0.5}NiO_4$, in which the checkerboard charge ordering is formed, was investigated by the inelastic neutron scattering. We found that the longitudinal modes show strong anomalies compared with La_2NiO_4. We argue the similarity and difference in the phonon anomalies between the present sample and the preceding works of different compositions.

$La_{2-x}Sr_xNiO_4$ is isostructural to one of the typical cupurate superconductors $La_{2-x}Sr_xCuO_4$. When $x = 0$, the compound is an antiferromagnetic insulator, where $S = 1$ spins on Ni^{2+} ($3d^8$) ions order antiferromagnetically. Similarly to the cupurates, one hole per Ni ion is introduced in the NiO_2 planes by the substitution of a La^{3+} ion with a Sr^{2+} ion. In the course of exploring the origin of the superconductivity in the cupurates, the role of phonons have been extensively studied, and it was revealed that the hole doping induces a strong anomaly in the Cu-O bond stretching mode [1]. The phonon anomaly was regarded as a result of the electron-lattice coupling associated with charge inhomogeneity in the CuO_2 planes [2]. In the nickelates, the electron-lattice coupling is considered to be larger than the cupurates, as the charge inhomogeneity appears in the more obvious form of static charge stripes along the direction diagonal to the Ni square lattice [3–6]. Then, phonons in the nickelates have been studied by inelastic neutron scattering (INS) to investigate whether the doped holes or charge stripes affect particularly the bond-stretching modes, which are the highest-energy longitudinal optical (LO) modes in this system. Pintschovius *et al.* reported that the LO modes in non-doped La_2NiO_4 show only weak dispersion at ~85 meV [7]. However, they found that the LO mode propagating along the [100] direction is susceptible to the non-stoichiometry, and observed similar softening to the cupurates in $La_{1.9}NiO_{3.93}$ [7, 8]. A study on powder samples of $La_{2-x}Sr_xNiO_4$ by McQueeney *et al.* showed that the ~85 meV band in the phonon density of states (DOS) splits into two sub-bands at ~75 meV and ~85 meV at $x \geq 1/3$ [9]. Tranquada *et al.* reported that in a single crystal of $La_{1.69}Sr_{0.31}NiO_4$, the LO mode along [100] shows the similar softening, while that along [110] shows a splitting of the same magnitude as the softening. However, the anomalies in the phonons are independent on the charge stripe wave vector in this compound, $\boldsymbol{q}_{CO} = (0.31, 0.31, 0)$ [10].

In the present study, we performed an INS study of the high-energy phonons in a single

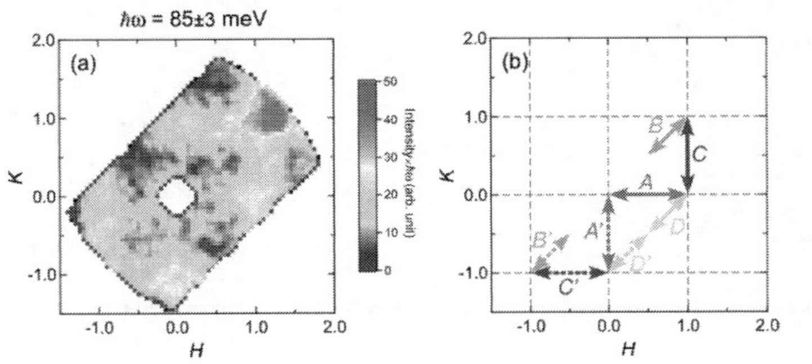

Figure 1. (a) Intensity map of the excitations in $La_{1.5}Sr_{0.5}NiO_4$ at ~5 K, which is cut at $\hbar\omega = 85 \pm 3$ meV and displayed on the H-K plane. The intensity is smoothed by a Gaussian filter. (b) Directions in the H-K plane along which the data in figure 2 are shown (see caption of figure 2 and text).

crystal of $La_{1.5}Sr_{0.5}NiO_4$. The wave vector of the charge stripes is proportional to x, and in this composition, the charge ordering results in a nearly checkerboard ordering of Ni^{2+} and Ni^{3+} sites: The checkerboard charge ordering with $\boldsymbol{q}_{\rm CO} = (1/2, 1/2, 0)$ is formed below ~480 K, and it is taken over by incommensurate charge order with $\boldsymbol{q}_{\rm CO} \sim (0.44, 0.44, 0)$ below ~180 K [6, 11]. We found that the dispersions of the phonon modes in $x = 0.5$ show anomalies similar to those observed in the $x = 0.31$ compound. However, we also found some difference between the two compositions.

Single crystals of $La_{1.5}Sr_{0.5}NiO_4$ were grown by the floating-zone method. The crystal structure is tetragonal with the space group $I4/mmm$ [12]. The lattice constants determined by powder x-ray diffraction are $a = 3.814$ Å and $c = 12.74$ Å. Four crystal rods, each of which is ~5 mm in diameter and ~30 mm in length, were assembled for the present work. The INS measurement was performed on the chopper spectrometer 4SEASONS at J-PARC [13]. The incident energy was $E_{\rm i} = 111$ meV with the energy resolution being 10 meV at the elastic scattering condition. We aligned the crystals so that the [001] axis is parallel to the incident beam and the [110] axis is in the horizontal plane. Due to this crystal orientation, the observed data may include the contributions from out-of-plane phonon modes. However, they are expected to be small, because the high-energy phonon modes observed in the current experiment arise mainly from in-plane polarized oxygen vibrations [7, 9]. We converted the raw data taken at ~5 K to a histogram of the intensity proportional to the dynamical structure factor $S(\boldsymbol{Q}, \hbar\omega)$, where $\boldsymbol{Q}$ and $\hbar\omega$ are momentum and energy transfers, using the software package Utsusemi [14]. In contrast to the previous INS studies using triple-axis spectrometers [7, 8, 10], scanning $Q_x = Ha^*$, $Q_y = Kb^*$, or $\hbar\omega$ simultaneously changes the value of $Q_z = Lc^*$ in the present study using the chopper spectrometer. Actually, for example, $(H, K) = (0, 0)$, $(1, 0)$, $(1, 1)$ correspond to $(H, K, L) = (0, 0, 7.66)$, $(1, 0, 8.48)$, $(1, 1, 9.43)$ at $\hbar\omega = 85$ meV, respectively. However, considering the layered structure of the sample, we ignore Q_z and express $\boldsymbol{Q}$ in terms of (H, K). Though this assumption is not strictly correct, it is supported by the weak Q_z dependence of the highest-energy modes throughout the Brillouin zone observed in $La_{1.9}NiO_{3.87}$ [8]. Since the obtained intensity decreases as a function of $\hbar\omega$, we further divided the intensity by $\hbar\omega$ for clarity in a high $\hbar\omega$ region. To correct the $\hbar\omega$-dependent background, we estimated it by fitting the background of the intensity in the region of $0.9 < H < 1.1$ and $-0.1 < K < 0.1$ to a quadratic function of $\hbar\omega$, then subtracted it from the data.

First, to survey the overall structure of the high-energy phonons, we investigated a Q map of the neutron scattering intensity. Figure 1(a) shows the intensity map at $\hbar\omega = 85$ meV on the H-K plane. This energy is almost equal to that of the bond-stretching mode in La_2NiO_4 [7]. A characteristic intensity modulation is observed in Fig. 1, which reflects some dispersions of the high-energy modes. There are strong spots at $(H, K) = (1, 1)$, $(\pm 1, 0)$, and $(0, \pm 1)$. In addition, the intensity shows streaks along the $(H, \pm 1)$ and $(\pm 1, K)$ lines. These spots and streaks suggest that the phonon mode along these lines shows a weak dispersion, while those along the other

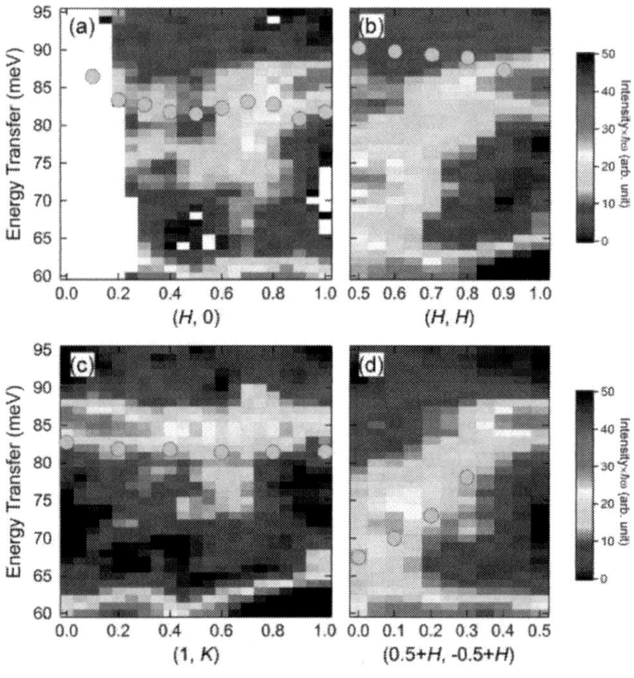

Figure 2. Intensity map of the high-energy excitations in $La_{1.5}Sr_{0.5}NiO_4$ at ~5 K on the Q-$\hbar\omega$ planes. The intensity is smoothed by a Gaussian filter. Each data is sliced along each Q direction with a width of H or $K = \pm 0.1$. The directions of the Q axis of (a), (b), (c), and (d) are $A + A'$, $B + B'$, $C + C'$, and $D + D'$ in figure 1(b), respectively (see text). Circles indicate the energies of phonons observed in La_2NiO_4 [7].

directions show stronger dispersions.

Next, to investigate the $\boldsymbol{Q}$ dependence of the high-energy phonons in more detail, we cut the data along the lines shown in figure 1(b). Figure 2 shows thus obtained intensity maps as functions of Q and $\hbar\omega$. Here, data in (a), (b), (c), and (d) correspond to the cuts along the lines indicated by A, B, C, and D in figure 1(b), respectively. To improve the statistics, we averaged the symmetrically identical data along A and A', B and B', C and C', and D and D'. Clear excitations are observed in the energy region between ~70 meV and ~90 meV where the in-plane modes of phonons were observed in the previous studies of nickelates and cupurates with the similar crystal structures [1, 2, 7–10], which should be an evidence of the observed excitations are caused by the phonons. Figures 2(a) and 2(b) reflect scattering from the longitudinal optical (LO) modes along [100] and [110], while figures 2(c) and 2(d) reflect that from the transverse optical (TO) modes along [100] and [110], respectively. We also show the energies of phonons observed in La_2NiO_4 along the equivalent directions (circles) [7].

The energies of the LO modes in La_2NiO_4 show weak Q dependences both along [100] and [110], as shown by circles in figures 2(a) and 2(b). In contrast, the scattering spectra of $La_{1.5}Sr_{0.5}NiO_4$ in figures 2(a) and 2(b) show quite different dispersions from La_2NiO_4. They show large Q dependences with the minimum energies of ~73 meV at $H = 0.5$ along both the directions. We should note that this minimum energy is almost equal to that of the sub-band in the phonon DOS appearing at $x > 1/3$ [9], which suggests these anomalies in the LO modes are the origin of the ~75 meV sub-band. On the other hand, for the TO modes in figures 2(c) and 2(d), the observed dispersions are almost similar to La_2NiO_4, though the energies are slightly higher. The hole doping and charge ordering do not affect the energies of the TO modes except for slight hardenings.

The observed anomalies in the LO modes as well as their minimum energies at ~73 meV are apparently similar to those observed in $x = 0.31$ [10]. In $x = 0.31$, though the energy of the LO phonon along [100] continuously softens as a function of Q, that along [110] shows a splitting into two modes. Moreover, the latter anomaly is independent on $\boldsymbol{q}_{CO}$. Tranquada *et al.* interpreted the splitting of the LO mode is caused by the local breathing motion of O ions about the hole-doped Ni sites (Ni^{3+} sites). The similarities between the phonon anomalies in $x = 0.5$ and those in $x = 0.31$ supports their idea that the doped holes cause local effects on the bond-stretching phonons independent on the hole concentration.

However, there is some difference between the two compounds. Contrary to $x = 0.31$, in $x = 0.5$, the dispersion of the LO mode along [110] shows a clear softening with its minimum at (0.5,0.5) [figure 2(b)], though there is very weak Q independent intensity at ~85 meV. In other words, the softening behavior is superior to the splitting behavior along [110] in this compound. On the other hand, the splitting behavior is more clearly observed in the data along [100] [figure 2(a)]. The difference between the present $x = 0.5$ sample and $x = 0.31$ may be related to the difference in types of the charge ordering. In the case of the checkerboard charge ordering in $x = 0.5$, the holes enter every two Ni sites, and Ni^{2+} sites and Ni^{3+} sites always share the in-plane O ions. Therefore, the vibrations of the shared in-plane O ions may become more coherent along the direction of the charge-ordering wave vector, [110], resulting in the softening of the dispersion.

In conclusion, we performed an INS study on $La_{1.5}Sr_{0.5}NiO_4$ to investigate the effects of hole doping on the high-energy phonon modes. We found that the longitudinal modes show clearly different dispersions compared with La_2NiO_4, though the transverse modes are quite similar to those in La_2NiO_4. In particular, the longitudinal mode along the direction diagonal to the Ni square lattice shows a clear softening, in contrast to $x = 0.31$ where the mode shows a splitting [10]. We interpret the difference in the phonon anomalies between $x = 0.5$ and $x = 0.31$ as the difference in the type of charge ordering.

Acknowledgments
The experiment on 4SEASONS was conducted under the project numbers 2012I0100 and 2012I0101.

References

[1] Reznik D 2010 *Advances in Condensed Matter Physics* **2010** 523549

[2] Reznik D, Pintschovius L, Ito M, Iikubo S, Sato M, Goka H, Fujita M, Yamada K, Gu G D and Tranquada J M 2006 *Nature* **440** 1170

[3] Chen C H, Cheong S W and Cooper A S 1993 *Phys. Rev. Lett.* **71** 2461

[4] Tranquada J M, Buttrey D J, Sachan V and Lorenzo J E 1994 *Phys. Rev. Lett.* **73** 1003

[5] Tranquada J M, Buttrey D J and Sachan V 1996 *Phys. Rev.* B **54** 12318

[6] Yoshizawa H, Kakeshita T, Kajimoto R, Tanabe T, Katsufuji T and Tokura Y 2000 *Phys. Rev.* B **61** R854

[7] Pintschovius L, Reichardt W, Braden M, Dhalenne G and Revcolevschi A 2001 *Phys. Rev.* B **64** 094510

[8] Pintschovius L, Bassat J M, Odier P, Gervais F, Chevrier G, Reichardt W and Gompf F 1989 *Phys. Rev.* B **40** 2229

[9] McQueeney R J, Sarrao J L and Osborn R 1998 *Phys. Rev.* B **60** 80

[10] Tranquada J M, Nakajima K, Braden M, Pintschovius L and McQueeney R J 2002 *Phys. Rev. Lett.* **88** 075505

[11] Kajimoto R, Ishizaka K, Yoshizawa H and Tokura Y 2003 *Phys. Rev.* B **67** 014511

[12] Millburn J E, Green M A, Neumann D A and Rosseinsky M J 1999 *J. Solid State Chem.* **145** 401

[13] Kajimoto R, Nakamura M, Inamura Y, Mizuno F, Nakajima K, Ohira-Kawamura S, Yokoo T, Nakatani T, Maruyama R, Soyama K, Shibata K, Suzuya K, Sato S, Aizawa K, Arai

M, Wakimoto S, Ishikado M, Shamoto S, Fujita M, Hiraka H, Ohoyama K, Yamada K and Lee C H 2010 *J. Phys. Soc. Jpn.* **80** SB025

[14] Inamura Y, Nakajima K, Kajimoto R, Nakatani T, Arai M, Otomo T, Suzuki J, So J Y and Park J G 2010 *Proc. 19th Meet. Int. Collaboration of Advanced Neutron Sources* (PSI-Proceedings 10-01)

Journal of Physics: Conference Series **502** (2014) 012057 doi:10.1088/1742-6596/502/1/012057

Quasielastic Neutron Scattering Study on Polymer Nanocomposites

T Masui[1], H Kishimoto[1], T Kikuchi[2], S Ohira-Kawamura[2], Y Inamura[2], T Koga[3], K Nakajima[2]
[1] Material Research and Development HQ, Sumitomo Rubber Industries, Ltd.
2-1-1 Tsutsui-cho, Chuo-ku, Kobe, Hyogo 651-0071, Japan
[2] J-PARC Center, Japan Atomic Energy Agency
Tokai, Ibaraki 319-1195, Japan
[3] Department of Materials Science and Engineering, Stony Brook University
Stony Brook, NY 11794-2275
E-mail: t-masui.az@srigroup.co.jp

Abstract. By using a quasielastic neutron scattering technique (disk-chopper-type spectrometer AMATERAS at J-PARC), we have investigated the microscopic dynamics of polybutadiene (PB) in which silica particles were embedded. The experimental results gave us both a damped vibration mode of PB and jumped diffusion motion of the segments. It was found that these dynamical features remain unaffected by the presence of the silica particle, while the residence time of the jumped diffusion motion becomes longer at the high filler loading.

1. Introduction

It is well known that the incorporation of nanofillers to a polymer improves its mechanical property. Thus, polymer nanocomposite have been widely used in the industrial field. Numerous experiments have been attemped to elucidate how the nanofillers improve the mechanical property. However, most previous studies focused on the macroscopic nature including rheological, calorimetric and dielectric characteristics. Thus, it is still unclear how such macroscopic nature relates to the microscopic nature of a polymer. In recent years, by employing a neutron spin echo technique that is a powerful tool to observe polymer dynamics directly, several experimental results to reveal the relation between a microscopic and a macroscopic view on the relaxation dynamics of polymer in the presence of fillers. It is reported that the increase in the filler concentration does not affect the basic segmental relaxation rate (Rouse rate) but decreases the diameter of tube (the reputation tube) [1].

A quasielastic neutron scattering (QENS) technique allows one to observe the dynamics of polymer chains directly with a time scale of $\sim 10^{-12}$ s, and numerous studies have been studied on pure polymers. Kanaya and co-workers showed a damped vibration mode and local conformation transition of polybutadiene (PB) using QENS [2]. Although some QENS studies on polymer nanocomposites were also reported [3], the question of how the microscopic dynamics is affected by the presence of fillers remains unrevealed. To answer the question we used silica nanofillers embedded in a PB matrix as a model and aimed to clarify the effect of the fillers on the dynamics of polybutadiene by means of QENS.

Journal of Physics: Conference Series **502** (2014) 012057 doi:10.1088/1742-6596/502/1/012057

2. Materials and Methods

2.1. Sample Preparation

Two kinds of silica (Si)-polybutadiene (PB) nanocomposites were used for this study. The sample characteristics are tabulated in Table.1. The PB used in this study was UBE UBEPOL BR150B(Ube Industries. Ltd.). Microstructure, molecular weight and distribution of polybutadiene is cis: trans: vinyl = 98: 1: 1 (wt.%), $M_w = 5.0\times 10^5$ and $M_w/M_n = 2.8$, respectively. Si pellets (Ultrasil VN3, Evonik) and a silane coupling agent (Si266, Degussa GmbH) were used. These samples were mixed with an internal mixer and then laminated to be 2.0 mm thickness with two roll mills. To avoid multiple scattering, these samples were further sliced to be 0.4 mm thickness for QENS experiments.

Table 1. Compositions of the samples used in this study (phr[a])

Sample	VF(%)[b]	PB	Silica	Si266[c]	ZnO	Stearic acid	Sulfur	Acc. CBS[d]	DPG
Φ_{15}	14.5	100	40	3.2	2.5	2.0	1.5	1.7	2.0
Φ_{24}	23.8	100	75	3.2	2.5	2.0	1.5	1.7	2.0

[a] Parts per one hundred rubber by weight. [b] volume fraction of silica. [c] Bis-triethozysilylpropyldisulfane.
[d] N-cyclohexyl-2-benzothiazole sulfenamide.

2.2. Quasi-Elastic Neutron Scattering (QENS)

QENS experiments were performed at a cold neutron disk-chopper-type spectrometer (AMATERAS, BL14) installed at Material and Life Science Experimental Facility (MLF) of J-PARC [4]. The energy resolution is about 0.12 meV. The experimental temperatures used were 1, 27, 47, and 88 °C.

2.3. Data Analysis

Dynamic structure factors $S(Q,\omega)$ obtained from the QENS experiments were first fitted with a model scattering function given by the sum of a δ-funcion and Lorentzian function. However, we turned out that the agreement between observed and calculated spectra was not so good. Therefore, we instead fitted the observed spectra with a funcion given by the sum of a δ-funcion and two Lorentzian functions expressed by the following equation:

$$S_{\mathrm{inc}}(Q,\omega) = C(Q,\omega)\{A_\delta \delta(\omega) + A_{\mathrm{n}}(Q)L(Q,\omega,\Gamma_{\mathrm{n}}) + A_{\mathrm{b}}(Q)L(Q,\omega,\Gamma_{\mathrm{b}})\} + B_{\mathrm{in}}(Q)\,, \quad (1)$$

where $C(Q, \omega)$ is a dynamic structure factor of the Si filled PB at T=-223 °C, $\delta(\omega)$ is a δ-function, $L(Q, \omega, \Gamma)$ is a Lorentz function, $B_{\mathrm{in}}(Q)$ is an inelastic background, A_{n} and Γ_{n} are prefactor and half width at half maximum (HWHM) of the narrower Lorentz function, and A_{b} and Γ_{b} are a prefactor and HWHM of the broader Lorentz function, respectively. The fitting was performed with the DAVE Peak Analysis (PAN) software developed by National Institute of Standard Technology(NIST) [5].

The mean-square displacement $\langle U^2 \rangle$ was evaluated through equation $I_{\mathrm{el}}(Q) = \exp[-\langle U^2 \rangle Q^2]$ where $I_{\mathrm{el}}(Q)$ is elastic scattering intensity.

3. Result and Disucussion

Figure 1 shows the dynamic structure factor $S(Q,\omega)$ of the sample Φ_{24} at 47 °C and $Q = 2.09$ Å^{-1}. The thick solid line calculated $S(Q,\omega)$ from equation (1). is in a good agreement with the experimental data.

Figure 2 shows the Q dependence of HWHM of the broader Lorentz component Γ_{b} at 47 °C. Γ_{b} is almost independent of Q. Thus, it is reasonable to suppose that the observed dynamics is a localized

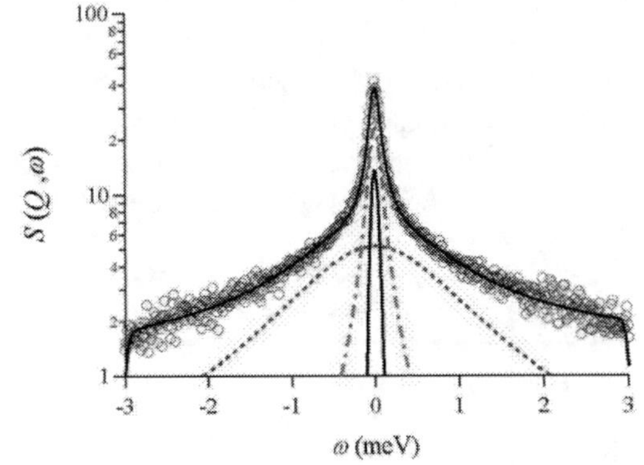

Figure 1. Quasielastic neutron-scattering spectra of sample Φ_{24} at Q=2.09 Å^{-1} and 47 °C. Thick solid line indicates fit result according to equation (1). Dotted line, dash-dot line and thin solid line indicate broader Lorentz, narrower Lorentz and δ-funcion.

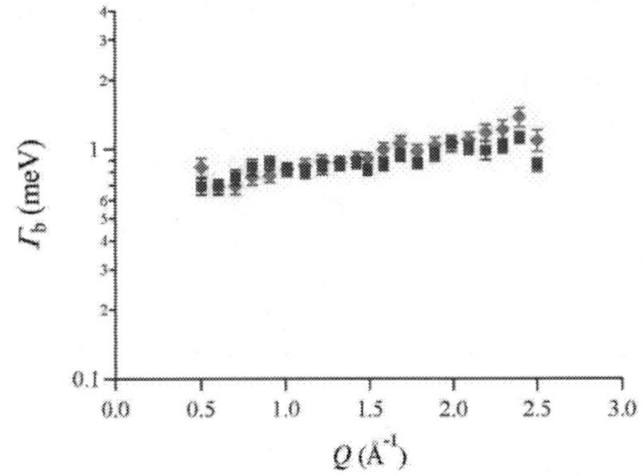

Figure 2. Q dependence of HWHM Γ_b of sample Φ_{24} at 47 °C.(◆) : Φ_{15}, (■) : Φ_{24}.

motion. The characteristic time τ derived from the relation of $\tau = h/\Gamma_b$ is estimated to be 5.5 × 10^{-12} s, where h is the plank constant.

Figure 3 shows the temperature denpendence of Γ_b at Q = 2.09 Å^{-1} plotted in the arrehenius fashion. The red and blue circles denote the Γ_b values for sample Φ_{15} and Φ_{24}, respectively. Both Γ_b values show a weak temperature denpendence and the estimated activation energy are 0.91 kcal/mol for sample Φ_{15} and 0.87 kcal/mol for Φ_{24}, respectively.

A QENS study on pure PB was reported by Kanaya and co-workers [2]. They showed a localized motion with an activation energy of about 0.5 kcal/mol. This motion was assinged to a damped vibrational motion of PB because the characteristic times of the motion was about 5.0 × 10^{-12} s at 45 °C, which is too fast for the conformational transition. The time range and activation energy of the observed motion in this study are the same order of those of the pure PB reported. Thus, the observed motion corresponds to the damped vibraional motion of the PB in the presence of the Si fillers.

Figure 4 shows the Q^2 dependence of HWHM of the narrow Lorentz function Γ_n at 47 °C. In the low Q region, Γ_n is proportional to Q^2 and becomes plateu at the high Q region. These characteristic Q-dependent features can be modeled as a jumped diffusion motion[6] expressed by equation (2):

$$\Gamma = h\frac{DQ^2 + (1-\exp(-\langle U^2\rangle Q^2)/\tau_0}{1+DQ^2\tau_0} \tag{2}$$

where τ_0, D and $<U^2>$ are a residence time, diffusion coefficient, and mean square displacement, respectively. In the low Q region, equation (2) becomes simple diffusion:

$$\Gamma = hDQ^2 \tag{3}$$

On the other hand, in the high Q region, equation (2) becomes a constant value:

$$\Gamma = \frac{h}{\tau_0} \tag{4}$$

The solid lines in Figure 4 represent the best fits of equation (2) to the experimental data. In this study, the energy resolution was about 0.12 meV so that it is difficult to estimate a diffusion constant D from the low Q region of Q < 1.0 Å^{-1}. Instead, we used the high Q region to obtain the residence time τ_0. The residence time τ_0 for the samples Φ_{15} and Φ_{24} are estimated to be 35.3 ps and 50.8 ps, respectively.

It was reported that pure PB shows a jumped diffusion motion with its residence time τ_0 of 81 ps and 48.6 ps for 23 °C and 60 °C, respectively [2]. By interpolating these reported data, the residence time τ_0 at 47 °C is estimated to be about 58 ps. Hence, it is obvious that the residence time of the Si filled PB is shorter than that of pure PB. This may be due to the molecular weight or microstructure

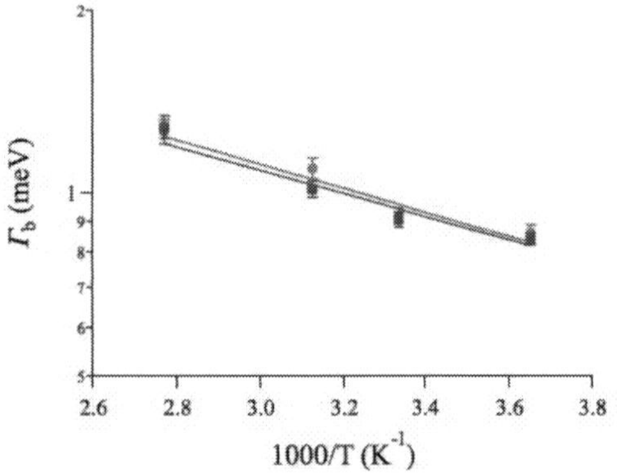

Figure 3. Temperature dependences of HWHM of the broad quasielastic components Γ_b at Q = 2.09 Å^{-1}. (◆) : Φ_{15}, (■) : Φ_{24}.

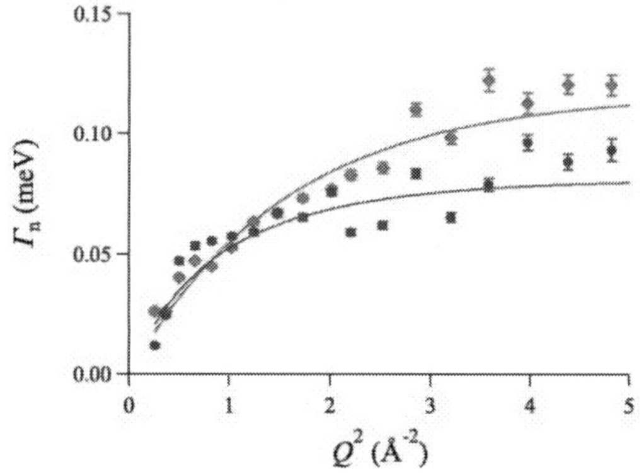

Figure 4. Q^2 dependences of HWHM of narrow quasielastic components Γ_n at 47 °C. (◆) : Φ_{15}, (■) : Φ_{24}.

difference in PB used in this study. Further studies are currently in progress to see the effects of the addition of Si particles to the PB. The important finding in this study is that the increase in the Si concentration makes the residence time τ_0 longer.

If we assume all the Si particles are well dispersed in the polymer matrix, the distances between the nearest neighbor Si particles are estimated to be 29.7 nm for samples Φ_{15} and 25.2 nm for Φ_{24}, respectively, based on the volume fraction of the Si and its mean diameter (= 17.2 nm). Given the fact that the radius of polymer gyration (R_g) of the PB is 12.9 nm, the polymer chains are confined between the Si particles and squeezed to be 48 % for the sample Φ_{15} and 31 % for the sample Φ_{24}, respectively. We postulate these confinement effects make the local conformation transition of the polymer slow down, resulting in the longer residence time τ_0 at the higher Si loading.

4. Conclusion

The QENS experiments allowed us to investigate the in-situ microscopic dynamics of PB filled with the Si fillers. We found that the increase in the Si concentration makes the residence time τ_0 longer.

5. Acknowledgment

The experiment in the MLF at J-PARC was performed with the approval of J-PARC (Proposal No. 2012A0137, 2012B0183).

6. References

[1] Schneider G S, Nusser K, Wilner L, Falus P and Richter D 2011 macromolecules **44** 5857
[2] Kanaya T, Kaji K and Inoue K 1991 *Macromolecules* **24** 1826
[3] Gagliardi S, Arrighi V, Ferguson R and Telling M T F 2001 Physica B **301** 110
[4] Nakajima K et al., 2011 *J. Phys. Soc. Jpn.* **80** SB028
[5] Azuah R T, Kneller L R, Qiu Y, Tregenna-Piggott P L W, Brown C M, Copley J R D and Dimeo R M 2009 *J. Res. Natl. Inst. Stan. Technol.* **114** 341
[6] Singwi K S and Sjolander A 1960 *Phys Rev.* **119** 863

Journal of Physics: Conference Series **502** (2014) 012058 doi:10.1088/1742-6596/502/1/012058

Adsorption of water to double-network polymers having a hierarchical structure

Taiki Tominaga[1], Shin-ichi Takata[1], Jun-ichi Suzuki[1,2], Kazuya Aizawa[1], Hideki Seto[3] and Masatoshi Arai[1]

[1]Materials and Life Science Division, J-PARC Center, Japan Atomic Energy Agency, 2-4 Shirakatashirane, Tokai, Ibaraki 319-1195, Japan

[2]Research Center for Neutron Science and Technology, Comprehensive Research Organization for Science and Society, 162-1 Shirakata, Tokai, Ibaraki 319-1106

[3]Japan Institute of Materials Structure Science, High Energy Accelerator Research Organization, Tsukuba 305-0801, Japan

E-mail: taiki.tominaga@j-parc.jp

Abstract. Double-network hydrogels (DN-gels) have cross-linked aqueous polymer networks which result in unique mechanical properties [1,2]. Although the structure of the hydrophilic polymer networks have been previously determined [3,4,5], it was not clear how water molecules are adsorbed to the polymer network. We prepared freeze-dried DN-gels (DN-polymers) made of polyacrylamide and sodium salt of poly(2-acrylamido-2-methylpropane sulfonic acid), and small-angle neutron scattering (SANS) experiments were conducted to determine the humidity dependence of the nanoscale structure. The SANS results show that water molecules adsorb on larger structures than the mesh size of polymer networks at low relative humidity (RH), and adsorb gradually on the segmental scale of polymers with increasing RH.

1. Introduction

Soft matter has a hierarchical structure on the scale of angstroms up to nanoscale and macroscale. Additionally it is known that the physical properties of soft matters are greatly influenced by water. Synthetic hydrogels are good materials to study the effect of water adsorption because they contain an abundance of water. In fact, most hydrogels, which have a three-dimensional polymer network, can adsorb more than 90 wt% of water. Nevertheless, most hydrogels are composed of 90 wt% water, which means only 10 wt% of the polymer network governs its mechanical properties. Although there are a number of studies on water in hydrogels, there is little experimental study on the water near the polymer network.

Double-network hydrogels (DN-gels) [1,2] are water-soluble synthetic polymer gels that have high mechanical strength equivalent to cartilage. Several studies on polymer networks of DN-gels have been conducted using small-angle neutron scattering (SANS) [3,4,5] so far. However, it was not determined how water molecules are adsorbed to the polymer network.

In this study, we performed SANS measurements to investigate the humidity dependence of the nanoscale structure of the freeze-dried DN-gels (DN-polymers) to clarify the mutual relationship between a polymer network and water.

Journal of Physics: Conference Series **502** (2014) 012058 doi:10.1088/1742-6596/502/1/012058

2. Experiments

2.1. Materials

The water-swollen DN-gels [1,2] were prepared with 2-acrylamido-2-methylpropanesulfonic acid (Toagosei Co., Ltd, Japan) and were neutralized by NaOH. 2-Acrylamide-2-methylpropanesulfonic acid sodium salt was used as the monomer for first network, purified acrylamide was used as the monomer for the second network and purified N,N′ -methylenebisacrylamide was used as the cross-linker for the first network with 2-oxoglutaricacid as the UV initiator (Wako Pure Chemical Industries, Ltd., Japan). The DN-gels were then freeze-dried by EYELA FDU-2100 (Tokyo Rikakikai Co., Ltd., Japan), which creates the DN-polymers [6]. The DN-polymer samples were placed in a dry state with a glove box evacuated by a turbo-pump (VPT-030, ULVAC Kiko, Inc., Japan) to remove light water. After the drying process, heavy water (435767-100G, Aldrich Chemical Company) vapours were introduced into the system for more than an hour to generate a certain relative humidity (RH). The humidity conditions were as follows: 0% (dry state), 23%, 40%, 68% and 79%. Measurements of all humidity conditions were prepared by the same sample, and we confirmed the reproducibility of SANS results shown below.

2.2. Small-Angle Neutron Scattering (SANS) measurements

The SANS measurements were performed with a small- and wide- angle neutron scattering instrument, TAIKAN (BL15), which is in the Materials and Life Science Experimental Facility (MLF) of the Japan Proton Accelerator Research Complex (J-PARC). The pinhole collimator set before the sample was 10 mm in diameter. All SANS measurements were performed at 25 °C. The wavelengths of the incident neutrons ranged from 0.07 nm to 0.76 nm, and the time of flight-method was used.

2.3. *Pore size determination*

The specific pore size distribution of the DN-polymers was determined with the BELSORP-mini II (BEL Japan, Inc.). Nitrogen gas adsorption–desorption isotherm analysis was performed at 77.4 K.

3. Results and Discussion

The SANS profiles of the DN-polymers under various RH conditions are shown in Fig. 1. The intensity below $q = 0.2$ nm^{-1} increases with increasing RH. The power law behaviour is observed in this q-range with an exponent of approximately −3.1 for all RH conditions. The samples studied here are dried, porous materials synthesized via radical polymerization and intrinsically have a random structure. Therefore, it is reasonable to assume that the SANS intensities originate from the nanoscale interfaces of polymer networks.

The pore size distribution is determined by the nitrogen gas adsorption–desorption measurements. Fig. 2 shows the BJH-plot of the DN-polymers, where d_p is the pore diameter in nanometers, and V_p is the pore volume in cm^{3}g^{-1}. From this plot, it is known that the number of pores larger than 30 nm in diameter is more than the number of nanoscale pores. This spatial scale corresponds to the q-range lower than $q = 0.2$ nm^{-1}. Thus, we could say that the origin of the SANS intensities in the region following the power-law behaviour are due to the density fluctuation of the pores.

To clarify the effect of water adsorption on the SANS profiles, the profiles measured between 23% RH and 79% RH were divided by the profile measured at 0% RH as shown in Fig. 3. The scaled intensity increases with increasing RH below $q \approx 0.8$ nm^{-1}. When the RH is less than 68%, the scaled intensity does not change above $q = 0.8$ nm^{-1}, but when it is more than 68%, the scaled intensity decreases with increasing RH. This threshold, $q \approx 0.8$ nm^{-1}, corresponds to the correlation length of the DN-polymers, which is approximately 8 nm [5].

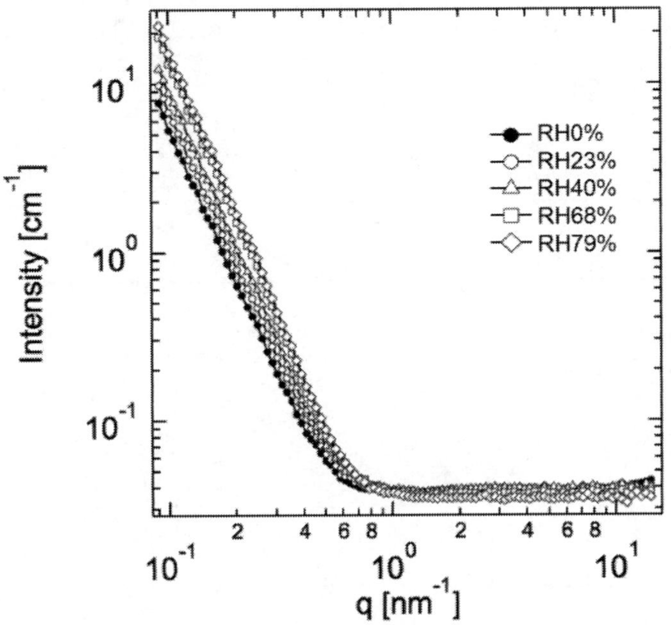

Figure 1. SANS profiles obtained from the DN-polymers under different humidity conditions. The relative humidity is 0% (dry state, closed circle), 23% (open circle), 40% (triangle), 68% (square) and 79% (diamond).

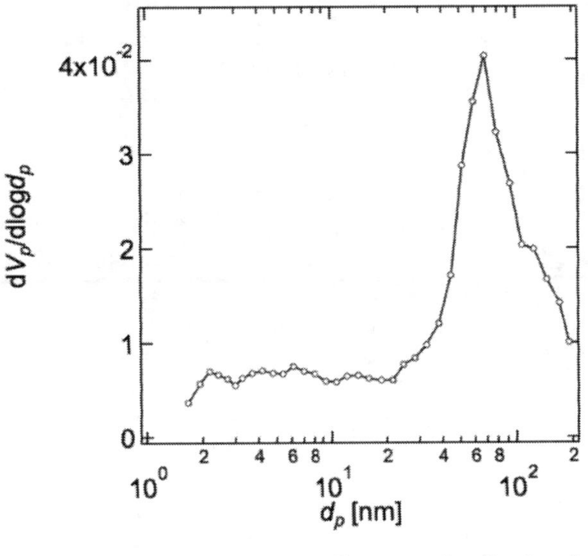

Figure 2. BJH-pore diameter distribution for DN-polymers.

The adsorption behaviour of water molecules in the polymer network of the DN-polymers could be interpreted as follows: There is no change in the polymer network structure larger than d (= $2\pi/q$) = 20 nm, which is comparable to the pore size below 79% RH. Below 79% RH, water molecules adsorb onto a larger polymer structure than a mesh size scale, which is greater than 8 nm, in the dry state of the DN-polymers. When the RH is 68%–79%, water molecules adsorb not only on the large polymer structure but also on internal structures such as polymer segments that are smaller than the mesh size. A detailed structure analysis of the SANS results is beyond the scope of this proceedings and will be presented in a forthcoming article with SAXS data.

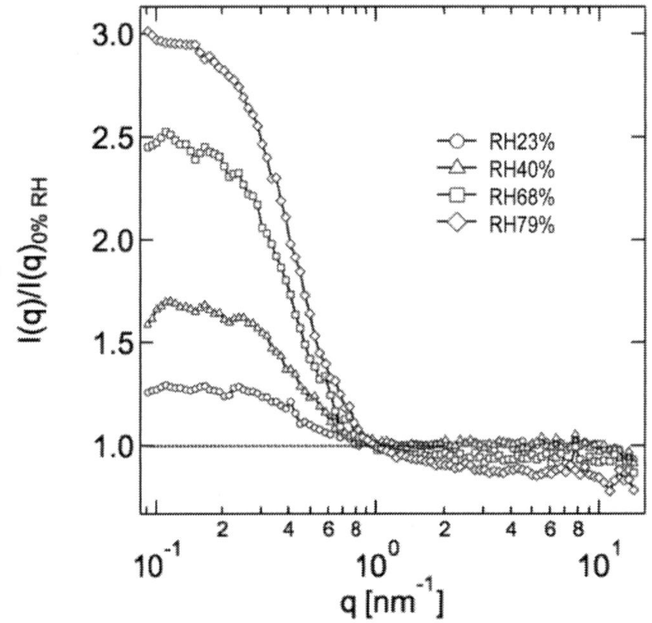

Figure 3. The relative SANS profile divided by the SANS profile measured at 0% RH. The relative humidity is 23% (open circle), 40% (triangle), 68% (square) and 79% (diamond).

Acknowledgments

This study was performed under the following project in the Materials and Life Science Experimental Facility: *Structure and dynamics of water molecules in soft matters* (2012P0401, 2013P0401). This work was supported by JSPS KAKENHI (Grant Number 25810131). The samples were produced by T.T. at the Molecular and System Life Science Unit, RIKEN ASI, Wako, Saitama, Japan with financial support from Toyota Motor Corporation.

Reference

[1] Gong J.P., Katsuyama K., Kurokawa T., and Osada Y. 2003 *Adv. Mater.* **15** 1155
[2] Gong J.P. 2012 *Soft Matter* **6** 2583
[3] Tominaga T., Tirumala V. R., Lin E. K., Gong J. P., Furukawa H., Osada Y., and Wu W-l. 2007 *Polymer* **48** 7449
[4] Tominaga T., Tirumala V. R., Lee S., Lin E. K., Gong J. P., and Wu W-l. 2007 *J. Phys. Chem B* **112** 3903
[5] Tirumala V. R, Tominaga T., Lee S., Butler P., Lin E. K., Gong J. P., and Wu W-l. 2008 *J. Phys. Chem. B* **112** 8024
[6] Tominaga T., Sano K-I., Kikuchi J., Mitomo H., Ijiro K., and Osada Y. 2012 *ACS Macro Lett.* **1** 432

AUTHOR INDEX

AUTHOR INDEX

AUTHOR INDEX